Remote Sensing of Above-Ground Biomass

Remote Sensing of Above-Ground Biomass

Special Issue Editors

Lalit Kumar
Onisimo Mutanga

MDPI • Basel • Beijing • Wuhan • Barcelona • Belgrade

MDPI

Special Issue Editors
Lalit Kumar
University of New England
Australia

Onisimo Mutanga
University of KwaZulu-Natal
South Africa

Editorial Office
MDPI
St. Alban-Anlage 66
4052 Basel, Switzerland

This is a reprint of articles from the Special Issue published online in the open access journal *Remote Sensing* (ISSN 2072-4292) from 2016 to 2017 (available at: https://www.mdpi.com/journal/remotesensing/special_issues/remotesensing_biomass)

For citation purposes, cite each article independently as indicated on the article page online and as indicated below:

LastName, A.A.; LastName, B.B.; LastName, C.C. Article Title. *Journal Name* **Year**, *Article Number*, Page Range.

ISBN 978-3-03921-209-5 (Pbk)
ISBN 978-3-03921-210-1 (PDF)

Cover image courtesy of Lalit Kumar and Onisimo Mutanga.

Contents

About the Special Issue Editors

Lalit Kumar, of the School of Environmental and Rural Science of the University of New England in Australia, maintains a research focus in the field of environmental modelling—specifically, the utilization of high spatial and high spectral resolution data for vegetation mapping and understanding light interactions at the plant canopy level. Most of his recent work has focused on the impacts of climate change in the coastal areas. He utilizes time series remote sensing data and global climate models to map historical changes and project future changes in coastal areas.

Onisimo Mutanga is a full professor and SARChI Chair of land use planning and management. He is known for his use of remote sensing to analyse the patterns and condition of vegetation (including agricultural crops) in the face of global and land use change. His work integrates ecology, biodiversity conservation and remote sensing to model the impacts of forest fragmentation, pests, disease, and invasive species on agricultural and natural ecosystems.

remote sensing

MDPI

Editorial

Remote Sensing of Above-Ground Biomass

Lalit Kumar [1,*] and Onisimo Mutanga [2]

[1] Ecosystem Management, School of Environmental and Rural Science, University of New England, Armidale NSW 2351, Australia
[2] School of Agricultural, Earth and Environmental Sciences, University of KwaZulu Natal, P. Bag X01 Scottsville, Pietermaritzburg 3209, South Africa; MutangaO@ukzn.ac.za
* Correspondence: lkumar@une.edu.au; Tel.: +612-6773-5239

Received: 8 September 2017; Accepted: 8 September 2017; Published: 10 September 2017

1. Importance of Above-Ground Biomass

Accurate measurement and mapping of biomass is a critical component of carbon stock quantification, climate change impact assessment, suitability and location of bio-energy processing plants, assessing fuel for forest fires, and assessing merchandisable timber. While above-ground biomass includes both live and dead plant material, most of the recent research effort on biomass estimation has focussed on the 'live' component (live trees) due to the prominence of this component. Accurate estimates of biomass is a prerequisite for better understanding of the impacts of deforestation and environmental degradation on climate change.

The Intergovernmental Panel on Climate Change (IPCC) [1] has listed five terrestrial ecosystem carbon pools involving biomass: above-ground biomass, below-ground biomass, litter, woody debris and soil organic matter. Of these five, above-ground biomass is the most visible, dominant, dynamic and important pool of the terrestrial ecosystem, constituting around 30% of the total terrestrial ecosystem carbon pool. Above-ground biomass estimation, and especially forest biomass, has received considerable attention over the last few decades because of increased awareness of climate warming and the role forest biomass plays in carbon sequestration and release of greenhouse gases due to deforestation.

Above-ground biomass estimates are the central basis for carbon inventories and most international negotiations in carbon trading schemes. Carbon trading markets require long-term information on carbon stocks, particularly on the above-ground 'live' biomass component as this is the most dynamic, changing and manipulable component of all the biomass pools. This is the 'merchantable' component of biomass.

Above-ground forest biomass accounts for between 70% to 90% of total forest biomass [2]. While soil organic matter holds two to three times more carbon than biomass on a global scale, much of the soil carbon is more protected and not easily oxidised [3]. On the other hand, above-ground biomass is in a continuous state of flux due to fire, logging, storms, landuse changes, etc., and thus contributes to atmospheric carbon fluxes to a much greater extent and so is of much greater interest. Due to this dynamism of above-ground biomass, it is necessary to monitor it continuously and not measure once and forget.

While detailed estimations of biomass is necessary for accurate carbon accounting, reliable estimation methods are few. Accurate estimates of stored carbon (biomass as dry weight is 50% carbon [4] and understanding sources and sinks can improve the accuracy of carbon flux models and thus lead to better projections of climate change and impacts. Initiatives such as Reducing Emissions from Deforestation and Forest Degradation (REDD) and REDD+ also rely heavily on above-ground biomass estimates [5,6]. REDD+ includes financing schemes and incentives with the aim of mitigating climate change by reducing deforestation and forest degradation through sustainable forest management and conservation, and enhancement of carbon stocks [7,8]. The countries that participate in the REDD+ scheme have a requirement to produce accurate estimates of their forest carbon stocks and changes.

Wildfire and fuel management are becoming an increasingly important part of forest management. Forest biomass, and especially crown biomass and the dry litter component, are important factors in any fire model. Traditionally crown biomass was a neglected component as there was much greater emphasis on the commercial component of the trees but with fire playing a more important role in environmental planning, this biomass component has gained prominence.

Biomass is also a highly abundant source of energy that is widely used around the world. Its attractiveness is that it is a renewable fuel. However, biomass resources are distributed over wide geographical regions and their qualities for energy production varies over space and time. Also, very often the resources are located far away from the centres where energy generation is required. Because of this link between distribution over space and time and centres of requirements, it is important to have accurate and consistent means of measurement methods for biomass to evaluate feasibility of biomass for energy generation.

2. Methods of Assessing Above-Ground Biomass

Above-ground biomass can be measured or estimated both destructively and non-destructively. In the destructive method, sometimes also known as the harvest method, the trees are actually cut down and weighed. Sometimes a selected sample of trees are harvested and estimations for the whole population are based on these, especially where there is uniformity in tree size, for example a pine plantation. The destructive method of biomass estimation is limited to a small area due to the destructive nature, time, expense and labour involved. It is also not suitable where there may be threatened flora and fauna.

The non-destructive methods include the estimation based on allometric equations or through remote imagery. Allometric equations have been developed through the use of tree dimensions [9–15], such as diameter at breast height (dbh) and tree height, however these are not very useful in heterogenous forests. Allometric equations are most useful in uniform forests or plantations with similar aged stands.

3. Role of Remote Sensing in Mapping Above-Ground Biomass

While biomass derived from field data measurements is the most accurate, it is not a practical approach for broad-scale assessments. This is where Remote Sensing has a key advantage. It can provide data over large areas at a fraction of the cost associated with extensive sampling and enables access to inaccessible places. Data from Remote Sensing satellites are available at various scales, from local to global, and from a number of different platforms. There are also different types of data, such as optical, radar and LiDAR, with each one having certain advantages over the others [15].

Optical Remote Sensing probably provides the best alternative to biomass estimation through field sampling due to its global coverage, repetitiveness and cost-effectiveness. Optical Remote Sensing data is available from a number of platforms, such as IKONOS, Quickbird, Worldview, SPOT, Sentinel, Landsat and MODIS. The spatial resolutions vary from less than one metre to hundreds of metres. Optical Remote Sensing data has been used by numerous researchers for biomass estimation [16–27].

Radar Remote Sensing has gained prominence for above-ground biomass estimation in recent years due to its cloud penetration ability as well as detailed vegetation structural information [15]. While airborne Synthetic Aperture Radar (SAR) systems have been operating for many years, space-borne systems such as Terra-SAR, ALOS and PALSAR have become available since 2000. This has enabled repetitiveness and cost-effectiveness. A large number of recent studies have explored the use of radar data for above-ground biomass estimation [28–44].

LiDAR is a relatively new technology that has found favour in biomass estimation. It has the ability to sample the vertical distribution of canopy and ground surfaces, providing detailed structural information about vegetation. This leads to more accurate estimations of basal area, crown size, tree height and stem volume. A number of studies have established strong correlations between LiDAR parameters and above-ground biomass [45–55].

4. Purpose of this Special Issue

Vegetation biomass plays a crucial role in understanding and monitoring ecosystem response and its contribution to the global carbon cycle. The recognition of forests as potential sinks of atmospheric carbon has resulted in numerous studies being conducted in estimating above-ground biomass or carbon stocks across varying scales. In addition, grassland biomass quantification is critical in understanding rangeland productivity as a resource for animal grazing. However, uncertainties in the Remote Sensing of AGB are high due to vegetation structural variations, heterogeneity of landscapes, seasonality and disproportionate data availability, among others. Recent developments in high resolution space-borne and air-borne satellite data have provided an opportunity to better estimate and map AGB across different spatial and temporal scales. The use of drones and UAVs has opened up avenues for super-fine resolution biomass estimation for targeted applications. Recent sensors, such as the Worldview series, now provide meter level spatial resolution while Sentinel and Landsat 8 provide free data for the whole world, opening up accessibility and more applications of Remote Sensing data, including for biomass estimation.

Remote sensing is a constantly evolving technology with new applications and methods being regularly introduced. This special issue was a call for the latest innovative methods and applications to map AGB at different scales. The range of topics included, but was not limited to, algorithm development and implementation, accuracy assessment, scaling issues (local-regional-global biomass mapping), integration of microwave (i.e., LiDAR) and optical sensors, forest biomass mapping, rangeland productivity and abundance (grass biomass, density, cover), bush encroachment biomass, seasonality and long term biomass monitoring, and climate change impacts and temporal monitoring.

5. Summary of Papers Published in this Special Issue

This special issue details results from a total of 15 papers that unpack the importance of Remote Sensing in biomass estimation and mapping across different spatial scales. The rich spectral, temporal and spatial information contained in satellite images has seen an improvement on productivity mapping under complex environmental conditions. Data sets used range from field spectrometers, multispectral and multi temporal images as well as microwave derived images. The data sets were combined with advanced machine learning algorithms and other state of the art processing techniques to reveal spatial and temporal biomass patterns.

A number of papers in this issue incorporated phenology in biomass estimation using Remote Sensing. Schucknecht et al. [56] used in-situ spatiotemporal biomass production and Remote Sensing to build a biomass model for the period 2001 to 2015. The phenology based seasonal NDVI was used as a proxy for biomass production and the model successfully predicted biomass at the end of the growing season. In a related study, NDVI Land Surface phenology derived from MODIS was used to model biomass in seasonal wetlands. The method was robust across different environmental conditions [57]. The use of high spatial resolution images in estimating carbon stocks across seasons was evaluated across two seasons in an abandoned agricultural land [58]. Pixel-scale vegetation indices derived from the dry season images yielded higher correlations with biomass than those derived from the wet season. The result confirms the saturation problem encountered using NDVI during biomass peak. Long term trends in biomass production are also critical in understanding the mechanism, direction and magnitude of climatic effects. Feng et al. [59] simulated potential productivity and actual productivity using Remote Sensing as well as climatic and anthropogenic data sets in Northern Tibetan Plateau between 1993 and 2011. Their results showed the importance of precipitation in regulating Net Primary Productivity.

Microwave Remote Sensing has also gained popularity in vegetation mapping in recent years. This issue presents three papers that used LiDAR and RADAR data to model vegetation biomass across different biomes. In a comparative study of modelling the biophysical attributes between deciduous and conifer forests, low density footprint LiDAR data successfully revealed spatial patterns of stem volume and total dry biomass [60]. Vaglio Laurin et al. [61] tested the potential of ALOS2 and NDVI to estimate above ground biomass. The objective was to solve the saturation problem experienced at low to moderate biomass levels when using JERS and ALOS SAR. The integrated SAR and NDVI improved estimation of biomass, while only a slight saturation was experienced at higher forest biomass levels. Pasture biomass was also modelled by integrating optical and X-Band radar data in an open savannah woodland [62]. The objective was to minimise the saturation problem experienced when using optical data as well as difficulties in separating dry matter from litter. TerraSAR-X (TSX) and image data from Landsat TM data from both dry and wet seasons could predict standing dry matter with high accuracy.

Optical multispectral data, with varying resolutions, was also used to estimate biomass. A multi-source satellite data approach was used to evaluate the potential of Remote Sensing in reducing biomass estimation error in an Alpine Meadow grassland [63]. Results showed that filtered MODIS NDVI data reduces biomass estimation errors and the error increases with the increasing spatial scale of investigation. Addition of Laser Altimeter data to Landsat TM in a multivariate modelling framework improved forest biomass predictions in Northern China [64]. Sibanda et al. [65] integrated texture metrics and the red edge derived indices from simulated Worldview-3 data to estimate biomass. Results showed the robustness of the model in reducing errors of prediction across all management treatments.

A total of 3 studies in this issue also applied hyperspectral data to improve biomass estimation. Jin et al. [66] used indices generated from the field spectroscopic data to calibrate the AquaCrop model for wheat yield and biomass estimation. In another study [67], fusion of Ultrasonic Sward Height and with narrow band normalized spectral index and simulated Worldview −2 data improved biomass prediction accuracy in grasslands with heterogeneous sward structure. Cheng et al. [68] showed the importance of dry matter indices in predicting biomass in canopy components of paddy rice as compared to chlorophyll indices, which saturate at high biomass levels.

Finally, this issue reports the utility of multi-angle data in improving biomass predictions. A study using multi angle CHRIS/PROBA data [69] showed that off nadir vegetation indices could predict forest biomass more accurately than the nadir derived indices. The result underscores the importance of Bidirectional Reflectance Distribution Function (BRDF) as a source of information than a source of uncertainty.

In summary, this Special Issue explored the role of Remote Sensing in estimating grassland, forest and woody biomass using a plethora of data and processing methods. Seasonality information was successfully built into biomass models with improved accuracies. The fusion of microwave and multispectral/hyperspectral data also reduced uncertainty errors in biomass estimation, especially in environments with complex canopy structure. Of critical importance is that the special issue highlighted methods and data sets that solves the problem of saturation in biomass estimation using the conventional vegetation indices [70]. The issue provides a platform for day to day methods and approaches to operationalize Remote Sensing in vegetation productivity management.

Acknowledgments: The authors thank the various contributors, reviewers and journal staff at Remote Sensing for making this Special Issue a success.

Author Contributions: The two authors contributed equally to all aspects of this paper.

Conflicts of Interest: The authors declare no conflict of interest.

References

1. Eggleston, H.S.; Buendia, L.; Miwa, K.; Ngara, T.; Tanabe, K. *IPCC Guidelines for National Greenhouse Gas Inventories Volume—IV Agriculture, Forestry and other Land-Use*; Institute of Global Environmental Strategies (IGES): Hayama, Japan, 2006.

2. Cairns, M.A.; Brown, S.; Helmer, E.H.; Baumgardner, G.A. Root biomass allocation in the world's upland forests. *Oecologia* **1997**, *111*, 1–11. [CrossRef] [PubMed]
3. Davidson, E.A.; Janssens, I.A. Temperature sensitivity of soil carbon decomposition and feedbacks to climate change. *Nature* **2006**, *440*, 165–173. [CrossRef] [PubMed]
4. Houghton, R.A.; Hall, F.; Goetz, S.J. Importance of biomass in the global carbon cycle. *J. Geophys. Res.* **2009**, *114*. [CrossRef]
5. Gibbs, H.K.; Brown, S.; Niles, J.O.; Foley, J.A. Monitoring and estimating tropical forest carbon stocks: Making REDD a reality. *Environ. Res. Lett.* **2007**, *2*, 1–13. [CrossRef]
6. Koch, B. Status and future of laser scanning, synthetic aperture radar and hyperspectral remote sensing data for forest biomass assessment. *ISPRS J. Photogramm. Remote Sens.* **2010**, *65*, 581–590. [CrossRef]
7. Angelsen, A.; Hofstad, O. Inputs to the Development of a National REDD Strategy in Tanzania. Available online: http://cf.tfcg.org/pubs/Angelsen2008REDD%20Tanzania%20rpt.pdf (accessed on 8 September 2017).
8. United Nations. Outcome of the Ad Hoc Working Group on longterm cooperative action under the convention. In Proceedings of the United Nations Framework Convention on Climate Change (2011), Durban, South Africa, 28 November–11 December 2011.
9. Nelson, B.W.; Mesquita, R.; Pereira, J.L.G.; Aquino de Sauza, S.G.; Batista, G.T.; Couto, L.B. Allometric regressions for improved estimate of secondary forest biomass in the Central Amazon. *For. Ecol. Manag.* **1999**, *117*, 149–167. [CrossRef]
10. Wang, X.C.; Ceulemans, R. Allometric relationships for below- and aboveground biomass of young Scots pines. *For. Ecol. Manag.* **2004**, *203*, 177–186.
11. Montès, N.; Gauquelin, T.; Badri, W.; Bertaudiere, V.; Zaoui, E.H. A non-destructive method for estimating above-ground forest biomass in threatened woodlands. *For. Ecol. Manag.* **2000**, *130*, 37–46. [CrossRef]
12. Navár, J. Allometric equations for tree species and carbon stocks for forests of Northwestern Mexico. *For. Ecol. Manag.* **2009**, *257*, 427–434. [CrossRef]
13. Brown, S.; Gillespie, A.R.; Lugo, A.E. Biomass estimation methods for tropical forests with applications to forest inventory data. *For. Sci.* **1989**, *35*, 881–902.
14. Basuki, T.M.; Van Laake, P.E.; Skidmore, A.K.; Hussin, Y.A. Allometric equations for estimating the above-ground biomass in tropical lowland dipterocarp forests. *For. Ecol. Manag.* **2009**, *257*, 1684–1694. [CrossRef]
15. Kumar, L.; Sinha, P.; Taylor, S.; Alqurashi, A.F. Review of the use of remote sensing for biomass estimation to support renewable energy generation. *J. Appl. Remote Sens.* **2015**, *9*. [CrossRef]
16. Thenkabail, P.S.; Stucky, N.; Griscom, B.W.; Ashton, M.S.; Diels, J.; van der Meer, B.; Enclona, E. Biomass estimations and carbon stock calculations in the oil palm plantations of African derived savannas using IKONOS data. *Int. J. Remote Sens.* **2004**, *25*, 5447–5472. [CrossRef]
17. Hyde, P.; Dubayah, R.; Walker, W.; Blair, J.B.; Hofton, M.; Hunsaker, C. Mapping forest structure for wildlife habitat analysis using multi-sensor (LiDAR, SAR/InSAR, ETM+, Quickbird) synergy. *Remote Sens. Environ.* **2006**, *102*, 63–73. [CrossRef]
18. Foody, G.M.; Boyd, D.S.; Cutler, M.E.J. Predictive relations of tropical forest biomass from Landsat TM data and their transferability between regions. *Remote Sens. Environ.* **2003**, *85*, 463–474. [CrossRef]
19. Steininger, M.K. Satellite estimation of tropical secondary forest above-ground biomass: Data from Brazil and Bolivia. *Int. J. Remote Sens.* **2000**, *21*, 1139–1157. [CrossRef]
20. Zheng, D.; Rademacher, J.; Chen, J.; Crow, T.; Bresee, M.; Le Moine, J.; Ryu, S. Estimating aboveground biomass using Landsat 7 ETM data across a managed landscape in northern Wisconsin, USA. *Remote Sens. Environ.* **2004**, *93*, 402–411. [CrossRef]
21. Baccini, A.; Friedl, M.A.; Woodcock, C.E.; Warbington, R. Forest biomass estimation over regional scales using multisource data. *Geophys. Res. Lett.* **2004**, *31*. [CrossRef]
22. Dong, J.; Kaufmann, R.K.; Myneni, R.B.; Tucker, C.J.; Kauppi, P.E.; Liski, J.; Buermann, W.; Alexeyev, V.; Hughes, M.K. Remote sensing estimates of boreal and temperate forest woody biomass: Carbon pools, sources, and sinks. *Remote Sens. Environ.* **2003**, *84*, 393–410. [CrossRef]
23. Calvão, T.; Palmeirim, J.M. Mapping Mediterranean scrub with satellite imagery: Biomass estimation and spectral behaviour. *Int. J. Remote Sens.* **2004**, *25*, 3113–3126. [CrossRef]

24. Lu, D. Aboveground biomass estimation using Landsat TM data in the Brazilian Amazon. *Int. J. Remote Sens.* **2005**, *26*, 2509–2525. [CrossRef]

25. Rahman, M.M.; Csaplovics, E.; Koch, B. An efficient regression strategy for extracting forest biomass information from satellite sensor data. *Int. J. Remote Sens.* **2005**, *26*, 1511–1519. [CrossRef]

26. Li, M.; Tan, Y.; Pan, J.; Peng, S. Modeling forest aboveground biomass by combining spectrum, textures and topographic features. *Front. For. China* **2008**, *3*, 10–15. [CrossRef]

27. Muukkonen, P.; Heiskanen, J. Estimating biomass for boreal forests using ASTER satellite data combined with standwise forest inventory data. *Remote Sens. Environ.* **2005**, *99*, 434–447. [CrossRef]

28. Ranson, K.J.; Sun, G. Mapping biomass of a northern forest using multifrequency SAR data. *IEEE Trans. Geosci. Electron.* **1994**, *32*, 388–396. [CrossRef]

29. Imhoff, M.L.; Johnson, P.; Holford, W.; Hyer, J.; May, L.; Lawrence, W.; Harcombe, P. BioSAR (TM): An inexpensive airborne VHF multiband SAR system for vegetation biomass measurement. *IEEE Trans. Geosci. Remote Sens.* **2000**, *38*, 1458–1462. [CrossRef]

30. Castel, T.; Guerra, F.; Caraglio, Y.; Houllier, F. Retrieval biomass of a large Venezuelan pine plantation using JERS-1 SAR data. Analysis of forest structure impact on radar signature. *Remote Sens. Environ.* **2002**, *79*, 30–41. [CrossRef]

31. Sun, G.; Ranson, K.J.; Kharuk, V.I. Radiometric slope correction for forest biomass estimation from SAR data in the Western Sayani Mountains, Siberia. *Remote Sens. Environ.* **2002**, *79*, 279–287. [CrossRef]

32. Santos, J.R.; Freitas, C.C.; Araujo, L.S.; Dutra, L.V.; Mura, J.C.; Gama, F.F.; Soler, L.S.; Sant'Anna, S.J.S. Airborne P-band SAR applied to the aboveground biomass studies in the Brazilian tropical rainforest. *Remote Sens. Environ.* **2003**, *87*, 482–493. [CrossRef]

33. Kasischke, E.S.; Melack, J.M.; Dobson, M.C. The use of imaging radars for ecological applications—A review. *Remote Sens. Environ.* **1997**, *59*, 141–156. [CrossRef]

34. Kasischke, E.S.; Goetz, S.; Hansen, M.C.; Ozdogan, M.; Rogan, J.; Ustin, S.L.; Woodcock, C.E. Temperate and boreal forests. In *Remote Sensing for Natural Resource Management and Environmental Monitoring*; Ustin, S.L., Ed.; John Wiley & Sons: Hoboken, NJ, USA, 2004; pp. 147–238.

35. Sarker, M.L.R.; Nichol, J.; Iz, H.B.; Ahmad, B.B.; Rahman, A.A. Forest biomass estimation using texture measurements of high resolution dual-polarization C-band SAR data. *IEEE Trans. Geosci. Remote Sens.* **2013**, *51*, 3371–3384. [CrossRef]

36. Buckley, J.R.; Smith, A.M. Monitoring grasslands with RADARSAT 2 quad-pol imagery. *IEEE Int. Geosci. Remote Sens. Symp.* **2010**, 3090–3093. [CrossRef]

37. Le Toan, T.; Beaudoin, A.; Riom, J.; Guyon, D. Relating forest biomass to SAR data. *IEEE Trans. Geosci. Remote Sens.* **1992**, *30*, 403–411. [CrossRef]

38. Santos, J.R.; Lacruz, M.S.P.; Araujo, L.S.; Keil, M. Savanna and tropical rainforest biomass estimation and spatialization using JERS-1 data. *Int. J. Remote Sens.* **2002**, *23*, 1217–1229. [CrossRef]

39. Balzter, H. Forest mapping and monitoring with interferometric synthetic aperture radar (InSAR). *Progr. Phys. Geogr.* **2001**, *25*, 159–177. [CrossRef]

40. Beaudoin, A.; Le Toan, T.; Goze, S.; Nezry, E.; Lopes, A.; Mougin, E.; Hsu, C.C.; Han, H.C.; Kong, J.A.; Shin, R.T. Retrieval of forest biomass from SAR data. *Int. J. Remote Sens.* **1994**, *15*, 2777–2796. [CrossRef]

41. Harrell, P.A.; Kasischke, E.S.; Bourgeau-Chavez, L.L.; Haney, E.M.; Christensen, N.L., Jr. Evaluation of approaches to estimating aboveground biomass in Southern pine forests using SIR-C data. *Remote Sens. Environ.* **1997**, *59*, 223–233. [CrossRef]

42. Le Toan, T.; Quegan, S.; Davidson, M.W.J.; Balzter, H.; Paillou, P.; Papathanassiou, K.; Plummer, S.; Rocca, F.; Saatchi, S.; Shugart, H.; et al. The BIOMASS mission: Mapping global forest biomass to better understand the terrestrial carbon cycle. *Remote Sens. Environ.* **2011**, *115*, 2850–2860. [CrossRef]

43. Lucas, R.M.; Mitchell, A.L.; Rosenqvist, A.; Proisy, C.; Melius, A.; Ticehurst, C. The potential of L-band SAR for quantifying mangrove characteristics and change: Case studies from the tropics. *Aquat. Conserv. Mar. Freshwater Ecosyst.* **2007**, *17*, 245–264. [CrossRef]

44. Solberg, S.; Astrup, R.; Gobakken, T.; Naesset, T.; Weydahl, D.J. Estimating spruce and pine biomass with interferometric X-band SAR. *Remote Sens. Environ.* **2010**, *114*, 2353–2360. [CrossRef]

45. Patenaude, G.; Hill, R.A.; Milne, R.; Gaveau, D.L.A.; Briggs, B.B.J.; Dawson, T.P. Quantifying forest above ground carbon content using LiDAR remote sensing. *Remote Sens. Environ.* **2004**, *93*, 368–380. [CrossRef]

46. Hall, S.A.; Burke, I.C.; Box, D.O.; Kaufmann, M.R.; Stoker, J.M. Estimating stand structure using discrete-return LiDAR: An example from low density, fire prone ponderosa pine forests. *For. Ecol. Manag.* **2005**, *208*, 189–209. [CrossRef]

47. Saremi, H.; Kumar, L.; Stone, C.; Melville, G.; Turner, R. Sub-compartment variation in tree height, stem diameter and stocking in a *Pinus. radiata* D. Don plantation examined using airborne LiDAR data. *Remote Sens.* **2014**, *6*, 7592–7609. [CrossRef]

48. Saremi, H.; Kumar, L.; Turner, R.; Stone, C. Airborne LiDAR derived canopy height model reveals a significant difference in radiata pine (*Pinus. radiata* D. Don) heights based on slope and aspect of sites. *Trees* **2014**, *28*, 733–744. [CrossRef]

49. Chen, Q.; Baldocchi, D.; Gong, P.; Kelly, M. Isolating individual trees in a savanna woodland using small footprint LiDAR data. *Photogramm. Eng. Remote Sens.* **2006**, *72*, 923–932. [CrossRef]

50. Lim, K.S.; Treitz, P.M. Estimation of above ground forest biomass from airborne discrete return laser scanner data using canopy-based quantile estimators. *Scand. J. For. Res.* **2004**, *19*, 558–570. [CrossRef]

51. Popescu, S.C. Estimating biomass of individual pine trees using airborne LiDAR. *Biomass Bioenergy* **2007**, *31*, 646–655. [CrossRef]

52. Næsset, E.; Økland, T. Estimating tree height and tree crown properties using airborne scanning laser in a boreal nature reserve. *Remote Sens. Environ.* **2002**, *79*, 105–115. [CrossRef]

53. Cohen, W.B.; Spies, T.A. Estimating structural attributes of Douglas-fir/western hemlock forest stands from Landsat and SPOT imagery. *Remote Sens. Environ.* **1992**, *41*, 1–17. [CrossRef]

54. Loos, R.; Niemann, O.; Visintini, F. Identification of partial canopies using first and last return LiDAR data. In Proceedings of the Our Common Borders—Safety, Security, and the Environment through Remote Sensing, Ottawa, ON, Canada, 28 October–1 November 2007.

55. García, M.; Riano, D.; Chuvieco, E.; Danson, M. Estimating biomass carbon stocks for a Mediterranean forest in central Spain using LiDAR height and intensity data. *Remote Sens. Environ.* **2010**, *114*, 816–830. [CrossRef]

56. Schucknecht, A.; Meroni, M.; Kayitakire, F.; Boureima, A. Phenology-based biomass estimation to support rangeland management in semi-arid environments. *Remote Sens.* **2017**, *9*, 463. [CrossRef]

57. Lumbierres, M.; Méndez, P.; Bustamante, J.; Soriguer, R.; Santamaría, L. Modeling biomass production in seasonal wetlands using MODIS NDVI land surface phenology. *Remote Sens.* **2017**, *9*, 392. [CrossRef]

58. Liu, N.; Harper, R.; Handcock, R.; Evans, B.; Sochacki, S.; Dell, B.; Walden, L.; Liu, S. Seasonal timing for estimating carbon mitigation in revegetation of abandoned agricultural land with high spatial resolution remote sensing. *Remote Sens.* **2017**, *9*, 545. [CrossRef]

59. Feng, Y.; Wu, J.; Zhang, J.; Zhang, X.; Song, C. Identifying the relative contributions of climate and grazing to both direction and magnitude of alpine grassland productivity dynamics from 1993 to 2011 on the Northern Tibetan Plateau. *Remote Sens.* **2017**, *9*, 136. [CrossRef]

60. Awaya, Y.; Takahashi, T. Evaluating the differences in modeling biophysical attributes between deciduous broadleaved and evergreen conifer forests using low-density small-footprint LiDAR data. *Remote Sens.* **2017**, *9*, 572. [CrossRef]

61. Vaglio Laurin, G.; Pirotti, F.; Callegari, M.; Chen, Q.; Cuozzo, G.; Lingua, E.; Notarnicola, C.; Papale, D. Potential of ALOS2 and NDVI to estimate forest above-ground biomass, and comparison with Lidar-derived estimates. *Remote Sens.* **2017**, *9*, 18. [CrossRef]

62. Schmidt, M.; Carter, J.; Stone, G.; O'Reagain, P. Integration of optical and X-band radar data for pasture biomass estimation in an open savannah woodland. *Remote Sens.* **2016**, *8*, 989. [CrossRef]

63. Meng, B.; Ge, J.; Liang, T.; Yang, S.; Gao, J.; Feng, Q.; Cui, X.; Huang, X.; Xie, H. Evaluation of remote sensing inversion error for the above-ground biomass of alpine meadow grassland based on multi-source satellite data. *Remote Sens.* **2017**, *9*, 372. [CrossRef]

64. Liu, K.; Wang, J.; Zeng, W.; Song, J. Comparison and evaluation of three methods for estimating forest above ground biomass using TM and GLAS data. *Remote Sens.* **2017**, *9*, 341. [CrossRef]

65. Sibanda, M.; Mutanga, O.; Rouget, M.; Kumar, L. estimating biomass of native grass grown under complex management treatments using WorldView-3 spectral derivatives. *Remote Sens.* **2017**, *9*, 55. [CrossRef]

66. Jin, X.; Kumar, L.; Li, Z.; Xu, X.; Yang, G.; Wang, J. Estimation of winter wheat biomass and yield by combining the AquaCrop model and field hyperspectral data. *Remote Sens.* **2016**, *8*, 972. [CrossRef]

67. Moeckel, T.; Safari, H.; Reddersen, B.; Fricke, T.; Wachendorf, M. Fusion of ultrasonic and spectral sensor data for improving the estimation of biomass in grasslands with heterogeneous sward structure. *Remote Sens.* **2017**, *9*, 98. [CrossRef]
68. Cheng, T.; Song, R.; Li, D.; Zhou, K.; Zheng, H.; Yao, X.; Tian, Y.; Cao, W.; Zhu, Y. Spectroscopic estimation of biomass in canopy components of paddy rice using dry matter and chlorophyll indices. *Remote Sens.* **2017**, *9*, 319. [CrossRef]
69. Wang, Q.; Pang, Y.; Li, Z.; Sun, G.; Chen, E.; Ni-Meister, W. The potential of forest biomass inversion based on vegetation indices using multi-angle CHRIS/PROBA data. *Remote Sens.* **2016**, *8*, 891. [CrossRef]
70. Mutanga, O.; Skidmore, A.K. Narrow band vegetation indices overcome the saturation problem in biomass estimation. *Int. J. Remote Sens.* **2004**, *25*, 3999–4014. [CrossRef]

remote sensing

MDPI

Article

The Potential of Forest Biomass Inversion Based on Vegetation Indices Using Multi-Angle CHRIS/PROBA Data

Qiang Wang [1,2,3,*], Yong Pang [4], Zengyuan Li [4], Guoqing Sun [5], Erxue Chen [4] and Wenge Ni-Meister [3]

[1] Harbin Institute of Technology, School of Electronics Information Engineering, Harbin 150001, China
[2] Department of Surveying Engineering, Heilongjiang Institute of Technology, Harbin 150040, China
[3] Department of Geography, Hunter College of CUNY, New York, NY 10065, USA; wnimeist@hunter.cuny.edu
[4] Institute of Forest Resource Information Techniques, Chinese Academy of Forestry, Beijing 100091, China; pangy@ifrit.ac.cn (Y.P.); lizy@caf.ac.cn (Z.L.); chenrx@caf.ac.cn (E.C.)
[5] Department of Geography, University of Maryland, College Park, MD 20742, USA; guoqing.sun@gmail.com
[*] Correspondence: wangqiang310108@aliyun.com; Tel.: +86-451-8802-8725

Academic Editors: Lalit Kumar, Onisimo Mutanga and Prasad S. Thenkabail
Received: 27 August 2016; Accepted: 19 October 2016; Published: 28 October 2016

Abstract: Multi-angle remote sensing can either be regarded as an added source of uncertainty for variable retrieval, or as a source of additional information, which enhances variable retrieval compared to traditional single-angle observation. However, the magnitude of these angular and band effects for forest structure parameters is difficult to quantify. We used the Discrete Anisotropic Radiative Transfer (DART) model and the Zelig model to simulate the forest canopy Bidirectional Reflectance Distribution Factor (BRDF) in order to build a look-up table, and eight vegetation indices were used to assess the relationship between BRDF and forest biomass in order to find the sensitive angles and bands. Further, the European Space Agency (ESA) mission, Compact High Resolution Imaging Spectrometer onboard the Project for On-board Autonomy (CHRIS-PROBA) and field sample measurements, were selected to test the angular and band effects on forest biomass retrieval. The results showed that the off-nadir vegetation indices could predict the forest biomass more accurately than the nadir. Additionally, we found that the viewing angle effect is more important, but the band effect could not be ignored, and the sensitive angles for extracting forest biomass are greater viewing angles, especially around the hot and dark spot directions. This work highlighted the combination of angles and bands, and found a new index based on the traditional vegetation index, Atmospherically Resistant Vegetation Index (ARVI), which is calculated by combining sensitive angles and sensitive bands, such as blue band 490 nm/$-55°$, green band 530 nm/$55°$, and the red band 697 nm/$55°$, and the new index was tested to improve the accuracy of forest biomass retrieval. This is a step forward in multi-angle remote sensing applications for mining the hidden relationship between BRDF and forest structure information, in order to increase the utilization efficiency of remote sensing data.

Keywords: multi-angle remote sensing; forest structure information; vegetation indices; forest biomass; Bidirectional Reflectance Distribution Factor

1. Introduction

Emissions from land surfaces are considered the most uncertain component of the global carbon cycle. Forest structure is an important factor in the estimation of energy and carbon fluxes between land and the atmosphere, and in the biodiversity of ecosystems. Forest structure is determined by several factors, including species composition and the three-dimensional distribution of leaves/needles, canopy size, tree height, and woody biomass [1]. Of primary importance is aboveground standing biomass, which represents an important constraint on process-based biogeochemical models, and can be used to validate these models; additionally, it is estimated in the field from basal area and canopy height using empirically-derived allometric functions [2,3]. It is important to accurately extract biomass using remote sensing data.

There are some indirect biomass measurements from remote sensors. Optical data provide the best global coverage and information on forest type, crown cover, Leaf Area Index LAI, etc.; however, they provide limited structural information. Synthetical Aperture Radar SAR provides volumetric scattering related to fresh biomass, but water contents, forest spatial structure, and terrain slopes cause errors and "saturation" [4]. Light Detection and Ranging (LiDAR) is an active remote-sensing laser technology, capable of providing detailed, spatially explicit, three-dimensional information on vegetation structure; however, regional or global repeat coverage will not be available in the near future, because its imaging method [5,6]. Extending LiDAR and field samples for regional or global coverage should make use of other image data.

Most traditional optical sensor observations are made near or normalized to the nadir, although they provide two-dimensional information of the horizontal extent of canopies, and allow us to measure vegetation cover types and density. They do not provide three-dimensional information on vegetation structure [7]. This requires the capability to remotely measure the vertical and spatial distribution of forest structural parameters, which are needed for more accurate inversion of aboveground standing biomass over regional, continental, and global scales. Compared with traditional nadir-viewed remote sensing, a multi-angle optical sensor can provide three-dimensional structural information of a forest through different directional observations [8]. The multi-angle information of the radiometric signal is often treated as noise, and is then removed through an angle normalization procedure [9]. However, canopy structure and disturbance information can be gained more accurately from multi-angle remote sensing [10,11]. There have been some studies demonstrating the utility of multiple angle measurements [12,13].

Some prediction methods, such as multivariable regression, neural networks, and nearest neighbor, were used to inverse forest biomass, of which input variables include spectral reflectance, vegetation indices, derived products (leaf area index, crown closure), etc. [14]. Many spectral vegetation indices (VIs) are designed to assess vegetation photosynthetic activities and biomass on the land surface [15], but the accuracy is low. It has already been demonstrated that VIs not only minimize, but, in fact, can also exaggerate the impacts of the solar zenith and view angle [16–18]. VIs do suffer from directionality, not only because of the reflectance anisotropy of surfaces, due to vegetation type and background contributions [19–21], but also because of the inherent viewing geometry of sensors, including canopy structure, tree height, stand density, shadowing, and local illumination, resulting from topography and sun position.

Some angle VIs, such as the hot spot–dark spot difference index (HDDI705), HDDI750, Hot spot–Dark Spot (HDS), and Normalized Difference between Hot spot and Dark spot (NDHD), were designed in order to characterize three-dimensional (3-D) vegetation structure [22,23]. In this study, the angle VIs were designed using all of the angles and bands of a look-up table (LUT), simulated by the Discrete Anisotropic Radiative Transfer (DART) model, to estimate forest biomass, and then the hidden relationship between the observation angles and bands and forest biomass was determined in order to get the sensitive angles and bands. Finally, Compact High Resolution Imaging Spectrometer (CHRIS) sensor data and field measurement were used to test the results, and the optimized angle VIs were built using the sensitive angle and band to extract forest biomass in order to increase the utilization efficiency of the multi-angle remote sensing data.

Remote Sens. **2016**, *8*, 891

2. Materials and Methods

2.1. Study Site

In this paper, the study area is located in Howland, Maine, USA (45.17°N–45.26°N, 68.65°W–68.81°W), as shown in Figure 1. The forest of Howland has tremendous ecological value and plays host to researchers as a vital site in research on how forests remove carbon dioxide from the atmosphere and store it in plant biomass. The natural stands in this northern hardwood boreal transitional forest consist of hemlock–spruce–fir, aspen–birch, and hemlock–hardwood mixtures. Common species include quaking aspen, paper birch, eastern hemlock, red spruce, balsam fir, and red maple. The regional features are relatively level, where the elevation ranges from 20 m to 158 m within an area covered by the Compact High Resolution Imaging Spectrometer onboard the Project for On-board Autonomy (CHRIS/PROBA) data used in this study. Additionally, almost 450 ha of the surrounding area consist of bogs and other wetlands. Generally, the soils throughout the forest are glacial tills, acidic in reaction, with a low fertility and high organic composition. The climate is chiefly cold, humid, and continental. Summer maximum temperatures of 30 °C are common, and winter minimums can reach −30 °C. The mean annual air temperature (1996–2010) at Howland tower is 6.7 °C and average annual rainfall (1950–2000) is 1050 mm.

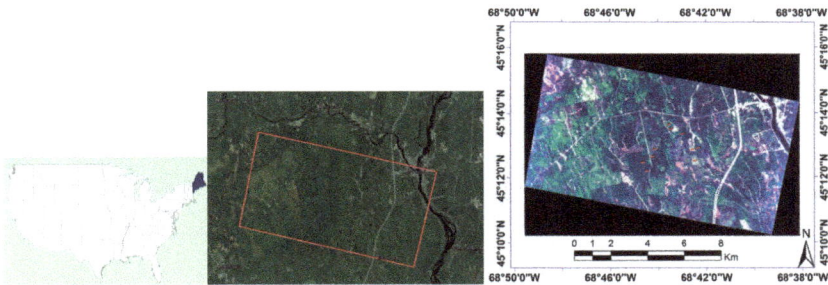

Figure 1. The spatial coverage of CHRIS data used in the Howland study area, and the geographical locations of field samplings (red square).

2.2. Field Samplings

Howland field measurements were conducted from 2009 to 2011 by the NASA Deformation, Ecosystem Structure and Dynamics of Ice (DESDynI) project. Twenty-four plots, measuring 1-ha (200 m × 50 m), were established in 2009 and 2010. Each plot was divided into neighboring 25 m × 25 m subplots, which can be aggregated into 96 subplots of 0.25-ha (50 m × 50 m). In total, 40 subplots were selected for testing the model after removing subplots outside of CHRIS image data, and the geographical locations are shown in Figure 1. In the paper the location of the sampling plots was measured using Real-time kinematic (RTK). For each subplot, Diameter at Breast Height (DBH, 1.3 m above ground), species, height of three highest trees, and typical tree crown information (crown width, live branch base height) were recorded. The biomass of subplots was calculated through the diameter-based allometric equations coming from the comprehensive report of USDA (United States Department of Agriculture) on North American forest given by Jenkins et al. [24]. Biomass was first calculated for each tree, and then total biomass was aggregated to subplot levels.

2.3. Satellite Data

The CHRIS sensor on PROBA provides spectral contiguous bands in the spectral wavelength range from 415 nm to 1050 nm, and a 17-m ground-sampling distance. PROBA is an experimental ESA space platform, which enables the sensor to capture images from five viewing angles. They were five CHRIS Fly-by Zenith Angles (FZA), namely: +36°, −36°, +55°, −55° and 0°. Five CHRIS modes were acquired for different observation tasks; CHRIS Mode 3 (Land) data were acquired over Howland. Data specifications are shown in Table 1, and the viewing geometry is shown in Figure 2. The sensor sweeps in the opposite direction, between two adjacent images, which results in the reversal of images of the Fly-by-Zenith Angle (FZA) ±36°, ±55° [25,26]; thus, it was rotated by the ENVI software. The swath effect of the images was eliminated using the official software, HDFclean, supplied by the European Space Agency (ESA). The geometric correction relied on ENVI orthorectification, considering the viewing geometry and geometric distortion due to the sensor, as well as the platform and topography, which used an ETM+ image from 23 August 2007, with a size of 28.5 × 28.5 m and a digital elevation model (DEM) of 1:50,000 [27].

Table 1. CHRIS specifications.

Study Area	Image Area	Sampling	Central Latitude and Longitude	Viewing Angles	Spectral Bands	Sun Zenith	Sun Azimuth
Howland	14 × 14 km (744 × 748 pixels)	17 × 17 m @ 662 km altitude	42°2′24″ 127°47′24″	5 nominal angles @ −55°, −36°, 0°, +36°, +55°	18 bands, 442~1019 nm with 6~11 nm width	44.3°	159°

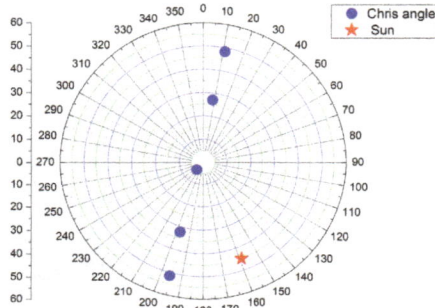

Figure 2. Polar plot of the CHRIS image-viewing geometry.

The atmospheric correction of the CHRIS radiation data was performed using the fast line-of-sight atmospheric analysis of spectral hypercubes (FLAASH) model of ENVI, a MODerate resolution atmospheric TRANsmission (MODTRAN4)-based approach to remove scattering and absorption effects of atmosphere constituents for nadir and non-nadir viewing instruments, which enables the processing of data from tiled sensors by considering for varying path lengths through the atmosphere and varying transmittances. One distinct characteristic of this approach is that FLAASH is able to correct paths scattered by radiation and other adjacent effects. The surface reflectance generated by FLAASH represents BRDF through band response functions, imaging time, solar and sensor positions, and the location of the study area [28,29].

2.4. The Building of a Look-Up Table

The Look Up Table (LUT) database should include all kinds of combinations of environmental conditions and forest structures. Firstly, the forest growth model, ZELIG, is used to generate forest scenes, which are used to drive the Discrete Anisotropic Radiative Transfer (DART) model, as described in References [30,31]. ZELIG is an individual tree simulator that simulates the establishment, annual diameter growth, and mortality of each tree on an array of model plots. Table 2 shows the input parameters related to the growth and environmental response of the dominant tree species used in the Howland simulation. Because of the various random processes in the forest growth model, the ZELIG model runs five times to generate forest stands from 0 to 500 years with increments of five years. Secondly, from the forest stands, the parameters used as inputs for the DART model were determined, including tree species, tree location, height below crown, height within crown, diameter at breast height (BDH), crown type, crown height, crown geometry parameters, and leaf are index (LAI). The BRDF was simulated using the DART model and then the biomass was calculated using tree height and BDH.

Table 2. Input parameters of growth and environmental response of dominant tree species used in the ZELIG simulation.

Species	A_{max}	D_{max}	H_{max}	G	DD_{min}	DD_{max}	Light	Drt	Nutri
aspen	150	60	2500	400	800	2300	5	2	2
birch	150	60	3000	400	1000	3000	5	2	2
spruce	300	110	3300	60	550	1800	1	5	3
fir	200	70	3000	50	500	1800	1	5	1

A_{max}: Maximum age, D_{max}: Maximum diameter at breast height (cm), H_{max}: Maximum height(m), G: Growth rate scaling coefficient, DD_{min} and DD_{max}: Minimum and maximum growing degree-days (5.56° base), Light: Shade-tolerance class (rank: 1 = very shade tolerant, 5 = very intolerant), Drt: Drought tolerance (rank: 1 = very drought-intolerant, 5 = very drough-tolerant), Nutri: Soil fertility response class (1 = nutrient stress intolerant, 3 = tolerant).

The DART model was developed by the Center for the Study of the BIOsphere from Space, and simulates the radiation transfer in any complex 3D scene, using an innovative multispectral approach (ray tracing, exact kernel, and discrete ordinate techniques) over the entire optical domain [32]. The results are accurate; however, they demand a considerable number of calculations. The DART model simulates natural 3D forested scenes, and accurately reflects the effects of different factors on BRDF, which is affirmed by the international radiation transfer model intercomparison (RAMI3). In the calculation of forest BRDF, 29 local incidence angles from −70° to 70°, with 5° intervals, are used. The solar zenith angle was calculated using information from the CHRIS data. The main parameters are presented in Table 3.

Table 3. Parameters of the look-up table.

Parameters	Range	Comments
Solar zenith	44.3°	The same as CHRIS data
View zenith	0°–70°	5° interval
Azimuth angle	0°, 180°	Principal and vertical plane
Bands	18 bands	The same as CHRIS data
Mean tree height	3.25–16.93 (m)	From Zelig model
Leaf area index	0.13–9.7	From Zelig model
Biomass	0.166–41.694 (ton/ha)	From Zelig model

2.5. Data Analysis

The following eight vegetation indices were selected for data analyses in order to represent forest structure and plant physiology, and are categorized into traditional VIs (1–6) and angle VIs (7–8) (Table 4). The equations and references are listed in Table 4. Because the reflectance properties of a land surface are anisotropic in nature, the vegetation indices are assumed to be sensitive to changing viewing angles, depending on the spectral bands used and the degree of surface anisotropy present in the observed scene [16]. Because forest surface anisotropy is affected by the canopy size, tree height, stands density, and shadow, there is a potential relationship between vegetation index and forest biomass. In the paper, the eight VIs were calculated using two types of reflectance, one from LUT and the other from CHRIS images. The pixels of the CHRIS images were selected according to the field measurement locations. CHRIS Land Mode 3 has 18 bands, and was categorized into four wavelength ranges: R_{BLUE} (442 nm, 490 nm), R_{GREEN} (530 nm, 551 nm, 570 nm), R_{RED} (631 nm, 661 nm, 672 nm, 697 nm, 703 nm, 709 nm, 742 nm, 748 nm), and R_{NIR} (781 nm, 872 nm, 895 nm, 905 nm, 1019 nm).

Firstly, in order to compare the off-nadir with the nadir, the six traditional VIs were calculated using the nadir angle reflectance from LUT, using the formula defined in Table 4, for example, the Simple Ratio Index (SRI) was calculated for $5 \times 8 = 40$ values. Then, the VIs were calculated using the 29 local incidence angle reflectances, the 29 angle reflectances can make up more than one value; for example, the SRI was calculated for $5 \times 8 \times 29 \times 29 = 33{,}640$ values, because there are five bands in the near infrared wavelength range and eight bands in the red wavelength range, and each band has 28 angles. The correlation coefficient (R) between the vegetation index and the biomass of LUT was calculated. RMSE (Root Mean Square Error) and rRMSE (Relative Root Mean Square Error) were calculated for each VI, rRMSE $= \mathrm{RMSE}/\overline{W_i}$, where $\overline{W_i}$ is average of $\hat{W_i}$ and $\hat{W_i}$ is the biomass estimated by a nonlinear regression model. According to the allometric functions and the nonlinear relationship between biomass and VIs, the following nonlinear regression model was used. $\ln Y = A + B \ln X$, where Y is biomass, X is VIs, and A and B are coefficients. The minimum RMSE of SRI was determined and the corresponding wavelengths and angles from the 40 and 33,640 values were determined. Then, the ground measure biomass and CHRIS image data were used to calculate the five VIs, using the same method, in order to test the advantages of multi-angle remote sensing. The VIs, built by a sensitive angle and band, were used to calculate the corresponding A and B, and then the model was used to inverse the biomass. However, the two angle VIs, HDS and NDHD, were calculated using the same band and different angles, as defined in Reference [22].

Secondly, the correlation coefficient between BRDF and the biomass was calculated, based on the LUT, and 18 curves were calculated.

Table 4. Eight Hyperion-derived vegetation indices used in the study.

Vegetation Index	Formula	Description	Reference
Traditional vegetation index			
SRI Simple Ratio Index	R_{NIR}/R_{RED}	Measure of green vegetation cover.	Tucker (1979) [33]
NDVI Normalized Difference Vegetation Index	$(R_{NIR} - R_{RED}) / (R_{NIR} + R_{RED})$	Measure of green vegetation cover.	Tucker (1979)
PVI perpendicular vegetation index	$\frac{(R_{NIR} - aR_{RED} - b)}{(\sqrt{1+a^2})}$ where a = 0.96916, b = 0.084726, and L = 0.5	To deduce information about soil surface conditions based on soil background line	Richardson and Everitt (1992) [34]
SAVI A soil-adjusted vegetation index	$\frac{(R_{NIR} - R_{RED})(1+0.5)}{(R_{NIR} + R_{RED} + 0.5)}$	Similar as NDVI while correcting for high soil reflectance	Huete (1988) [35]
EVI Enhanced Vegetation Index	$2.5 \left(\frac{(R_{NIR} - R_{RED})}{(R_{NIR} + 6R_{RED} - 7.5R_{BLUE} + 1)} \right)$	More sensitive to plant canopy differences and reduce the influence of atmospheric conditions	Huete et al. (2002) [36]
ARVI Atmospherically Resistant Vegetation Index	$\frac{(R_{NIR} - (2R_{RED} - R_{BLUE}))}{(R_{NIR} + (2R_{RED} - R_{BLUE}))}$	Similar as NDVI while being less sensitive to aerosol effects	Kaufman and Tanre (1992) [37]
Angle vegetation index			
HDS Hot spot–Dark Spot index	$(R_{HS} - R_{DS})/R_{DS}$	Measure of plant canopy structure information	Chen et al., (2003) [20]
NDHD Normalized Difference between Hot spot and Dark spot	$(R_{HS} - R_{DS}) / (R_{HS} + R_{DS})$	Measure of plant canopy structure information while reduce the influence of leaf optical properties	Chen et al., (2003)

Thirdly, the VIs were calculated using the 29 angles from $-70°$ to $70°$ with $5°$ intervals and 18 bands were used to find the sensitive angles and bands. For example, SRI was calculated to be $18 \times 18 \times 29 \times 29 = 272{,}484$ different values, R_{RED} was not limited in the red wavelength range, and 18 bands $\times$ 29 angles = 522 values were determined, this was also the case for R_{NIR}. The R correlation between each vegetation index and biomass was established and RMSE and rRMSE were calculated. The minimum RMSE of SRI was found, and the corresponding wavelength and angle from the 272,484 values was determined. The other VIs were processed using the same method, and then ground measure biomass and CHRIS image data were used to calculate the six VIs, using the same method. However, the two angle VIs, HDS and NDHD, have the same results as SRI and NDVI if calculated without the band limit, thus, the two VIs were not calculated again. This will help us to understand the viewing angle and the direction effects on biomass.

3. Results

3.1. Comparison Nadir Angle with Off-Nadir Angle

In Table 5, every traditional vegetation index was calculated using the nadir angle and the off-nadir angle, as noted in the Comment column. The "Red/angle" represents the optimal red wavelength and angle when the RMSE is minimal. The results show that the off-nadir RMSE and rRMSE were reduced by an average of about 20% when compared with the nadir, and the greatest reduction is 64% for NDVI and 61% for ARVI. The R square was improved from 0.398 to 0.943 for ARVI. For the other two angle vegetation indices, HDS and NDHD, the RMSE and rRMSE were also very small, although not the minimum values. The off-nadir RMSE of eight VIs were smaller than the nadir, which shows that multi-angle remote sensing can reflect the three-dimensional structural information of forests and improve the accuracy of biomass retrieval, relative to the traditional single angle.

Table 5. Comparison of nadir with off-nadir using the LUT.

	R Square	Red/Angle	Near-Infrared/Angle	Blue/Angle	RMSE	rRMSE	Comment
SRI	0.880	709/70	742/45		27.732	0.236	off-nadir
	0.327	709/0	905/0		46.644	0.419	nadir
NDVI	0.956	709/−65	748/45		16.850	0.142	off-nadir
	0.450	709/0	872/0		45.122	0.340	nadir
PVI	0.823	709/−70	742/45		34.141	0.287	off-nadir
	0.045	709/0	872/0		52.681	0.505	nadir
SAVI	0.834	709/−70	742/40		35.452	0.295	off-nadir
	0.114	709/0	872/0		50.922	0.481	nadir
EVI	0.654	709/−55	742/55	442/−55	40.783	0.349	off-nadir
	0.131	709/0	872/0	442/0	50.434	0.474	nadir
ARVI	0.943	703/−70	742/45	490/−65	18.611	0.157	off-nadir
	0.398	709/0	872/0	490/0	45.304	0.404	nadir
HDS	0.931	709/−40 709/45			21.070	0.176	off-nadir
NDHD	0.893	709/40 709/10			24.275	0.203	off-nadir

In order to prove the results, ground measure and CHRIS image data were used to find the optimal wavelength and angle, for both the nadir and off-nadir. In Table 6, the results show that, compared with the nadir, the off-nadir RMSE and rRMSE were reduced by an average of about 10%, and the maximum decrease is 23% for ARVI. The R square was also improved, from 0.612 to 0.697 for ARVI. The off-nadir RMSE of eight VIs were smaller than the nadir, and the result prove that multi-angle remote sensing can provide more structural information on forests than nadir angle observation.

In addition to the above-mentioned results, from Tables 5 and 6 we can see the optimal wavelength focused on some bands, for example, in Table 5, the optimal red wavelengths are 709 nm and 703 nm, the optimal near-infrared wavelengths are 742 nm and 748 nm, and the optimal blue wavelengths are 490 nm and 442 nm. In Table 6, the optimal red wavelengths are 709 nm and 672 nm, the optimal near-infrared wavelengths are 742 nm and 748 nm, and the optimal blue wavelength is 490 nm. The results show that the structural information of forests is more sensitive to angle information. In order to test the results, the correlation coefficient between each band reflectance and biomass was calculated and are shown in Figure 3.

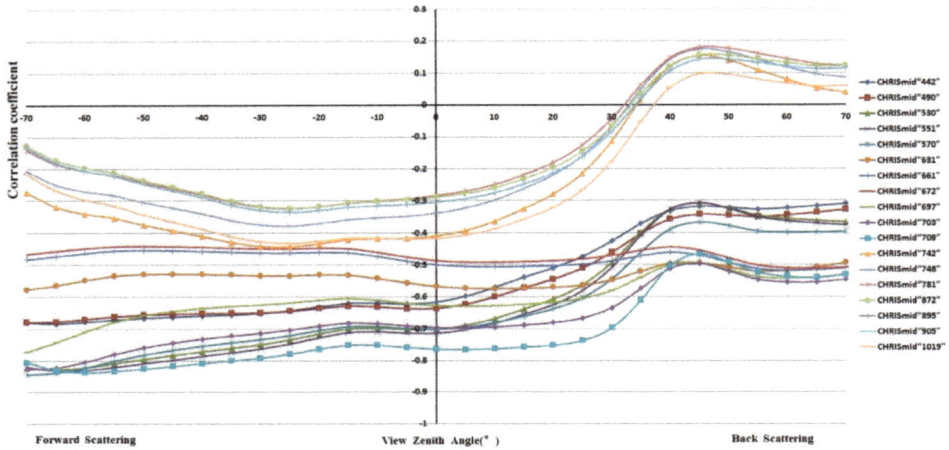

Figure 3. Correlation coefficients between all band reflectances and biomass.

Table 6. View-illumination effects for the nadir with the off-nadir angle, using ground measures and CHRIS data.

	R Square	Red/Angle	Near-Infrared/Angle	Blue/Angle	RMSE	rRMSE	Comment
SRI	0.712	672/−36	748/−36		54.518	0.482	off-nadir
	0.651	672/0	742/0		57.929	0.534	nadir
NDVI	0.643	672/−36	742/−55		55.926	0.535	off-nadir
	0.551	697/0	742/0		61.570	0.599	nadir
PVI	0.312	709/55	742/0		76.047	0.761	off-nadir
	0.232	709/0	742/0		79.776	0.843	nadir
SAVI	0.643	672/−36	742/−55		55.932	0.535	off-nadir
	0.551	697/0	742/0		61.573	0.599	nadir
EVI	0.724	672/−36	742/−36	490/−36	47.148	0.434	off-nadir
	0.677	672/0	742/0	490/0	52.919	0.489	nadir
ARVI	0.697	672/−36	742/−55	490/55	45.427	0.421	off-nadir
	0.612	672/0	742/0	490/0	58.663	0.565	nadir
HDS	0.572	672/−36 672/55			57.018	0.527	off-nadir
NDHD	0.618	672/36 672/−36			59.368	0.542	off-nadir

Note: Wavelength unit is nanometer, angle unit is degrees.

3.2. Band Effects on Biomass

From Tables 5 and 6, we can see that the structural information of forests is more sensitive to angle information compared with band information. In order to test the results, the correlation coefficient between each band reflectance and biomass was calculated and are shown in Figure 3. In Figure 3, the horizontal axis is the view zenith angle and the vertical axis is the correlation coefficient between each band reflectance and biomass; the 18 curves represent the 18 bands of a CHRIS image. According to the curve shape, the 18 bands were separated into four groups. The curves in the near-infrared band group, of which the wavelengths are more than 709 nm, have a similar shape. The blue bands, 442 nm and 490 nm, have a similar shape, the green bands, 530 nm, 551 nm, and 709 nm have a similar shape, and the red bands, 631 nm, 661 nm, 972 nm, 697 nm, 703 nm, and 709 nm also have a similar shape. There is a great difference between the near infrared group and the other three groups, and, in the visible band group, the blue, green, and red band curves also have small differences. The correlations are much lower than the VIs; however, the highest correlations appear in the forward scattering direction (shadowed canopy) within the red band range. Because of the greater red reflectance dependence on its formulation when compared to the other indices, EVI showed sensitivity to view angle, view direction, and solar illumination. EVI has the highest R^2, as shown in Table 6.

There were greater reflectance differences in the CHRIS viewing directions in the NIR for Howland Forest (Figure 4a). Despite the differences in the magnitude of reflectance, results were generally consistent with the DART modeled spectra (Figure 4b). In both figures, the reflectance increased from the forward scattering to the backscattering, as expected, and the reflectance of each band within the four groups varied with angle, and is the same as is shown in Figure 4b, which can be tested to show that each group has a similar shape.

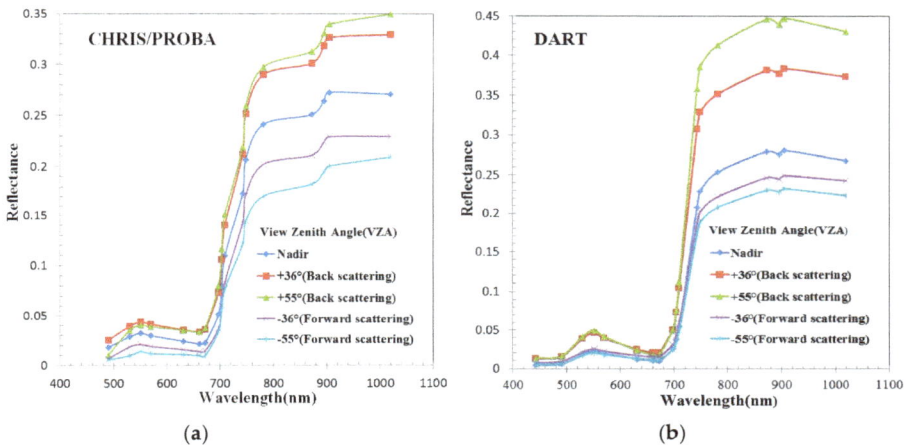

Figure 4. (a) Measured CHRIS/PROBA reflectance spectra; (b) simulated DART reflectance spectra of Howland for CHRIS/PROBA view zenith angles and directions.

3.3. View-Illumination Effects on Biomass

The above-mentioned results highlight that view-illumination is the main factor effect BRDF value in the change of forest biomass, thus, the sensitive angle should been found in order to improve the accuracy of forest biomass retrieval. Then, six VIs were calculated using all angles and all bands, without limiting the band range, in considering the influence of the band and the results are shown in Tables 7 and 8. In Table 7, LUT data were used, and, in Table 8, ground measure and CHRIS image data were used.

Table 7. Optimal bands and angles using LUT.

	R Square	Red/Angle	Near-Infrared/Angle	Blue/Angle	Green/Angle	RMSE	rRMSE
SRI	0.958	709/−70	781/45			15.940	0.135
NDVI	0.958		748/45		551/−65	16.432	0.139
PVI	0.945	709/−45		442/20		17.375	0.147
SAVI	0.910			490/−55	570/−50	18.734	0.158
EVI	0.941	709/45 709/−70			551/40	16.003	0.135
ARVI	0.964		781/45 781/60		551/−65	14.654	0.133

Note: Wavelength unit is nanometers, angle unit is degrees.

In order to compare Table 5 to Table 7, the RMSE and rRMSE of each VI were reduced, comparing the off-nadir with a limited band range with the off-nadir without limiting the band range. The greatest values are from 45.493 to 17.375 for PVI. The R square of each vegetation index was greater than 0.9. The results also show that multi-angle remote sensing can improve the accuracy of forest biomass retrieval. Table 7 also shows that the backward angles, 20°, 40°, 45°, and 60°, and the forward angles, −45°, −50°, −55°, −65°, and −70°, were found, and the most common backward angle value was 45° and the forward angles were –70° and −65°. These angles are the "hot" spot and "dark" spot angles, because the sun zenith angle is 44.3°. In addition, the vegetation index with the smallest RMSE is still ARVI, the angles of which are 45°, 60°, and −65°. In Table 5, the backward angles are 10°, 40°, 45°, and 70°, and the forward angles are −40°, −60°, −65°, and −70°, and the most common backward angle value is also 45°, and the forward angles are −70° and −65°. The present results implicate that the angles around the "hot" spot and the "dark" spot contain the main information about forest structure, and there was a maximum correlation relationship between them and forest biomass in Howland, where the tree species belong to coniferous forests and the forest structure is simple. However, in this paper, the results show that "hot" spots and "dark" spots should be combined in order to get better results and higher correlations with forest biomass. It is possible that the signal angle has a lower correlation with forest biomass.

The two angle VIs, HDS and NDHD, have the same results as SRI and NDVI if they are calculated without the band limit. In Table 7, HDS and NDHD were calculated using the different bands and different angles, and they have the same values as SRI and NDVI, respectively. The RMSE and rRMSE values were smaller than those calculated using same bands and different angles, which also show that the combination of sensitive angles and sensitive bands is suitable for forest biomass retrieval. The above-mentioned results were tested using a CHRIS image and field measured biomass, as shown in Table 8; the backward angles were 36° and 55°, which are around the "hot" spot angle, and the forward angle was –55°, which is around the "dark" spot angle, are the optimal angles.

Table 8. Optimal bands and angles using ground measure and a CHRIS image.

	R Square	Red/Angle	Near-Infrared/Angle	Blue/Angle	Green/Angle	RMSE	rRMSE
SRI	0.724	697/36			570/36	44.878	0.406
NDVI	0.716	697/36			570/36	45.336	0.413
PVI	0.790		781/55 1019/55			45.605	0.404
SAVI	0.716	697/36			570/36	45.325	0.413
EVI	0.867	697/55		490/−55	530/55	32.219	0.277
ARVI	0.852	697/55		490/−55	530/55	32.114	0.279

Note: Wavelength unit is nanometer, angle unit is degrees.

3.4. Biomass Estimation

From the scatter diagrams and biomass thematic maps of the six VIs in Figure 5, and the tables shown above, we found that ARVI is the best vegetation index to retrieval forest biomass from Howland. Because the points in the ARVI scatter diagrams are close to the dotted line, and the biomass thematic maps, are better for reflecting the actual forest distribution. The results are also shown in Tables 4–7, and, in the four tables, the RMSEs and rRMSEs of ARVI are always minimal. However, the biomass inversion results tend to be over-estimate at lower values and under-estimate at higher one. It is important to note the reflectance values estimated from orbital data are explained by LAI and other structural parameters, but, as a tree grows, it accumulates biomass but LAI and others important structural data do not increase. Younger trees present higher LAI and lower biomass. When they are adult or mature, they present low LAI and high biomass. The VIs values are affected by LAI.

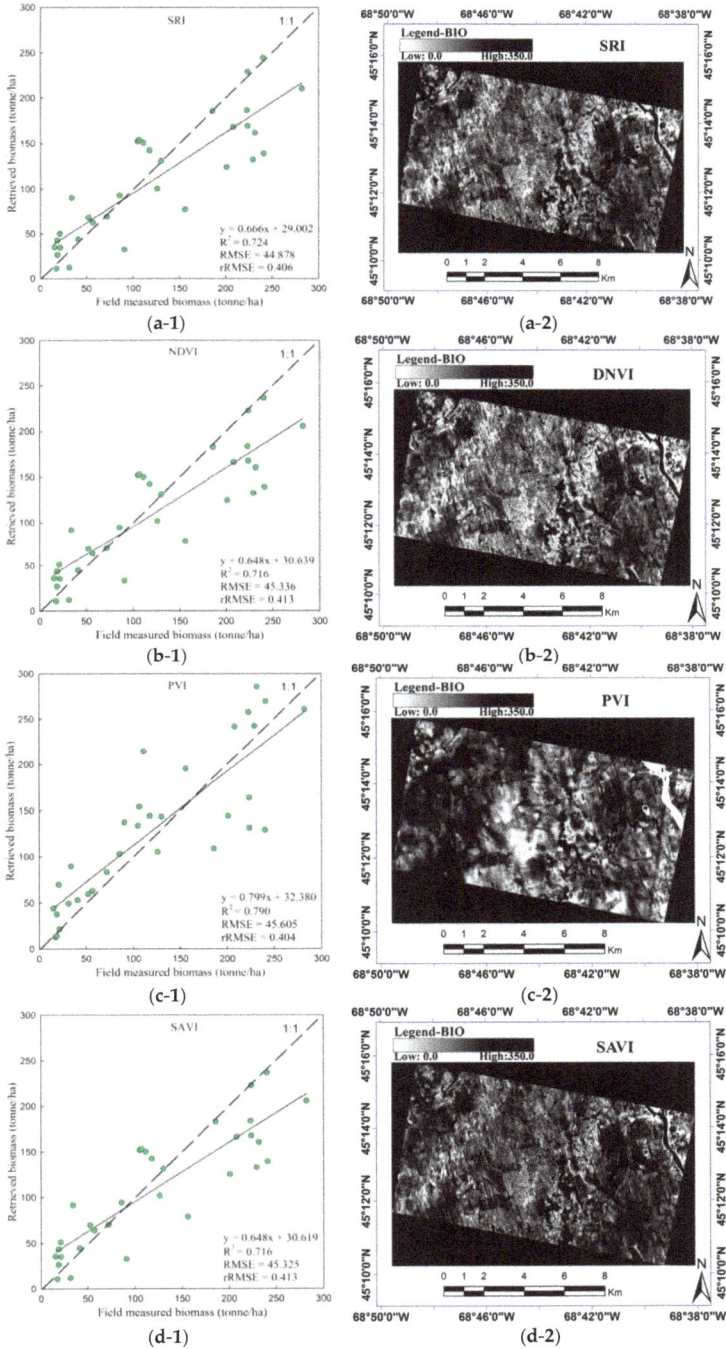

Figure 5. *Cont.*

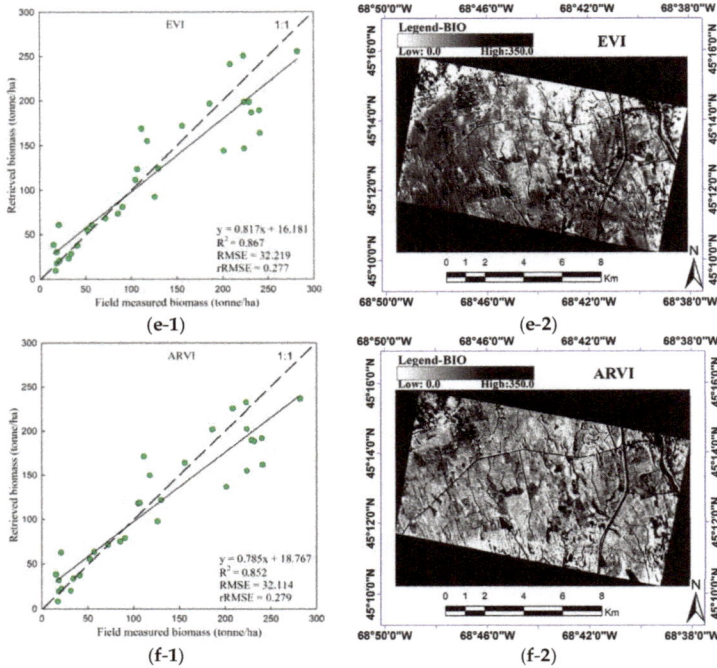

Figure 5. The performance of biomass inversion using six VIs, combining all angles with all bands in Table 8. (**a-1**) a scatter diagram comparing the measured biomass with the inversion biomass using SRI, and (**a-2**) the inversion biomass thematic map using SRI; (**b-1,b-2**) are for NDVI; (**c-1,c-2**) are for PVI; (**d-1,d-2**) are for SAVI; (**e-1,e-2**) are for EVI; (**f-1,f-2**) are for ARVI.

4. Discussion

Multi-angle remote sensing can provide more structural information regarding forests than nadir-angle observation. Compared with nadir angle observation, multi-angle can acquire some two-dimension images and then three-dimensional information can be extracted. The nadir image only provides one, two-dimension image, which loses the three-dimensional information.

Multiple scattering and shadowing are the main reasons that cause differences between the near-infrared and visible bands. The reflectance of leaf in NIR mainly depend on the inside structure of a cell, because the refraction index of a cell wall is high, which causes a high upward radiation energy. Because the chlorophyll in a leaf can perform photosynthesis, the reflectance is low in the red band [38]. Multiple scattering within canopies, among canopies, as well as between canopies and the background were increased [39]. A slight reflectance was observed in foliage and soil in the visible band, thus, the multiple scattering within canopies, among canopies, and between canopies and the background could be small.

However, in the visible band group, the blue, green, and red band curves also have small differences. As the effects of multiple scattering are the same for the visible band, there are other confounding factors affecting the relationship between BRDF and biomass, in addition to forest structure information. In the process of vegetation growth, not only does the forest three-dimensional structure change, but, also, the chemical composition and physical structure inside the tree components change, which cause blue, green, and red band curves to be of different shapes. This change also exists in the near infrared band. Some of chemical compositions include chlorophyll or carotenoid pigment levels [40]. The results show that the sensitive angle should be the main factor; at the same time, the effects of the bands should also be taken into account when the forest structure information is extracted using multi-angle remote-sensing data. However, considering orbital imagery, visible bands are also strongly influenced by shadowing. It is difficult to separate the influence of pigments and multiple shadowing, and the results aim to show that there is a primary relationship between angle and forest structures, such as biomass; however, there is a relationship between band and chemical composition that also vary in effects relative to biomass. Thus, the hope is to find a sensitive angle and suitable band in order to extract forest biomass.

The two VIs, PVI and SAVI, did not present well, because they were mainly influenced by soil background, and at the high angle of the viewed part of the ground is too small, near zero. Because view-illumination effects vary with canopy structure, angle is the key factor affecting the anisotropy of VIs determined from hyperspectral and multi-angle CHRIS/PROBA data. However, this is not the only factor that drives the VIs in Howland, in which the presence of within-canopy photosynthetic vegetation should be considered. Thus, there are other confounding factors, other than view-illumination effects, affecting the relationship between VIs and biomass. Some of them were reviewed by Middleton et al. (2011), and include chlorophyll and carotenoid pigments levels, and should be presented by different band reflectance [41].

The present results suggest that the angles around "hot" spots and "dark" spots contain the main forest structure information, and there were maximum correlation relationships between them and forest biomass in Howland. We also found that the angle values are greater, in several cases, and the optimal band/angle, with a minimum RMSE, occurs with a very high view geometry (70°). Thus, large view angles are suitable for extracting forest biomass. At a high view angle, more of the canopy can be seen and the viewed shadow and illumination canopy can include all of the vegetation information; however, the aerial proportions of the viewed ground are small. It may be reasonable to expect that the observed proportions of Photosynthetic Vegetation (PV) and Non-Photosynthetic Vegetation (NPV) depend on the viewing angle. For instance, it is likely that, at greater viewing angles, a lower proportion of PV and a greater proportion of NPV contribute to the observed canopy reflectance [16]. As the tree species of Howland are coniferous, the structure is simple. However, NPV includes canopy branches and twigs, which primarily make up canopy biomass, and there is a good relationship between canopy biomass and forest biomass [42].

Because vegetation and soil are composed of a complex non-Lambertian system, BRDF is a function of many variables, which include sensor viewing direction, solar radiation direction, geometric parameters (LAI, leaf angle distribution (LAD), canopy size, canopy spatial distribution and nelson foliage distribution, etc.), optical parameters (reflectance and transmittance of vegetation composition), etc. There is a direct relationship between some of the variables and forest biomass, which are called forest structural parameters. However, there is an indirect relationship between some of the variables and forest biomass, which are called forest non-structural parameters. Some variables are even unpredictable, such as LAD, of which the values are altered and does not only depend on forest scene, species, and growing season, but also change due to wind, plant diseases, and man-made factors. Obviously, single-band reflectance is greatly different due to changes to any single factor; thus, we need to use two or more wavelengths in order to reduce the influence on BRDF. Subtracting the different angle reflectance can highlight the structure information of a forest canopy, and dividing can reduce the influence of non-forest structure on BRDF. Therefore, ARVI, calculated by combining sensitive angles with sensitive bands, was suitable for retrieving forest biomass, and the RMSE value is minimal, which evolves based on NDVI and can reduce the influence of aerosol. The result show that the vegetation index, using three different angles, can extract the most forest structure information in Howland, and, further, it will improve the accuracy of forest biomass retrieval.

5. Conclusions

In this paper, we prove that multi-angle remote sensing can extract the biomass information of a forest via observation from different directions, and improves the accuracy of biomass retrieval compared with traditional single-angle remote sensing. The paper investigated how to determine the sensitive angle and band in order to build optimal angle vegetation indices for forest biomass retrieval. We have used ZELIG and DART models to simulate the BRDF of a forest canopy, and built a LUT for determining the sensitive angle and band; further, multi-angle CHRIS/PROBA data and field-measured biomass were used to test the results. We found that the reflectance around hot spots and dark spots include the main biomass information of a forest canopy, because the greater angle viewing is sensitive to non-photosynthetic leaf activity in a canopy (e.g., branches, trunks), which make up canopy biomass, and also has good relationship with forest biomass [43]. Crown size is considered to be one of the most important traits that affect radial tree growth, and because there is a relationship between canopy biomass and forest biomass [44]. The greater angle also reduces the influence of noise reflectance from the ground. The results also show that this mainly does not happen in the cold spot direction compared with hot spot direction; the hot spots included the main biomass information. However, the "hot" spot and "dark" spot should be combined in order to get better results and higher correlations with forest biomass. In addition, the signal angle may have a lower correlation with forest biomass. At a high view angle, the greater part of the canopy can be seen, and the viewed shadow and illumination canopy can include most of the vegetation information; however, the aerial proportions of the viewed ground are small.

Hence, the difference in the reflectance around a hot spot at different wavelength is normalized against that of a dark spot at different wavelength, and this accentuates the importance of canopy geometry on the new angle indices, and also takes into account of the influence of leaf optical properties on forest biomass. During the process of vegetation growth, as the biomass changes, not only does the forest's biophysical structure change, but also the forest's biochemical composition changes, which leads to changes in the reflectance of forest composition.

Finally, the authors presented a new and optimal angle vegetation index to retrieve the forest biomass from bidirectional signatures, based on the traditional vegetation index, ARVI, which is made of three bands with three directions. The results show that more than two bands can highlight forest biomass information and reduce the influence of other non-structural parameters on BRDF. The potential of forest biomass retrieval, based on angle vegetation indices using CHRIS/PROBA data, investigate how to take advantage of the implied information of BRDF; in particular, the two paramount directional signatures, which are around maximum (hot spot) and minimum (dark spot) reflectances. In this respect, the feasibility of retrieving the structural properties of vegetation from different satellite observations is a challenge. The POLarization and Directionality of the Earth's Reflectance onboard the Advanced Earth Observing Satellite (POLDER/ADEOS) mission fosters such an investigation, as it was shown that the angle effect is a major feature of BRDF in terrestrial biomes [45].

Acknowledgments: The research undertaken for this paper was funded by National Program on Key Basic Research Project-973 Program(Grant No. 2013CB733404),the National Natural Science Foundation of China (Grant No. 41471311, 41201435), Spatial Geography Information Laboratory Open Fund(Grant No. KJKF-12-02).

Author Contributions: Qiang Wang and Yong Pang conceived and designed the idea of paper; Zengyuan Li and Guoqing Sun performed the experiments; Erxue Chen, Wenge Ni-Meister contributed analysis methods and tools; Qiang Wang analyzed the data and wrote the paper.

Conflicts of Interest: The authors declare no conflict of interest.

References

1. Kimes, D.S.; Ranson, K.J.; Sun, G.Q.; Blair, J.B. Predicting lidar measured forest vertical structure from multi-angle spectral data. *Remote Sens. Environ.* **2006**, *100*, 503–511. [CrossRef]
2. Jenkins, J.C.; Chojnacky, D.; Heath, L.S.; Birdsey, R.A. *Comprehensive Database of Diameter-Based Biomass Regressions for North American Tree Species*; General Technical Report NE-319; U.S. Department of Agriculture, Forest Service, Northeastern Research Station: Newtown Square, PA, USA, 2004; p. 45.
3. Tritton, L.M.; Hornbeck, J.W. *Biomass Equations for Major Tree Species of the Northeast*; General Technical Report NE-69; U.S. Department of Agriculture, Forest Service, Northeastern Forest Experiment Station: Newtown Square, PA, USA, 1982; p. 46.
4. Ni, W.J.; Sun, G.Q.; Ranson, K.J.; Zhang, Z.Y.; He, Y.T.; Huang, W.L.; Guo, Z.F. Model based analysis of the influence of forest structures on the scattering phase center at L-band. *IEEE Trans. Geosci. Remote Sens.* **2014**, *52*, 3937–3946.
5. Pang, Y.; Li, Z.Y. Inversion of biomass components of the temperate forest using airborne Lidar technology in Xiaoxing'an Mountains, Northeastern of China. *Chin. J. Plant Ecol.* **2012**, *36*, 1095–1105. [CrossRef]
6. Lefsky, M.A.; Cohen, W.B.; Parker, G.G.; Harding, D.J. Lidar remote sensing for ecosystem studies. *Bioscience* **2002**, *52*, 19–30. [CrossRef]
7. Wulder, M. Optical remote sensing techniques for the assessment of forest inventory and biophysical parameters. *Proc. Phys. Geogr.* **1998**, *22*, 449–476. [CrossRef]
8. Wang, Q.; Pang, Y.; Li, Z.Y.; Chen, E.X.; Sun, G.Q.; Tan, B.X. Improvement and application of the conifer forest multiangular hybrid GORT Model-MGeoSAIL. *IEEE Trans. Geosci. Remote Sens.* **2013**, *51*, 5047–5059. [CrossRef]
9. Leroy, M.; Roujean, J.L. Sun and view angle corrections on reflectances derived from NOAA/AVHRR data. *IEEE Trans. Geosci. Remote Sens.* **1994**, *32*, 684–697. [CrossRef]
10. Asner, G.P. Contributions of multi-view angle remote sensing to land surface and biogeochemical research. *Remote Sens. Rev.* **2000**, *18*, 137–162. [CrossRef]
11. Asner, G.P.; Braswell, B.H.; Schimel, D.S.; Wessman, C.A. Ecological research needs from multi-angle remote sensing data. *Remote Sens. Environ.* **1991**, *63*, 155–165. [CrossRef]
12. Diner, D.J.; Asner, G.P.; Davies, R.; Knyazikhin, Y.; Muller, J.P.; Nolin, A.W.; Pinty, B.; Schaaf, C.B.; Stroeve, J. New directions in Earth observing: scientific application of multi-angle remote sensing. *Bull. Am. Meteorol. Soc.* **1999**, *80*, 2209–2228. [CrossRef]
13. Sandmeier, S.; Deering, D.W. Structure analysis and classification of boreal forests using hyper-spectral BRDF data from ASAS. *Remote Sens. Environ.* **1999**, *69*, 281–295. [CrossRef]
14. Kimes, D.S.; Nelson, R.F.; Manry, M.T.; Fung, A. Attributes of neural networks for extracting continuous vegetation variables from optical and radar measurements. *Int. J. Remote Sens.* **1998**, *19*, 2639–2663. [CrossRef]
15. Myneni, R.B.; Maggion, S.; Iaquinto, J.; Privette, J.L.; Gobron, N.; Pinty, B.; Kimes, D.S.; Verstraete, M.M.; Williams, D.L. Optical remote-sensing of vegetation-modeling, caveats, and algorithms. *Remote Sens. Environ.* **1995**, *51*, 169–188. [CrossRef]
16. Verrelst, J.; Schaepman, M.E.; Koetz, B.; Kneubühlerb, M. Angular sensitivity analysis of vegetation indices derived from CHRIS/PROBA data. *Remote Sens. Environ.* **2008**, *112*, 2341–2353. [CrossRef]
17. Jackson, R.D.; Teillet, P.M.; Slater, P.N.; Fedosejevs, G.; Jasinski, M.F.; Aase, J.K.; Moran, M.S. Bidirectional measurements of surface reflectance for view angle corrections of oblique imagery. *Remote Sens. Environ.* **1990**, *32*, 189–202. [CrossRef]
18. Pinter, P.J.; Zipoli, G.; Maracchi, G.; Reginato, R.J. Influence of topography and sensor view angles on NIR/red ratio and greenness vegetation indices of wheat. *Int. J. Remote Sens.* **1987**, *8*, 953–957. [CrossRef]
19. Roberts, D.A.; Roth, K.L.; Perroy, R.L. Chapter 14. Hyperspectral vegetation indices. In *Hyperspectral Remote Sensing of Vegetation*; Thenkabail, P.S., Lyon, J.G., Huete, A., Eds.; CRC Press, Taylor and Francis Group: Boca Raton, FL, USA, 2011; pp. 309–327.
20. Garbulsky, M.F.; Penuelas, J.; Gamon, J.; Inoue, Y.; Filella, I. The Photochemical Reflectance Index (PRI) and the remote sensing of leaf, canopy and ecosystem radiation use efficiencies: A review and meta-analysis. *Remote Sens. Environ.* **2011**, *115*, 281–297. [CrossRef]

21. Huete, A.R.; Kim, Y.; Ratana, P.; Didan, K.; Shimabukuro, Y.E.; Miura, T. Chapter 11. Assessment of phenologic variability in Amazon tropical rainforests using hyperspectral Hyperion and MODIS satellite data. In *Hyperspectral Remote Sensing of Tropical and Sub-tropical Forests*; Kalacska, M., Anchez-Azofeifa, G.A., Eds.; CRC Press, Taylor and Francis Group: Boca Raton, FL, USA, 2008; pp. 233–259.

22. Chen, J.M.; Liu, J.; Leblanc, S.G.; Lacaze, R.; Roujean, J.L. Multi-angular optical remote sensing for assessing vegetation structure and carbon absorption. *Remote Sens. Environ.* **2003**, *84*, 516–525. [CrossRef]

23. Wu, C.Y.; Niu, Z.; Wang, J.D.; Gao, S.; Huang, W.J. Predicting leaf area index in wheat using angular vegetation indices derived from in situ canopy measurements. *Can. J. Remote Sens.* **2010**, *36*, 301–312. [CrossRef]

24. Jenkins, J.C.; Chojnacky, D.C.; Heath, L.S.; Birdsey, R.A. National-scale biomass estimators for United States tree species. *Forest Sci.* **2003**, *49*, 12–35.

25. Barnsley, J.M.; Settle, J.J.; Cutter, M.A.; Lobb, D.R. The PROBA/CHRIS mission: A low-cost smallsat for hyperspectral multi-angle observations of the earth surface and atmosphere. *IEEE Trans. Geosci. Remote Sens.* **2004**, *42*, 1512–1520. [CrossRef]

26. Guanter, L.; Alonso, L.; Moreno, J. A method for the surface reflectance retrieval from PROBA/CHRIS data over land: Application to ESA SPARC campaigns. *IEEE Trans. Geosci. Remote Sens.* **2005**, *43*, 2907–2917. [CrossRef]

27. Kneubühler, M.; Kötz, B.; Richter, R.; Schaepman, M.; Ltten, K. Geimetric and radionmetric pre-processing of CHRIS/PROBA over mountainous terrain. In Proceedings of the 3rd CHRIS/PROBA Workshop, Frascati, Italy, 21–23 March 2005; pp. 59–64.

28. Verrelst, J.; Clevers, J.G.P.; Schaepman, M.E. Merging the Minnaert-k parameter with spectral unmixing to map forest heterogeneity with CHRIS/PROBA data. *IEEE Trans. Geosci. Remote Sens.* **2010**, *48*, 4014–4022.

29. Chan, J.C.W.; Ma, J.; van de Voorde, T.; Canters, F. Preliminary results of superresolution-enhanced angular hyperspectral (CHRIS/PROBA) images for land-cover classification. *IEEE Trans. Geosci. Remote Sens.* **2011**, *8*, 1011–1015. [CrossRef]

30. Urban, D.L.; Bonan, G.B.; Smith, T.M.; Shugart, H.H. Spatial applications of gap models. *Forest Ecol. Manag.* **1991**, *42*, 95–110. [CrossRef]

31. Gastellu-Etchegorry, J.P. DART User Manual. 2015. Available online: http://www.cesbio.upstlse.fr/dart/license/documentationsDart/DART_User_Manual.pdf (accessed on 27 October 2016).

32. Gastellu-Etchegorry, J.P.; Martin, E.; Gascon, F. DART: A 3-D model for simulating satellite images and studying surface radiation budget. *Int. J. Remote Sens.* **2004**, *25*, 73–96. [CrossRef]

33. Tucker, C.J. Red and photographic infrared linear combinations for monitoring vegetation. *Remote Sens. Environ.* **1979**, *8*, 127–150. [CrossRef]

34. Richardson, A.J.; Everitt, J.H. Using spectral vegetation indices to estimate rangeland productivity. *Geocarto Int.* **1992**, *7*, 63–77. [CrossRef]

35. Huete, A.R. A Soil-Adjusted Vegetation Index (SAVI). *Remote Sens. Environ.* **1988**, *25*, 295–309. [CrossRef]

36. Huete, A.R.; Didan, K.; Miura, T.; Rodriguez, E.P.; Gao, X.; Ferreira, L.G. Overview of the radiometric and biophysical performance of the MODIS vegetation indices. *Remote Sens. Environ.* **2002**, *83*, 195–213. [CrossRef]

37. Kaufman, Y.J.; Tanre, D. Atmospherically resistant vegetation index (ARVI) for EOS-MODIS. *IEEE Trans. Geosci. Remote Sens.* **1992**, *30*, 261–270. [CrossRef]

38. Fang, X.Q.; Zhang, W.C. The application of remotely sensed data to the estimation of the leaf area index. *Remote Sens. Land Res.* **2003**, *57*, 58–62.

39. Chen, J.M.; Leblanc, S.G. Multiple-scattering scheme useful for geometric optical modeling. *IEEE Trans. Geosci. Remote Sens.* **2001**, *39*, 1061–1071. [CrossRef]

40. Galvao, L.S.; Breunig, F.M.; dos Santos, J.R.; de Moura, Y.M. View-illumination effects on hyperspectral vegetation indices in the Amazonian tropical forest. *Int. J. Appl. Earth Obs. Geoinf.* **2013**, *21*, 291–300. [CrossRef]

41. Middleton, E.M.; Huemmrich, K.F.; Cheng, Y.; Margolis, H.A. Chapter 12. Spectral bioindicators of photosynthetic efficiency and vegetation stress. In *Hyperspectral Remote Sensing of Vegetation*; Thenkabail, P.S., Lyon, J.G., Huete, A., Eds.; CRC Press, Taylor and Francis Group: Boca Raton, FL, USA, 2011; pp. 265–288.

42. Mäkelä, A.; Valentine, H.T. Crown ration influences allometric scaling in trees. *Ecology* **2006**, *87*, 2967–2972. [CrossRef]

43. Vauhkonen, J.; Holopainen, M.; Kankare, V.; Vastaranta, M.; Viitala, R. Geometrically explicit description of forest canopy based on 3D triangulations of airborne laser scanning data. *Remote Sens. Environ.* **2016**, *173*, 248–257. [CrossRef]

44. Fichtner, A.; Sturm, K.; Rickert, C.; Oheimb, G.; Härdtle, W. Crown size-growth relationships of European beech (*Fagus sylvatica* L.) are driven by the interplay of disturbance intensity and inter-specific competition. *Forest Ecol. Manag.* **2013**, *302*, 178–184. [CrossRef]

45. Lacazea, R.; Chen, J.M.; Roujeana, J.L.; Leblanc, S.G. Retrieval of vegetation clumping index using hot spot signatures measured by POLDER instrument. *Remote Sens. Environ.* **2002**, *79*, 84–95. [CrossRef]

remote sensing

MDPI

Article

Estimation of Winter Wheat Biomass and Yield by Combining the AquaCrop Model and Field Hyperspectral Data

Xiuliang Jin [1,*], Lalit Kumar [2], Zhenhai Li [3,4], Xingang Xu [3,4], Guijun Yang [3,4] and Jihua Wang [4]

[1]   Key Laboratory of Wetland Ecology and Environment, Northeast Institute of Geography and Agroecology, Chinese Academy of Sciences, Changchun 130102, China
[2]   Ecosystem Management, School of Environmental and Rural Science, University of New England, Armidale, NSW 2351, Australia; lkumar@une.edu.au
[3]   Beijing Research Center for Information Technology in Agriculture, Beijing Academy of Agriculture and Forestry Sciences, Beijing 100097, China; lizh323@126.com (Z.L.); xxg2007@aliyun.com (X.X.); guijun.yang@163.com (G.Y.)
[4]   National Engineering Research Center for Information Technology in Agriculture, Beijing 100097, China; wangjh@nercita.org.cn
*   Correspondence: jinxiuliang@iga.ac.cn or jinxiuxiuliang@126.com; Tel.: +86-156-5250-0901

Academic Editors: Onisimo Mutanga, Nicolas Baghdadi, Clement Atzberger and Prasad S. Thenkabail
Received: 9 August 2016; Accepted: 18 November 2016; Published: 24 November 2016

Abstract: Knowledge of spatial and temporal variations in crop growth is important for crop management and stable crop production for the food security of a country. A combination of crop growth models and remote sensing data is a useful method for monitoring crop growth status and estimating crop yield. The objective of this study was to use spectral-based biomass values generated from spectral indices to calibrate the AquaCrop model using the particle swarm optimization (PSO) algorithm to improve biomass and yield estimations. Spectral reflectance and concurrent biomass and yield were measured at the Xiaotangshan experimental site in Beijing, China, during four winter wheat-growing seasons. The results showed that all of the measured spectral indices were correlated with biomass to varying degrees. The normalized difference matter index (NDMI) was the best spectral index for estimating biomass, with the coefficient of determination (R^2), root mean square error (RMSE), and relative RMSE (RRMSE) values of 0.77, 1.80 ton/ha, and 25.75%, respectively. The data assimilation method ($R^2 = 0.83$, RMSE = 1.65 ton/ha, and RRMSE = 23.60%) achieved the most accurate biomass estimations compared with the spectral index method. The estimated yield was in good agreement with the measured yield ($R^2 = 0.82$, RMSE = 0.55 ton/ha, and RRMSE = 8.77%). This study offers a new method for agricultural resource management through consistent assessments of winter wheat biomass and yield based on the AquaCrop model and remote sensing data.

Keywords: biomass; yield; AquaCrop model; spectral index; particle swarm optimization; winter wheat

1. Introduction

Wheat is an important food source for the rapidly increasing population in China [1]. The attention paid to national food security and sustainable agricultural development has increased over recent years, with increased concern for the improvement of field wheat management. Therefore, it is important to estimate wheat growth status and predict wheat yield in a timely and accurate way [2]. The integration of crop models and remote sensing data has become a useful method for monitoring crop growth status and crop yield based on data assimilation over extensive regions [3,4].

In most cases, researches have developed remote sensing and crop models used in their respective study areas [5]. Crop models simulate crop physiological growth status using mathematical formulas [6]. They have been used to analyze the influences of climate, soil conditions, and management strategies on agronomic parameters (e.g., canopy aboveground biomass, LAI, and grain yield) [7]. The Agricultural Model Intercomparison and Improvement Project (AgMIP) recently reviewed 27 wheat models from around the world [8]. This review showed that poor performance may be obtained when a crop model is applied over a large region due to uncertainties in the spatial distribution of soil properties, initial model conditions, crop parameters, and field management practices, resulting in biased simulations [9,10]. Large amounts of high-quality data have been used to improve the calibration and parameterization of crop models, thereby increasing the simulation accuracy of crop models on a regional scale.

Rapid development of remote sensing technology has facilitated the acquisition of crop growth information with high temporal and spatial resolutions [5,11–16]. Previous studies have indicated that combining crop models and remote sensing data can be used to improve the accuracy of crop yield estimates [17–25]. Curnel et al. [17] evaluated the feasibility of assimilating wheat leaf area index (LAI) derived from remote sensing data into the World Food Studies' (WOFOST) crop growth model using a recalibration-based assimilation method; the results indicated that remote sensing data can be used to improve yield estimations. Dente et al. [19] assimilated LAI from Environment Satellite (ENVISAT) Advanced Synthetic Aperture Radar (ASAR) and Medium Resolution Imaging Spectrometer (MERIS) data into the Crop Estimation through Resource and Environment Synthesis-Wheat (CERES-Wheat) model at a catchment scale using a variational assimilation algorithm; the results suggested that this approach minimizes the difference between simulated and remotely-sensed LAI and achieves high estimation accuracy. Soil moisture data from the Advanced Microwave Scanning Radiometer-EOS (AMSR-E) and LAI from the Moderate Resolution Imaging Spectroradiometer-LAI (MODIS-LAI) were assimilated into the Decision Support System for Agro-technology Transfer-Cropping System Model (DSSAT-CSM)-Maize using an Ensemble Kalman Filter algorithm, and simulated yield was more accurate when both LAI and soil moisture were used [22]. The ensemble-based four-dimensional variational method was used to assimilate HJ-1A/B satellite data into the CERES-Wheat model, and estimates of winter wheat yield in field plots were reported (R^2 = 0.73; RMSE = 319 kg/ha) [23]. Huang et al. [25] assimilated time series of LAI data with a 30-m spatial resolution into the WOFOST model with a Kalman Filter (KF) algorithm and reported more accurate estimates of regional winter wheat yield compared with more traditional approaches.

The integration of crop models and remotely sensed data using optimization algorithms (data assimilation methods) is becoming an effective and potential method for monitoring crop growth status and estimating crop yields, as it overcomes certain defects and combines the advantages of individual methods [5,26–30]. The data assimilation method can be used to reduce uncertainty in crop models to ensure that the simulated state variables (e.g., LAI and biomass) are in agreement with the measured state variables from remote sensing data. Several assimilation schemes have been developed [3,12,31–33]. Delecolle et al. [34] divided schemes into three categories. The first is the forcing method, in which state variables in crop models are directly substituted by remote sensing variables. This method is easy to use, but it relies on the calibrated parameters of crop models and the accuracy of remote sensing data [5,12]. The second is the calibration method, in which the initial parameters of crop models are recalibrated, based on the relationship between the remote sensing state variables and the simulated state variables [5,10,34]. In recent years, the calibration method has gained more attention, as it has greatly benefited from several intelligent optimization algorithms. The main shortcoming of the calibration method is that a great deal of computation time is required [3,5,30]. The third is the updating method, in which the simulated state variables are continuously renewed whenever remote sensing state variables are available. It is more flexible than the forcing method, but the remote sensing data must be of a higher accuracy than those of the simulated state variables, and this method heavily relies on the selection of the remote sensing data [4,33,35].

The combination of remote sensing data and light-driven or carbon-driven models has been widely studied, but few studies have focused on water-driven models for estimating the biomass and yield of crops. Therefore, in this study we focused on the AquaCrop model, a water-driven crop model, which was recently introduced to optimize crop water management strategies and improve crop yield in irrigation regions [36]. Winter wheat is the main crop grown in the North China Plain (NCP), and thus an important food source in China. Increased industrial and domestic water use has resulted in reduced water availability for irrigation of winter wheat crops. Therefore, improving water resource management in this region is crucial for increasing winter wheat yield. The main goal of this study was to improve estimates of winter wheat biomass and yield by assimilating field spectroscopic data into the AquaCrop model with a Particle Swarm Optimization (PSO) algorithm. The biomass and yield of winter wheat were used to optimize field irrigation management strategies and then to increase water use efficiency under different planting dates and irrigation management strategies. The specific objectives of this study were: (1) to select the best spectral indices from hyperspectral data for estimating winter wheat biomass; (2) to calibrate the AquaCrop model with biomass estimates derived from these indices using the PSO algorithm for improving accuracy of biomass and yield estimates; and (3) to evaluate the performance of the data assimilation method in estimating wheat biomass and yield.

2. Methodology

2.1. Description of the Study Site

Field experiments were carried out during the 2008/2009, 2009/2010, 2010/2011, and 2011/2012 growing seasons, at the Xiaotangshan experimental site (40°10′31″~40°11′18″N, 116°26′10″~116°27′05″E), Beijing, PR China. The soil type in the study site is fine-loamy. Beijing is characterized by a typical continental climate. The maximum temperature is 26.1 °C in summer, and the minimum temperature is −4.7 °C in winter. For the experimental period, the average annual precipitation was 650 mm and the frost-free period was 180 days on average [37].

2.2. Experimental Setup

Table 1 shows the winter wheat planting dates and cultivars. The area of each plot was 100 m², in 2008, 2009, and 2010, and 300 m² in 2011. A two-way factorial arrangement of treatments (winter wheat cultivar and planting date) in a randomized complete block design with three replicates was used in this experiment. Weed control, pest management, and fertilizer application were performed according to the local standard practices for wheat production.

Table 1. Winter wheat cultivars and planting dates in 2008, 2009, 2010, and 2011.

Winter Wheat Cultivars	Planting Date
Nongda195, Jingdong8, Jing9428	28 September, 7 October, and 20 October 2008
Nongda195, Jingdong13, Jing9428	25 September, 5 October, and 15 October 2009
Nongda195, Yannong19, Jing9428	25 September, 5 October, and 15 October 2010
Nongda211, Zhongmai175, Jingdong8, Jing9843	25 September 2011

Note: There were three winter wheat cultivars, and each had three planting dates per year, in 2008, 2009, and 2010. In 2011, four cultivars were planted on the same date.

2.3. Data Acquisition

2.3.1. Meteorological Data Collection

The local Xiaotangshan meteorological station was used to obtain meteorological data. Daily relative humidity, rainfall, total sunshine hours, wind speed, and maximum, minimum, and mean temperatures were recorded directly at the Xiaotangshan experimental site. The Food and Agriculture Organization Penman–Monteith method was used to calculate the reference evapotranspiration (ET_o) [38].

2.3.2. Measurement of Canopy Reflectance

Spectral measurements of winter wheat were taken at different growth stages. The specific growth stages and dates are presented in Table 2. All canopy spectral measurements were taken at a nadir orientation, 1.0 m above the canopy, under clear sky condition between 10:00 and 14:00 Beijing local time, using an ASD Field Spec Pro Spectrometer (Analytical Spectral Devices, Boulder, CO, USA). The spectrometer was fitted with a 25° field of view optical fiber, operating in the 350–2500 nm spectral region. The scanned area of the ASD sensor was about 0.70 m^2. A 40 × 40 cm BaSO$_4$ calibration panel was used for calculating the black and baseline reflectance. Spectral measurements taken at all four sites in each plot were averaged to represent the canopy reflectance of each plot to reduce the possible effects due to field conditions. Vegetation radiance measurements were averaged from 10 scans at an optimized combination time at each site, and a dark current correction was conducted before each measurement. For each plot, a total of 40 spectra data points were obtained. Panel radiance measurements were taken twice, before and after the canopy spectral measurements.

Table 2. Spectral reflectance measurement dates for 2009, 2010, 2011, and 2012.

Wheat Growth Stages	Measurement Dates			
	2009	2010	2011	2012
Jointing	16 April 28 April	23 April	18 April	13 April 28 April
Heading	6 May	6 May	7 May	10 May
Anthesis	12 May	19 May	17 May	21 May
Grain filling	26 May 10 June	1 June 7 June 12 June		30 May

2.3.3. Biomass and Yield Data Collection

The aboveground biomass, at the measuring positions of canopy spectral reflectance data, were obtained 5–6 times using random sampling of a 0.25-m^2 area, in 2009–2012, with four replicates from each plot. A 4 × 0.25 m^2 area for each plot was deemed sufficient, based on previous results [3]. All samples were heated to 105 °C, then oven dried at 70 °C to a constant weight, and their final dry weights were recorded.

The grain yields of each plot with three replicates for each treatment were obtained by randomly sampling a 1.5-m^2 area. Finally, selected grain was dried and weighed on an electronic scale (±0.01 g).

2.3.4. Selection of Spectral Indices and Biomass Estimation from Spectral Indices

Fifteen spectral indices from the literature [13,16,39–50], determined to be good candidates for estimating biomass, were selected for the entire winter wheat growing season, based on 2009, 2010, and 2011 field data (Table 3). To refine the relationships between spectral indices and biomass, linear and nonlinear regression relationships between each of the spectral indices and biomass were determined based on field data from all growth stages during 2009, 2010, and 2011 (n = 135, calibration dataset). Field data taken in 2012 (n = 20, validation dataset) was used to validate the estimation accuracy of the models. Since the four winter wheat cultivars exhibited larger differences in 2012, resulting in greater variation in the biomass of the four winter wheat cultivars, the dataset from 2012 was selected to validate the estimation accuracy of the models. To determine the most sensitive spectral indices, we compared the coefficient of determination (R^2), root mean square error (RMSE), and relative RMSE (RRMSE) values of the different models. Best-fitting regression equations were used for estimating winter wheat biomass. In addition, the homoscedasticity values (F) of the estimated and measured biomass were calculated using Levene's test [51].

Table 3. Summary of spectral indices studied.

Spectral Index	Name	Formula	References
WI I	Water index (970, 900)	R_{970}/R_{900}	[39]
WI II	Water index (1300, 1450)	R_{1300}/R_{1450}	[40]
NDII	Normalized difference infrared index	$(R_{850} - R_{1650})/(R_{850} + R_{1650})$	[41]
NDMI	Normalized difference matter index	$(R_{1649} - R_{1722})/(R_{1649} + R_{1722})$	[42]
TBWI	Three band water index	$(R_{973} - R_{1720})/R_{1447}$	[43]
EVI	Enhanced vegetation index	$2.5 \times (R_{800} - R_{660})/(1 + R_{800} + 2.4 \times R_{660})$	[44]
TCARI	Transformed chlorophyll absorption in reflectance index	$3 \times ((R_{700} - R_{670}) - 0.2 \times (R_{700} - R_{550}) \times (R_{700}/R_{670}))$	[23]
OSAVI	Optimized soil-adjusted vegetation index	$1.16 \times (R_{800} - R_{670})/(R_{800} + R_{670} + 0.16)$	[45]
TCARI/OSAVI	Combined Index II	TCARI/OSAVI	[16]
MTCI	MERIS terrestrial chlorophyll index	$(R_{750} - R_{710})/(R_{710} - R_{680})$	[46]
$CI_{red\ edge}$	Red edge model	$(R_{750}/R_{720}) - 1$	[47]
NDVI	Normalized difference vegetation index	$(R_{800} - R_{670})/(R_{800} + R_{670})$	[48]
DCNI I	Double-peak canopy nitrogen index I	$(R_{750} - R_{700})/(R_{700} - R_{670})/(R_{750} - R_{670} + 0.09)$	[49]
OSAVI × $CI_{red\ edge}$	Combined Index I	OSAVI × $CI_{red\ edge}$	[13]
WDRVI	Wide dynamic range vegetation index	$WDRVI = (\alpha \times R_{800} - R_{670})/(\alpha \times R_{800} + R_{670})$ $\alpha = 0.1$	[50]

Note: R_i denotes reflectance at band i (nanometer).

2.4. Description of the AquaCrop and ACsaV40 (AquaCrop Plug-In) Models

2.4.1. Description of the AquaCrop Model

The AquaCrop model was reported by the FAO in 2009, and detailed descriptions are reported in Steduto et al. [36], Raes et al. [52], and Jin et al. [37]. It computes daily crop transpiration and soil evaporation. The model subsequently estimates yield based on daily crop transpiration [36].

2.4.2. Description of the ACsaV40 (AquaCrop Plug-In) Model

The AquaCrop plug-in program, ACsaV40, was created to simultaneously run large amounts of data without a user interface [53]. It facilitates external and practical applications of AquaCrop. The input parameters of ACsaV40 are sorted in a text file, which can be created using the AquaCrop model, or by manually replacing the values of each variable with new values in the existing text files [54]. ACsaV40 runs the successive project files, and the simulated results of each project file are reserved in an output file, which includes the simulation period, stress factors, canopy cover, biomass, crop yield, and so on [53].

2.5. Assimilation of the AquaCrop Model and Remote Sensing Data Using the Particle Swarm Optimization (PSO) Algorithm

Particle swarm optimization (PSO) is a comparatively simple principle that can be easily combined into crop models with high calculation efficiency and few input parameters. Compared with various optimization algorithms, PSO is easier to apply in a practical study and has the advantages of a high precision and rapid convergence [55]. It has received widespread attention among scientists who have demonstrated its superiority in solving practical problems. In addition, PSO has the capability of parallel computing. Therefore, we used PSO to carry out the assimilation of remote sensing data into the AquaCrop model. PSO is based on the assumption of a group consisting of m (25 groups in this study) particles with certain speeds, without quality and size, in a d-dimensional search space. Each particle can modify its position and velocity based on both the best point in the current generation (p_{id}) and the best point of all particles in the swarm (p_{gd}). In this study, estimated biomass was used to optimize the crop parameters used in the AquaCrop model to obtain the optimal simulated biomass based on the fit of the cost function. The corresponding optimal yield is produced when the optimal simulated biomass is achieved. The PSO assimilation method for the The AquaCrop model and remote sensing data are presented in Figure 1. The specific steps to execute the PSO are as follows:

(1) The velocity and position (initial value) of each particle are determined. The adjusted parameters include eight crop parameters (*cgc, ccx, cdc, eme, num, psen, pstoshp,* and *rootdep*) [56]. Specific information and ranges for these parameters are listed in Table 4.

(2) ACsaV40 is executed with the required data using MATLAB (version 2007, MathWorks, Natick, MA, USA), and simulated biomass (*BIOs*) is obtained.

(3) Regression relationships between spectral indices and measured biomass are analyzed, and the best regression model is determined to estimate biomass (*BIOe*).

(4) A cost function is constructed according to the relationship *BIOs* and *BIOe*, reflecting the difference between *BIOs* and *BIOe*. The fit of the cost function determines whether the optimization algorithm had achieved the optimal input parameters.

(5) The values of p_{id} and p_{gd} are searched in each iteration.

(6) The position and velocity of each particle are updated based on p_{id} and p_{gd}. The values of C_1 and C_2 are set as 2, and random values between 0 and 1 are assigned to ξ and η [57].

(7) If the iteration target (100 generations) is not reached, the updated positions are replaced and the previous step is repeated.

(8) If the final iteration is achieved, the values of *BIOs* and corresponding simulated YIELDs are produced.

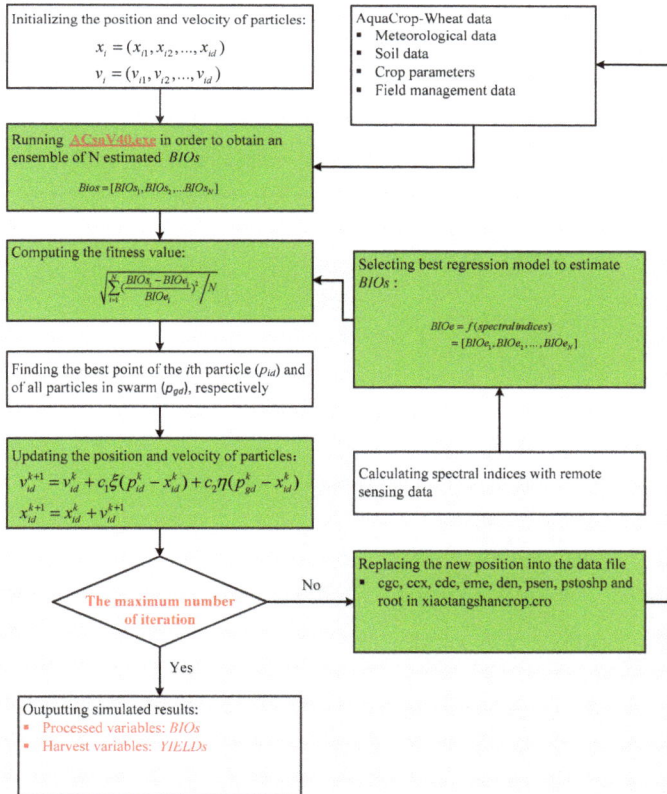

Figure 1. Flowchart of the Particle Swarm Optimization (PSO) assimilation method for the AquaCrop model and remote sensing data.

Table 4. Initial values and ranges of calibration parameters, or initial data, of the AquaCrop model.

Variables	Values	Ranges
Canopy growth coefficient (*cgc*)	0.06	0.05–0.07
Maximum canopy cover in fraction soil cover (*ccx*)	0.65	0.82–0.99
Canopy decline coefficient (*cdc*)	0.05	0.04–0.07
Growth degree day: from sowing to emergence (*eme*)	175	100–250
Number of plants per hectare (*num*)	4,500,000	2,500,000–5,500,000
Soil water depletion factor for canopy senescence (*psen*)	0.65	0.55–0.75
Shape factor for water stress coefficient for stomatal control (*pstoshp*)	2.5	1.5–3.5
Growth degree day: from sowing to maximum rooting depth (*rootdep*)	1500	1200–1800

3. Results

3.1. Biomass Estimation

The regression relationships between biomass and spectral indices are provided in Table 5. The lowest and highest R^2 values (0.25 and 0.84) were obtained for DCNI I and NDMI, respectively. The order of the spectral indices was WI I, WI II, NDII, NDMI, TBWI, EVI, TCARI, OSAVI, TCARI/OSAVI, MTCI, $CI_{red\ edge}$, NDVI, DCNI I, OSAVI $\times$ $CI_{red\ edge}$, and WDRVI. Of the R^2 values, two were above 0.70, five were equal to or above 0.60, and eight were below 0.6. All spectral indices were fitted to power regression equations, with the exception of NDMI, DCNI I, and WDRVI, which were fitted to exponential regression equations (Table 5). The result showed that the assumption of homoscedasticity is met, based on the calculated F values between the estimated and measured biomass (Table 5).

Table 5. Correlations between biomass and spectral indices of winter wheat (*n* = 135).

Vegetation Index	Regression Equations	R^2	F	RMSE (Ton/Ha)	RRMSE (%)
WI I	$y = 1.169x^{-13.5}$	0.72 **	0.85	2.05	29.33
WI II	$y = 0.264x^{2.149}$	0.56 **	0.62	2.98	42.63
NDII	$y = 29.40x^{1.620}$	0.67 **	0.80	2.24	32.04
NDMI	$y = 0.883e^{70.06x}$	0.77 **	0.94	1.80	25.75
TBWI	$y = 2.3x^{1.040}$	0.52 **	0.80	3.41	48.78
EVI	$y = 15.13x^{1.660}$	0.61 **	0.73	2.92	41.77
TCARI	$y = 28.11x^{0.847}$	0.38 **	0.45	3.87	55.36
OSAVI	$y = 22.73x^{3.352}$	0.60 **	0.81	2.89	41.34
TCARI/OSAVI	$y = 1.165x^{0.361}$	0.30 **	0.34	4.08	58.37
MTCI	$y = 0.451x^{1.838}$	0.63 **	0.77	2.48	35.48
$CI_{red\ edge}$	$y = 3.767x^{1.750}$	0.68 **	0.80	2.18	31.19
NDVI	$y = 15.33x^{4.835}$	0.59 **	0.73	2.95	42.20
DCNI I	$y = 2.626e^{0.458x}$	0.25 **	0.27	4.42	63.23
OSAVI $\times$ $CI_{red\ edge}$	$y = 3.704x^{0.619}$	0.58 **	0.78	2.96	42.34
WDRVI	$y = 4.947e^{2.533x}$	0.54 **	0.71	3.03	43.35

Note: *n* = number of data pairs; x represents the spectral index; and y represents biomass. In addition, x and e represents power and exponential function in regression equations, respectively. Probability levels of 0.05 and 0.01 are indicated by * and **, respectively; F represents the homoscedasticity value in Levene's test. If the associated probability for the F test is larger than 0.05, the assumption of homoscedasticity is met.

The correlation between biomass and NDMI was highest compared with the other spectral indices, and the corresponding RMSE and RRMSE values for measured (*BIOm*) and estimated (*BIOe*) biomass were 1.80 ton/ha and 25.75%, respectively, which were lower than the values for the other indices (Table 5 and Figure 2). Therefore, NDMI was selected to estimate winter wheat biomass.

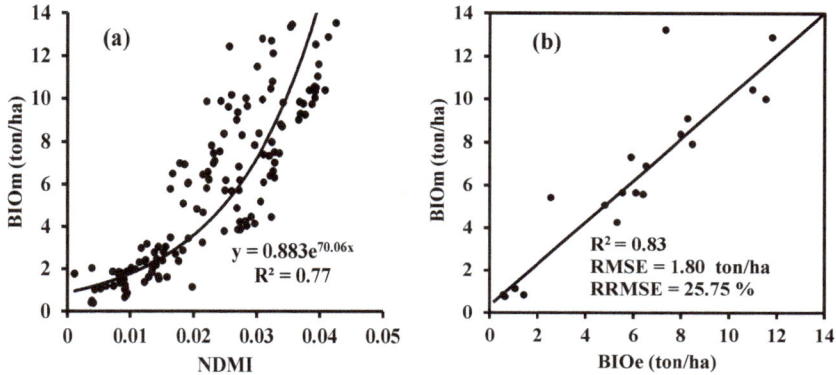

Figure 2. Regression model between biomass and normalized difference matter index (NDMI) (**a**); and model validation (**b**).

3.2. Data Assimilation for Biomass Estimation

The value of *BIOe* derived from the NDMI exponential regression equation was used as a variable to calibrate the AquaCrop model using the PSO algorithm. The results are presented in Figure 3 and Table 6, and the statistical regression equations are shown in Table 6. The *BIOs* was consistent with the *BIOm* across four years with different winter wheat cultivars, sowing dates, and irrigation management strategies, and the corresponding R^2 and RMSE values were 0.83 and 1.65 ton/ha, respectively. The estimation accuracies of our experiments varied between years. The R^2 and RMSE values were 0.81 and 1.69 ton/ha for 2008/2009, 0.82 and 1.67 ton/ha for 2009/2010, 0.81 and 1.56 ton/ha for 2010/2011, and 0.87 and 1.72 ton/ha for 2011/2012. The RRMSE values ranged from 23.60% to 30.65%. The deviation between the *BIOs* and *BIOm* in 2008/2009 was larger than that for the other years. Strong relationships between *BIOs* and *BIOm* were found, although biomass was often overestimated when the measured values exceeded 2 ton/ha (Figure 3). However, biomass was underestimated when the measured values were less than 2 ton/ha. The value of F was from 0.87 to 0.96 between the simulated and measured biomass (Table 6). The results show that the assumption of homoscedasticity was met.

Table 6. Equations for regressions between data assimilation biomass (*BIOs*) and field measurement biomass (*BIOm*) of winter wheat for the four experiments.

Year	n	Regression Equations	R^2	F	RMSE (Ton/Ha)	RRMSE (%)
2009	54	y = 0.847x − 0.114	0.81	0.96	1.69	26.68
2010	54	y = 0.853x + 0.847	0.82	0.78	1.67	24.58
2011	27	y = 0.754x − 0.198	0.81	0.84	1.56	30.65
2012	20	y = 0.863x + 0.135	0.87	0.83	1.72	25.94
2009–2012	155	y = 0.872x + 0.310	0.82	0.82	1.70	26.57
C/V [a]	135/20 [b]	y = 0.865x + 0.066	0.83	0.85	1.65	23.60

Note: [a] C represents the calibration dataset (2009–2011, n = 135), and V represents the validation dataset (2012, n = 20). The calibration dataset was used to refine the linear regression relationships between the data assimilation biomass (*BIOs*) and field measurement biomass (*BIOm*) across three years of experiments. The validation dataset taken in 2012 was used to validate the estimation accuracy of the linear regression equation based on 2009, 2010, and 2011; [b] R^2 was calculated from 135 calibration datasets, and RMSE was calculated from 20 validation datasets; x represents simulated biomass; y represents measured biomass; and F represents the homoscedasticity value for the Levene's test. If the associated probability for the F test is larger than 0.05, the assumption of homoscedasticity is met.

We compared *BIOs* with *BIOe* using the spectral index method. The data assimilation method (R^2 = 0.83 and RMSE = 1.65 ton/ha, Table 6) achieved better biomass estimations than the spectral index method (R^2 = 0.77 and RMSE = 1.80 ton/ha, Table 5).

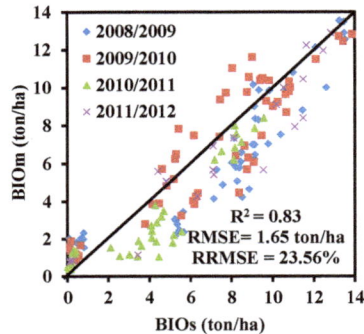

Figure 3. Comparison of data assimilation biomass (*BIOs*) and field measurement biomass (*BIOm*) values in winter wheat across the four experiments.

3.3. Data Assimilation for Yield

The yield of winter wheat was obtained after the data assimilation. The relationship between the measured and simulated yields is shown in Figure 4 and Table 7. There was a significant relationship between simulated (YIELDs) and measured (YIELDm) yield across all four years (R^2 and RMSE values of 0.82 and 0.55 ton/ha, respectively) (Table 7). YIELDs varied between the four growing seasons. The R^2 and RMSE values for YIELDs and YIELDm were 0.79 and 0.51 ton/ha in 2008/2009, 0.83 and 0.57 ton/ha in 2009/2010, 0.81 and 0.52 ton/ha in 2010/2011, and 0.89 and 0.61 ton/ha in 2011/2012, respectively. The value of RRMSE ranged from 8.77% to 10.69%. There was a wider range of YIELDs values in 2008/2011 than in 2011/2012 because of the different sowing treatments (Figure 4). A good relationship between YIELDs and YIELDm was also found. Yield was often overestimated when the YIELDm was higher than 5 ton/ha (Figure 4) and underestimated when the YIELDm was less than 5 ton/ha. Table 7 shows that the F values ranged from 0.72 to 0.87. The results demonstrated that the assumption of homoscedasticity was met.

Table 7. Regression equations between data assimilation yield (YIELDs) and field measurement yield (YIELDm) values of winter wheat across the four experiments.

Year	*n*	Regression Equations	R^2	F	RMSE (Ton/Ha)	RRMSE (%)
2009	9	y = 0.406x + 3.225	0.79	0.72	0.51	9.42
2010	9	y = 0.482x + 2.775	0.83	0.75	0.57	10.69
2011	9	y = 0.481x + 2.785	0.81	0.76	0.52	9.29
2012	4	y = 0.583x + 2.245	0.89	0.83	0.61	9.79
2009–2012	31	y = 0.490x + 2.768	0.85	0.80	0.57	10.27
C/V [a]	27/4 [b]	y = 0.460x + 2.911	0.82	0.76	0.55	8.77

Note: [a] C represents the calibration dataset (2009–2011, *n* = 27), and V represents the validation dataset (2012, *n* = 4). The calibration dataset was used to refine the linear regression relationships between the data assimilation yield (YIELDs) and field measurement yield (YIELDm) across three years of experiments. The validation dataset taken in 2012 was used to validate the estimation accuracy of the linear regression equation based on data from 2009, 2010, and 2011; [b] R^2 was calculated from 27 calibration datasets, and RMSE was calculated from 20 validated datasets; x represents simulated biomass; y represents measured biomass; and F represents the homoscedasticity value for the Levene's test. If the associated probability for the F test is larger than 0.05, the assumption of homoscedasticity is met.

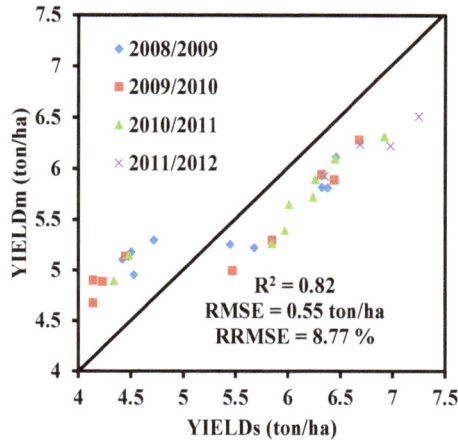

Figure 4. Comparison of data assimilation yield (YIELDs) and field measurement yield (YIELDm) values in winter wheat across the four experiments.

4. Discussion

Spectral data and concurrent biomass and yield were acquired during four winter wheat growing seasons. Fifteen spectral indices were related to biomass (Table 5); this is because red edge (670–780 nm) and near infrared (short NIR, 800-1100 nm) data contain useful information regarding vegetation biomass [13,39,44,50]. In particular, NDMI was found to be highly correlated with biomass, with R^2 and RMSE values of 0.77 and 1.80 ton/ha, respectively. NDMI does not contain red edge or short NIR data because absorption at these wavelengths is strongly influenced by chlorophyll content and canopy structure, which reduce the signal compared with that of dry matter. However, NDMI contains data at 1649 and 1722 nm, which are more sensitive to changes in dry matter [42]. These data were combined to establish NDMI, which includes signals from dry matter. For this reason, NDMI was more highly related with biomass than the other spectral indices and achieved more accurate biomass estimations. In this study, the linear and nonlinear regression relationships between each spectral index and biomass were analyzed to select the best-fitting regression equations. The results show that some models were fitted using power regression, and others fitted using exponential regression (Table 5). The difference between two regressions may have a close relationship with each spectral index and biomass dataset.

The model's initial variables (*num* and *eme*) and crop parameters (*cgc, ccx, cdc, eme, psen,* and *rootdep*) were calibrated by combining biomass retrieved from spectral indices and the AquaCrop model via the PSO assimilation algorithm, thereby achieving optimal biomass estimations. The simulated biomass values were consistent with the measured values. These findings are consistent with those of Soddu et al. [58]. Heng et al. [59] showed that the AquaCrop model is used to better simulate biomass when irrigation is adequate. Our results suggest that the AquaCrop model could be used to simulate winter wheat biomass. The data assimilation method, based on the PSO algorithm, achieved better biomass estimations than the spectral index method (Tables 5 and 6). The main reasons are as follows: (i) The AquaCrop model can be used to simulate dry biomass accumulation on the basis of a plant's physiological processes, and the effects of field management strategies and weather [36,37,59,60]; and (ii) the data assimilation method was used to minimize errors between the observed values from field spectroscopic data and the simulated values from the AquaCrop model, and the errors in the remote sensing data were reduced during data assimilation [10]. Typically, biomass simulated with the data assimilation method was overestimated when the measured values exceeded 2 ton/ha, but was underestimated when the measured values were less than 2 ton/ha

(Figure 3). This explains why the regression equations between NDMI and biomass were similar. Therefore, the integration of spectral indices into the AquaCrop model, using the PSO assimilation algorithm, is a useful tool for winter wheat biomass estimation.

Winter wheat grain yield was simulated according to the optimized values of the initial variables and calibrated crop parameters using the PSO data assimilation algorithm. A good relationship between the measured and simulated yields was found across all four years (R^2 = 0.82 and RMSE = 0.55 ton/ha). However, the RRMSE for yield was lower than that for biomass (Tables 6 and 7), mainly because the latest biomass measurements were taken at the grain filling stage (12 June) rather than at maturity and biomass simulated with the AquaCrop model becomes more accurate with the development of winter wheat [37,61]. Our results are in agreement with those of Wang et al. [61] and Jin et al. [37]. The AquaCrop model considers the effects of interannual variations in weather and field management strategies, as well as interactions between the two, on wheat growth status; therefore, it was used to analyze the nonlinear interannual variability in crop grain yield [36]. The results suggest that the AquaCrop model is an effective tool for deriving crop management strategies, and can be used to simulate biomass and grain yield of winter wheat. First, biomass retrieved from spectral indices is used to calibrate crop biomass simulated with the AquaCrop model. If crop biomass is accurately simulated, it can be used to simulate yield. The simulated yield is finally obtained directly from the AquaCrop model after data assimilation. Simulated grain yield is a useful measurement for informed decision-making regarding national food security issues. However, it is more important to obtain crop growth status information and then to improve field crop management for improving grain yield to ensure national food security, In short, the dynamic simulated biomass of wheat is used to enhance wheat management and decision-making, and then to ensure wheat yield.

The data assimilation accuracy of biomass and grain yield was acceptable according to the R^2, RSME, and RRMSE values (Tables 6 and 7). The results of Dente et al. [19] and Jiang et al. [23] indicated that assimilating remote sensing data (ENVISAT ASAR, MERIS, and HJ-1A/B satellites images) into the CERES-Wheat model with optimization algorithms (variational assimilation algorithm and Ensemble-Based Four-Dimensional Variational algorithm) can improve the estimation accuracy of wheat yield. Huang et al. [25] recently suggested that combining the WOFOST model and remote sensing data (MODIS and Landsat TM images) with a KF algorithm also increases the estimation accuracy of wheat yield. Our results are in agreement with the results of these studies and demonstrate that the combination of the AquaCrop model and spectral indices with a PSO algorithm can be used to enhance the estimation accuracy of winter wheat yield. A good relationship between the simulated and measured yields was found (Figure 4); however, the relationship between measured and simulated biomass was not reliable during each growth stage (Figure 3). This can be attributed to the influence of a large difference in the biomass measurement date on biomass simulation [37], which then introduced uncertainties into the process of data assimilation. However, the yield simulations were consistent during all crop growth stages. Therefore, the data assimilation method can improve crop yield estimations because the AquaCrop model considers the effects of management strategies and environmental factors on winter wheat growth status, based on a plant's physiological processes. Our results suggest that integrating remote-sensing data into the AquaCrop model is a feasible method for estimating winter wheat biomass and yield.

In this study, the hyperspectral data that were obtained were ground-based data. To improve our model for estimating biomass and yield in winter wheat, and to make it more practical, it is important to estimate the accuracy and stability of the model using hyperspectral satellite data. The current Landsat and Sentinel-2 satellites provide high spatial resolution imagery data (10–60 m) with relatively short revisit periods. Based on this, Landsat and Sentinel-2 sensors have the potential for improved estimates of biomass and yield in winter wheat at regional scales. With the development of unmanned aerial vehicles (UAV), the combination of UAV and hyperspectral imaging data should allow for the timely estimation of the growth status of crops, with high spatial resolution image data at the field and farm scales, in the future. In this study, we only carried out experiments at a single-site, and obtained good results over four years. The method used in this study is transferrable to other sites. The main insights from this study are as follows: (i) The crop parameters of the AquaCrop model for different crops are parameterized to better-simulate different crop biomass and yields, during all growth stages under different environmental conditions and experimental sites; (ii) different crops should be accurately classified using high temporal and spatial resolution image data when this method is applied to regional scales; (iii) PSO will further enhance the advantages of a parallel algorithm to quickly obtain estimated results at regional scales; (iv) corresponding field crop management strategies (such as water and fertilizer management) can then be carried out, based on the estimated crop biomass, resulting in improved crop yields at regional scales; and (v) in addition, this method can be combined with higher temporal and spatial resolution image data and the AquaCrop model to improve field crop management, and then to enhance crop yield at the sub-field and sub-farm scales in the future. The positive results obtained here were based on single-site experiments over four years, however, further experiments should be carried out to adjust crop parameters of the AquaCrop model under water and fertilizer stress treatments to maintain the stability of the simulated results. The effect of the soil parameter variations on the simulated results in the AquaCrop model should be further investigated to better apply it at regional scales. Further studies are needed to verify these results for different crops, and in different ecological areas, as this study was limited to winter wheat in Beijing, China.

5. Conclusions

In this study, the PSO data assimilation algorithm was used to assimilate field spectroscopic data into the AquaCrop model to improve the estimation accuracy of winter wheat yield under different planting dates and irrigation management strategies. The conclusions are as follows: (i) Several spectral indices were highly correlated with biomass in winter wheat. The exponential regression equation between the normalized difference matter index (NDMI) and biomass was the best model for estimating biomass, with R^2 and RMSE values of 0.77 and 1.80 ton/ha, respectively; (ii) The data assimilation method ($R^2 = 0.83$ and RMSE = 1.65 ton/ha) achieved more accurate biomass estimations than the spectral index method; (iii) Yield simulated with the data assimilation method was consistent with measured yield across all four years (R^2 and RMSE values of 0.82 and 0.55 ton/ha, respectively). In summary, the results indicated that the data assimilation method is an effective method for estimating biomass and yield of winter wheat. The results provide a guideline for optimizing irrigation management strategies for winter wheat in this region.

Acknowledgments: This study was supported by the Natural Science Foundation of China (41601369, 41471285, 41471351, 41301375, 41271345,), the Beijing Natural Science Foundation (4141001), the Special Funds for Technology Innovation Capacity Building, sponsored by the Beijing Academy of Agriculture and Forestry Sciences (KJCX20140417, KJCX20150409), and Yangzhou University Excellent Doctoral Foundation. We are grateful to staff for the collection of field data.

Author Contributions: Xiuliang Jin, Xingang Xu, Guijun Yang, and Jihua Wang conceived of and designed the study; Xiuliang Jin made substantial contributions to the acquisition, analysis, and interpretation of the data; Xiuliang Jin, Zhenhai Li, Xingang Xu, Guijun Yang, and Jihua Wang performed the experiments; Xiuliang Jin and Lalit Kumar discussed the basic structure of the manuscript; and Xiuliang Jin finished the first draft; Lalit Kumar reviewed and edited the draft. All authors have read and approved the submitted manuscript, have agreed to be listed as authors, and have accepted the final version for publication.

Conflicts of Interest: The authors declare no conflicts of interest.

References

1. Zhang, H.C. Thoughts on cultivation techniques for high quality of wheat in China and its processing. *Jiangsu Agric. Sci.* **2000**, *5*, 2–6. (In Chinese)
2. Lobell, D.B.; Asner, G.P.; Ortiz-Monasterio, J.I.; Benning, T.L. Remote sensing of regional crop production in the Yaqui Valley, Mexico: Estimates and uncertainties. *Agric. Ecosyst. Environ.* **2003**, *94*, 205–220. [CrossRef]

3. Fang, H.; Liang, S.; Hoogenboom, G.; Teasdale, J.; Cavigelli, M. Corn-yield estimation through assimilation of remotely sensed data into the CSM-CERES-Maize model. *Int. J. Remote Sens.* **2008**, *29*, 3011–3032. [CrossRef]
4. Thorp, K.R.; Wang, G.; West, A.L.; Moran, M.S.; Bronson, K.F.; White, J.W.; Mon, J. Estimating crop biophysical properties from remote sensing data by inverting linked radiative transfer and ecophysiological models. *Remote Sens. Environ.* **2012**, *124*, 224–233. [CrossRef]
5. Morel, J.; Begue, A.; Todoroff, P.; Martine, J.; Lebourgeois, V.; Petit, M. Coupling a sugarcane crop model with the remotely sensed time series of fIPAR to optimise the yield estimation. *Eur. J. Agron.* **2014**, *61*, 60–68. [CrossRef]
6. Curry, R.B. Dynamic simulation of plant growth, I. Development of amodel. *Trans. ASABE* **1971**, *14*, 946–959. [CrossRef]
7. Launay, M.; Guerif, M. Assimilating remote sensing data into a crop model to improve predictive performance for spatial applications. *Agric. Ecosyst. Environ.* **2005**, *111*, 321–339. [CrossRef]
8. Asseng, S.; Ewert, F.; Rosenzweig, C.; Jones, J.W.; Hatfield, J.L.; Ruane, A.C.; Boote, K.J.; Thorburn, P.J.; Rötter, R.P.; Cammarano, D. Uncertainty in simulating wheat yields under climate change. *Nat. Clim. Chang.* **2013**, *3*, 827–832. [CrossRef]
9. Hansen, J.W.; Jones, J.W. Scaling-up crop models for climate variability applications. *Agric. Syst.* **2000**, *65*, 43–72. [CrossRef]
10. Dorigo, W.A.; Zurita-Milla, R.; de Wit, A.J.; Brazile, J.; Singh, R.; Schaepman, M.E. A review on reflective remote sensing and data assimilation techniques for enhanced agroecosystem modeling. *Int. J. Appl. Earth Obs. Geoinf.* **2007**, *9*, 165–193. [CrossRef]
11. Bacour, C.; Baret, F.; Béal, D.; Weiss, M.; pavageau, K. Neural network estimation of LAI, fAPAR, fCover and LAI × Cab from top of canopy MERIS eflectance data: Principles and validation. *Remote Sens. Environ.* **2006**, *105*, 313–325. [CrossRef]
12. Baret, F.; Hagolle, O.; Geiger, B.; Bicheron, P.; Miras, B.; Huc, M.; Berthelot, B.; Nino, F.; Weiss, M.; Samain, O.; et al. LAI, fAPAR and fCOVER cyclopes global products derived from vegetation. Part 1: Principles of the algorithm. *Remote Sens. Environ.* **2007**, *110*, 275–286. [CrossRef]
13. Jin, X.L.; Diao, W.Y.; Xiao, C.H.; Wang, F.Y.; Chen, B.; Wang, K.R.; Li, S.K. Estimation of wheat agronomic parameters using new spectral indices. *PLoS ONE* **2013**, *8*, e72736. [CrossRef] [PubMed]
14. Jin, X.; Yang, G.; Xu, X.; Yang, H.; Feng, H.; Li, Z.; Shen, J.; Lan, Y.; Zhao, C. Combined Multi-Temporal Optical and Radar Parameters for Estimating LAI and Biomass in Winter Wheat Using HJ and RADARSAR-2 Data. *Remote Sens.* **2015**, *7*, 13251–13272. [CrossRef]
15. Jin, X.L.; Wang, K.R.; Xiao, C.H.; Diao, W.Y.; Wang, F.Y.; Chen, B.; Li, S.K. Comparison of two methods for estimation of leaf total chlorophyll content using remote sensing in wheat. *Field Crops Res.* **2012**, *135*, 24–29. [CrossRef]
16. Haboudane, D.; Miller, J.R.; Pattey, E.; Zarco-Tejada, P.J.; Strachan, I.B. Hyperspectral vegetation indices and novel algorithms for predicting green LAI of crop canopies: Modeling and validation in the context of precision agriculture. *Remote Sens. Environ.* **2004**, *90*, 337–352. [CrossRef]
17. Curnel, Y.; de WitA, J.W.; Duveiller, G.; Defourny, P. Potential performances of remotely sensed LAI assimilation in WOFOST model based on an OSS Experiment. *Agric. For. Meteorol.* **2011**, *151*, 1843–1855. [CrossRef]
18. Jarlan, L.; Mangiarotti, S.; Mougin, E.; Mazzega, P.; Hiernaux, P.; Le Dantec, V. Assimilation of SPOT/VEGETATION NDVI into a sahelian vegetation dynamics model. *Remote Sens. Environ.* **2008**, *112*, 1381–1394. [CrossRef]
19. Dente, L.; Satlino, G.; Mattia, F.; Rinaldi, M. Assimilation of leaf area index derived from ASAR and MERIS data into CERES-Wheat model to map wheat yield. *Remote Sens. Environ.* **2008**, *112*, 1395–1407. [CrossRef]
20. Duveiller, G.; Defourny, P. A conceptual framework to define the spatial resolution requirements for agricultural monitoring using remote sensing. *Remote Sens. Environ.* **2010**, *114*, 2637–2650. [CrossRef]
21. Lewis, P.; Gomez-Dans, J.; Kaminski, T.; Settle, J.; Quaife, T.; Gobron, N.; Styles, J.; Berger, M. An earth observation land data assimilation system (EO-LDAS). *Remote Sens. Environ.* **2012**, *120*, 219–235. [CrossRef]
22. Ines, A.V.M.; Das, N.N.; Hansen, J.W.; Njoku, E.G. Assimilation of remotely sensed soil moisture and vegetation with a crop simulation model for maize yield prediction. *Remote Sens. Environ.* **2013**, *138*, 149–164. [CrossRef]

23. Jiang, Z.; Chen, Z.; Chen, J.; Ren, J.; Li, Z.; Sun, L. The Estimation of Regional Crop Yield Using Ensemble-Based Four-Dimensional Variational Data Assimilation. *Remote Sens.* **2014**, *6*, 2664–2681. [CrossRef]
24. Liu, F.; Liu, X.; Ding, C.; Wu, L. The dynamic simulation of rice growth parameters under cadmium stress with the assimilation of multi-period spectral indices and crop model. *Field Crops Res.* **2015**, *183*, 225–234. [CrossRef]
25. Huang, J.; Sedano, F.; Huang, Y.; Ma, H.; Li, X.; Liang, S.; Tian, L.; Zhang, X.; Fan, J.; Wu, W. Assimilating a synthetic Kalman filter leaf area index series into the WOFOST model to improve regional winter wheat yield estimation. *Agric. For. Meteorol.* **2016**, *216*, 188–202. [CrossRef]
26. Abou-Ismail, O.; Huang, J.; Wang, R. Rice yield estimation by integrating remote sensing with rice growth simulation model. *Pedosphere* **2004**, *14*, 519–526.
27. Mo, X.; Liu, S.; Lin, Z.; Xu, Y.; Xiang, Y.; McVicar, T.R. Prediction of crop yield, water consumption and water use efficiency with a SVAT-crop growth model using remotely sensed data on the North China Plain. *Ecol. Model.* **2005**, *183*, 301–322. [CrossRef]
28. Thorp, K.R.; Hunsaker, D.J.; French, A.N. Assimilation leaf area index estimates from remote sensing into the simulations of a cropping systems model. *Trans. ASABE* **2010**, *53*, 251–262. [CrossRef]
29. Fang, H.; Liang, S.; Hoogenboom, G. Integration of MODIS LAI and vegetation index products with the CSM–CERES–Maize model for corn yield estimation. *Int. J. Remote Sens.* **2011**, *32*, 1039–1065. [CrossRef]
30. Li, Y.; Zhou, Q.G.; Zhou, J.; Zhang, G.F.; Chen, C.; Wang, J. Assimilating remote sensing information into a coupled hydrology-crop growth model to estimate regional maize yield in arid regions. *Ecol. Model.* **2014**, *291*, 15–27. [CrossRef]
31. Maas, S.J. Use of remotely-sensed information in agricultural crop growth models. *Ecol. Model.* **1988**, *41*, 247–268. [CrossRef]
32. Plummer, S.E. Perspectives on combining ecological process models and remotely sensed data. *Ecol. Model.* **2000**, *129*, 169–186. [CrossRef]
33. Jongschaap, R.E. Run-time calibration of simulation models by integrating remote sensing estimates of leaf area index and canopy nitrogen. *Eur. J. Agron.* **2006**, *24*, 316–324. [CrossRef]
34. Delecolle, R.; Maas, S.J.; Guerif, M.; Baret, F. Remote sensing and crop production models-present trends. *ISPRS J. Photogramm.* **1992**, *47*, 145–161. [CrossRef]
35. Liang, S.; Li, X.; Xie, X.H. *Land Surface Observation, Modeling and Data Assimilation*; Higher Education Press: Beijing, China, 2013.
36. Steduto, P.; Hsiao, T.C.; Raes, D.; Fereres, E. AquaCrop-The FAO crop model to simulate yield response to water. I. Concepts. *Agron. J.* **2009**, *101*, 426–437. [CrossRef]
37. Jin, X.L.; Feng, H.K.; Li, Z.H.; Song, S.N.; Zhu, X.K.; Song, X.Y.; Yang, G.J.; Xu, X.G.; Guo, W.S. Assessment of the AquaCrop model for use in simulation of irrigated winter wheat canopy cover, biomass, and grain yield in the North China Plain. *PLoS ONE* **2014**, *9*, e86938. [CrossRef] [PubMed]
38. Allen, R.G.; Pereira, L.S.; Raes, D.; Smith, M. *Crop Evapotranspiration-Guidelines for Computing Crop Water Requirements*; FAO Irrigation and Drainage Paper 56; FAO: Rome, Italy, 1998.
39. Penuelas, J.; Filella, I.; Biel, C.; Serrano, L.; Save, R. The reflectance at the 950–970 nm region as an indicator of plant water status. *Int. J. Remote Sens.* **1993**, *14*, 1887–1905. [CrossRef]
40. Seelig, H.D.; Hoehn, A.; Stodieck, L.S.; Klaus, D.M.; Adams, W.W.; Emery, W.J. The assessment of leaf water content using leaf reflectance ratios in the visible, near-, and short-wave-infrared. *Int. J. Remote Sens.* **2008**, *29*, 3701–3713. [CrossRef]
41. Hunt, E.R., Jr.; Rock, B.N. Detection of changes in leaf water content using near-and middle-infrared reflectances. *Remote Sens. Environ.* **1989**, *30*, 43–54.
42. Wang, L.; Qu, J.J.; Hao, X.; Hunt, E.R. Estimating dry matter content from spectral reflectance for green leaves of different species. *Int. J. Remote Sens.* **2011**, *32*, 7097–7109. [CrossRef]
43. Jin, X.L.; Xu, X.G.; Song, X.Y.; Li, Z.H.; Wang, J.H.; Guo, W.S. Estimation of leaf water content in winter wheat using grey relational analysis-partial least squares modeling with hyperspectral data. *Agron. J.* **2013**, *105*, 1385–1392. [CrossRef]
44. Jiang, Z.; Huete, A.R.; Didan, K.; Miura, T. Development of a two-band enhanced vegetation index without a blue band. *Remote Sens. Environ.* **2008**, *112*, 3833–3845. [CrossRef]
45. Rondeaux, G.; Steven, M.; Baret, F. Optimization of soil-adjusted vegetation indices. *Remote Sens. Environ.* **1996**, *55*, 95–107. [CrossRef]

46. Dash, J.; Curran, P.J. The MERIS terrestrial chlorophyll index. *Int. J. Remote Sens.* **2004**, *25*, 5403–5413. [CrossRef]

47. Gitelson, A.A.; Viña, A.; Ciganda, V.; Rundquist, D.C. Remote estimation of canopy chlorophyll content in crops. *Geophys. Res. Lett.* **2005**, *32*. [CrossRef]

48. Rouse, J.W.; Haas, R.H.; Schell, J.A.; Deering, D.W.; Harlan, J.C. *Monitoring the Vernal Advancement of Retrogradation (Green Wave Effect) of Natural Vegetation*; Type III, Final Report; NASA: Washington, DC, USA, 1974; pp. 1–371.

49. Jin, X.L.; Yang, G.J.; Li, Z.H.; Feng, H.K.; Xu, X.G. New combined spectral index to improve total leaf chlorophyll content estimation in cotton. *IEEE J. Sel. Top. Appl. Earth Obs. Remote Sens.* **2014**, *7*, 4589–4600. [CrossRef]

50. Gitelson, A.A. Wide dynamic range vegetation index for remote quantification of characteristics of vegetation. *J. Plant Physiol.* **2004**, *161*, 165–173. [CrossRef] [PubMed]

51. Brownlee, K.A. *Statistical Theory and Methodology in Science and Engineering*; John Wiley & Sons: New York, NY, USA, 1956.

52. Raes, D.; Steduto, P.; Hsiao, T.C.; Fereres, E. AquaCrop-The FAO Crop Model to Simulate Yield Response to Water: Reference Manual Annexes. Available online: http://www.fao.org/nr/water/aquacrop.html (accessed on 22 October 2009).

53. Raes, D.; Steduto, P.; Hsiao, T.C.; Fereres, E. *Reference Manual: AquaCrop Plug-in Program (Version 4.0)*; FAO: Rome, Italy, 2012.

54. Lorite, I.J.; García-Vila, M.; Santos, C.; Ruiz-Ramos, M.; Fereres, E. AquaData and AquaGIS: Two computer utilities for temporal and spatial simulations of water-limited yield with AquaCrop. *Comput. Electron. Agric.* **2013**, *96*, 227–237. [CrossRef]

55. Kenndy, J.; Eberhart, R. Particle swarm optimization. In Proceedings of IEEE International Conference on Neural Networks, Perth, Australia, 27 November–1 December 1995; pp. 1942–1948.

56. Vanuytrecht, E.; Raes, D.; Willems, P. Global sensitivity analysis of yield output from the water productivity model. *Environ. Model. Softw.* **2014**, *51*, 323–332. [CrossRef]

57. Wang, H.; Zhu, Y.; Li, W.; Cao, W.; Tian, Y. Integrating remotely sensed leaf area index and leaf nitrogen accumulation with RiceGrow model based on particle swarm optimization algorithm for rice grain yield assessment. *J. Appl. Remote Sens.* **2014**, *8*. [CrossRef]

58. Soddu, A.; Deidda, R.; Marrocu, M.; Meloni, R.; Paniconi, C.; Ludwig, R.; Soddea, M.; Mascarob, G.; Perrab, E. Climate variability and durum wheat adaptation using the AquaCrop model in southern Sardinia. *Procedia Environ. Sci.* **2013**, *19*, 830–835. [CrossRef]

59. Heng, L.K.; Hsiao, T.C.; Evett, S.; Howell, T.; Steduto, P. Validating the FAO AquaCrop model for irrigated and water deficient field maize. *Agron. J.* **2009**, *101*, 488–498. [CrossRef]

60. Jin, X.L.; Yang, G.J.; Li, Z.H.; Xu, X.G.; Wang, J.H.; Lan, Y.B. Estimation of water productivity in winter wheat using the AquaCrop model with field hyperspectral data. *Precis. Agric.* **2016**. [CrossRef]

61. Wang, X.X.; Wang, Q.J.; Fan, J.; Fu, Q.P. Evaluation of the AquaCrop model for simulating the impact of water deficits and different irrigation regimes on the biomass and yield of winter wheat grown on China's Loess Plateau. *Agric. Water Manag.* **2013**, *129*, 95–104.

remote sensing

MDPI

Article

Integration of Optical and X-Band Radar Data for Pasture Biomass Estimation in an Open Savannah Woodland

Michael Schmidt [1,*], John Carter [1], Grant Stone [1] and Peter O'Reagain [2]

[1] Queensland Department of Science, Information Technology and Innovation, GPO BOX 5078, Brisbane QLD 4102, Australia; john.carter@dsiti.qld.gov.au (J.C.); grant.stone@dsiti.qld.gov.au (G.S.)

[2] Department of Agriculture and Fisheries, P.O. Box 976, Charters Towers, QLD 4820, Australia; peter.O'Reagain@daf.qld.gov.au

* Correspondence: michael.schmidt@dsiti.qld.gov.au; Tel.: +61-7-3170-5675

Academic Editors: Lalit Kumar, Onisimo Mutanga, Xiaofeng Li and Prasad S. Thenkabail
Received: 2 September 2016; Accepted: 25 November 2016; Published: 1 December 2016

Abstract: Pasture biomass is an important quantity globally in livestock industries, carbon balances, and bushfire management. Quantitative estimates of pasture biomass or total standing dry matter (TSDM) at the field scale are much desired by land managers for land-resource management, forage budgeting, and conservation purposes. Estimates from optical satellite imagery alone tend to saturate in the cover-to-mass relationship and fail to differentiate standing dry matter from litter. X-band radar imagery was added to complement optical imagery with a structural component to improve TSDM estimates in rangelands. High quality paddock-scale field data from a northeastern Australian cattle grazing trial were used to establish a statistical TSDM model by integrating optical satellite image data from the Landsat sensor with observations from the TerraSAR-X (TSX) radar satellite. Data from the dry season of 2014 and the wet season of 2015 resulted in models with adjusted r^2 of 0.81 in the dry season and 0.74 in the wet season. The respective models had a mean standard error of 332 kg/ha and 240 kg/ha. The wet and dry season conditions were different, largely due to changed overstorey vegetation conditions, but not greatly in a pasture 'growth' sense. A more robust combined-season model was established with an adjusted r^2 of 0.76 and a mean standard error of 358 kg/ha. A clear improvement in the model performance could be demonstrated when integrating HH polarised TSX imagery with optical satellite image products.

Keywords: TerraSAR-X; Landsat; pasture biomass; Wambiana grazing trial; foliage projective cover; fractional vegetation cover

1. Introduction

Savannahs cover approximately 20% of the Earth's land surface and are characterised as a grassland ecosystem, with trees being sufficiently widely spaced so that the canopy does not close [1]. The understorey herbaceous layer consists primarily of grasses [1] which are a major contributor to the carbon balance. To a large extent these areas are extensively grazed by native, domestic, and feral herbivores—supporting conservation, tourism, and pastoral activities [2]. Pastures play an important role in rangeland ecology, ecosystem services, and livestock-related industries [2]. Physical sampling of pasture biomass over large areas is not generally considered feasible in rangeland and savannah systems; it is not possible to collect and collate sufficient field data to adequately inform land managers and provide sufficient input for pasture biomass modelling [3]. A major issue is the estimation of pasture biomass for livestock forage budgeting and conservation purposes [4]. Spatially explicit seasonal pasture biomass estimates could assist land managers and a host of other stakeholders to make assessments relating to livestock production and land resource management.

Pasture biomass estimates from space have been actively pursued over time with varying approaches, initially relating vegetation cover to biomass [5]. Vegetation indices of optical satellite imagery such as NDVI (Normalised Differenced Vegetation Index) or EVI (Enhanced Vegetation Index) focus on the green vegetation component. A general relationship between vegetative ground cover and pasture biomass exists for low ground cover areas, but when the ground cover is close to 100% the cover-to-mass relationship saturates and reliable estimates are not possible even at low biomass levels [6–8]. Investigating this relationship, Hobbs [5] related four vegetation indices to field data and found a breakdown of biomass levels >1000 kg/ha. The authors of [9] had success in relating bulk-green pasture biomass to NDVI for their study site in New Zealand, but also found that pasture biomass estimates based on NDVI were saturating. This has been confirmed in other studies, for example, Kawamura et al. [10], who compared NDVI and EVI from MODIS (Moderate Resolution Imaging Spectroradiometer) and AVHRR (Advanced Very High Resolution Radiometer) and found a saturation in the relationship when NDVI and EVI values were approximately 0.9 and 0.8, respectively. In most savannah systems this approach has significant limitations for pasture biomass estimation due to the presence of senescent dry grass and tree cover. To overcome this direct limitation e.g., Edirisinhe et al. [7] and Holm [11] used time series of optical MODIS and AVHRR imagery for a quantitative pasture biomass assessment using cumulative NDVI data. The author of [12] has related MODIS BRDF parameters to pasture biomass, with some success. Other approaches rely on physically based models incorporating remotely sensed raster images and meteorological data to model and forecast pasture biomass [4]. It is likely that cover-to-mass relationships are variable with pasture composition and the degree of grazing necessitating location specific calibration of satellite indices.

Synthetic Aperture Radar (SAR) imagery from active sensors can add to the cover signal as the backscatter data provide structural information of the surface [13,14]. In a review on biomass mapping with optical and SAR imagery, Kumar et al. [15] include a section on grasslands and reported that very few studies so far are using SAR for pasture biomass estimation. SAR imagery is a two-dimensional representation of the surface backscatter from an active sensor [16]. A range of SAR imaging systems exists in different wavelength, most notably: P-band (30–100 cm), L-band (15–30 cm), S-band (15–30 cm), C-band (3.75–7.5 cm), or X-band (2.4–3.75 cm). The sideways-looking radar pulses or chirps are emitted and recorded in either horizontal or vertical polarisation. The phase and backscatter information, converted to complex data, are stored in image bands with four potential polarisation combinations (quad-pol): HH, VV, VH, and HV (the first letter indicated the emitted and the second the received polarisation). The received backscatter from the surface is dependent on (a) sensor parameters, such as wavelength, polarisation, look angle, and resolution; and (b) scene parameters, such as surface roughness, local terrain, dielectric properties, target density, and distribution [16]. The features of interest should be on the order of magnitude of the wavelength, e.g., for pasture monitoring P-Band imagery with a 60-cm wavelength may not interact with the grass plants directly, while with lower wavelength, such as X-band, interactions with the grass stems are more likely [14]. Different scattering mechanisms occur when the emitted photons interact with the surface (direct single scattering, direct ground reflection, double bounce, etc.). In complex structures, such as tree crowns, multiple scattering (in the lower wavelength) and a change in polarisation are common [16].

The authors of [17] applied C-band RADARSAT-2 HH imagery in an explorative study to assess grassland spatial heterogeneity and concluded that it is possible to map pasture biomass with SAR imagery. The authors of [18] mapped pasture types in Western Australia and found that C-band SAR data alone were not effective for pasture discrimination, but the combination with optical imagery showed more discriminative power than either dataset alone. The authors of [19] used a time series of optical and RADARSAT-2 quad-pol imagery for grassland/crop differentiation with support vector machines and found that SAR imagery (via parameters from poliametric decomposition) resulted in better classification accuracies than optical imagery (0.98 compared to 0.81) for their study site in Brittany. The authors of [20] have analysed a time series of X-band TerrarSAR-X (TSX) and

COSMO-SkyMed imagery in combination with Landsat and Spot-4 optical data to monitor irrigated grasslands. They compared the in situ data of vegetation properties with satellite imagery and concluded that X-band data are sensitive to variations in moisture irrespective of the grass cover, though the potential for X-band data for monitoring grassland growth is very limited. Their study achieved better results when monitoring soil moisture variations with X-band imagery. The authors of [21] explored dual-polarisation ALOS PALSAR and TSX imagery (HH, VV) in combination with Landsat imagery and extensive field data for pasture biomass estimation. Their results showed some promise relating TSX-derived alpha entropy [21] to pasture biomass, but the statistical relationship with field data was inconclusive. The authors of [14] have used multi-temporal HH-polarised imagery from C-band ENVISAT ASAR, L-band ALOS PALSAR, and X-band COSMO-SkyMed for pasture monitoring in southern Australia. SAR backscatter data were correlated with vegetation indices derived from optical MODIS, Landsat, and SPOT 5 imagery and report on the feasibility of pasture biomass estimation with SAR imagery, particularly with X and C- band data in the early growing season. The authors of [22] performed a robust regression estimation of pasture biomass with TSX (HH, HV) in New Zealand, with a time series of imagery between February 2008 and April 2009 and associated field observations. Their model revealed a regression-based biomass model with a mean residual error of 317 kg/ha.

So far, to the knowledge of the authors, no reliable pasture biomass monitoring system in savannah ecosystems based on satellite imagery has been published.

In our approach we have focused on the question of whether satellite imagery can be used to establish a pasture biomass model; and if TSX X-band data with 3.1 cm wavelength have sufficient interaction with grass species to add to pasture biomass estimation in comparison with biophysical image products from optical imagery alone.

2. Materials and Methods

2.1. Field Data

On-ground estimates of pasture biomass can be acquired by destructive and non-destructive (e.g., visual) means. Destructive methods involve laborious cutting and drying of a large number of samples of a known metric such as a quadrat (e.g., 0.25 m^2), where the dried biomass values are scaled up to estimate a larger area [23]. Purely visual estimates of pasture biomass and larger areas often have large errors and are generally variable between different operators. The BOTANAL methodology [24] employed in this study enables multiple users to traverse large transect lines, where quadrats are visually estimated at given distances and calibration curves are applied to scale up to larger (e.g., field-scale areas). This results in more accurate and less subjective estimates. The estimates are calibrated to destructive measurements of dried grass to represent TSDM (Total Standing Dry Matter). Appendix A lists Australian pastoralism terminology used in the text.

The Wambiana grazing trial site is located southwest of the township of Charters Towers in anopen woodland. The landscape of the trial is characterised by three main tree species of Reid River box (*Eucalyptus brownii*), brigalow (*Acacia harpophylla*), and silver leaf ironbark (*Eucalyptus melanophloia*) (Figure 1). Mature tree heights are typically 12–15 m, with a foliage projective cover [25] ranging from 5%–20% across the paddock by land type combinations. The tree species are evergreen, but can suffer partial defoliation in drought, with the possibility of some variation in the canopy between the wet and dry seasons. Measure leaf size ranges for dominant tree species [26,27] are: *E. brownii* (8–15 cm × 2–4 cm); *E. melanophloia* (5–9 cm × 2–3 cm); and *A. harpophylla* (10–20 cm × 0.7–1.6 cm), which are likely to interact with the 3.1 cm X-band wave lengths. An understorey of non-edible native shrubs (currant bush; *Carissa ovata*) and false sandalwood (*Eremophila mitchellii*) are present, with Carissa covering 25%–30% of the box land type [28].

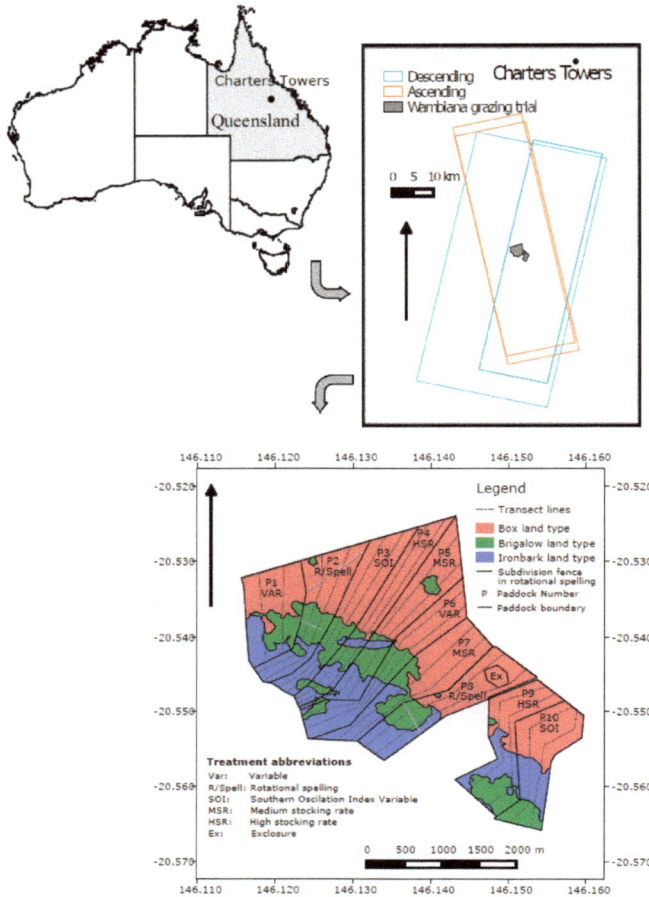

Figure 1. Location of the Wambiana grazing trial (in red, upper left panel) and the TSX overpasses in 2014 (magenta) and 2015 (cyan). The bottom panel shows the layout of experimental plots, associated grazing strategies, and approximate transect lines across the three major vegetation associations.

The dominant understorey native grasses include desert bluegrass (*Bothriochloa ewartiana*) and black speargrass (*Heteropogon contortus*), which are preferred for grazing; along with less desired grasses such as wiregrass (*Aristida* spp.) and wandarrie grass (*Eriachne mucronata*). These grasses are largely erect tussock grasses, generally less than 50 cm high with leaves less than 1 cm wide, although in recent years Indian couch (*Bothriochloa pertusa*), a more prostrate, thin-stemmed, introduced grass, has become increasingly present, particularly in heavily grazed treatments. The grasses generally change from green to dry as the season progresses and leaf disappears faster than stem material (with grazing and detachment). At higher stocking rates, grass tussocks at the end of the dry season are dominated by stem material (5–30 cm tall and 1–2 mm in diameter), generally from the least productive species. In prolonged drier periods (i.e., drought), there may be little standing material of any grass species present, with only grass crowns visible. The length-to-width ratio of the elements of the grass sward suggests that X-band radar should interact. Tussock density of the productive grasses can vary from about 5 tussocks/m^2 to 1.5 tussocks/m^2 in heavily grazed areas.

The soil types at the site are described as texture contrast (sodosols) associated with the Box trees, heavy clays (vertosols) aligned to the brigalow trees, and yellow-brown earths (kandosols) for the ironbark trees. The site is generally flat, with the heavy clay soils being "micro-gilgaied" (i.e., depressions) with vertical scales of a few centimetres.

The trial offers a time series of TSDM observations systematically and consistently acquired since 1997 to estimate land type and paddock scale TSDM. Five different grazing strategies with two replicates each have been tested on 10 paddocks, each 100 ha in size [29], for assessing sustainable and profitable land management. The trial has clearly demonstrated the productive benefits of improved grazing management, in a manner and scale of direct relevance to the grazing industry of northern Australia. A key outcome of the trial is that a loss of land condition under heavy stocking compromised productivity, profitability, and the local environment [28]. TSDM field data are acquired for the grazing trial at paddock scale in May and October each year. Two parallel transects per paddock were established, with vegetation and soil parameters recorded by experienced field operators approximately every 50 m along these transects (Figure 1). TSDM estimates were made using calibrated visual observations (BOTANAL method) [24], in association with 0.5 m × 0.5 m quadrats used to harvest pasture TSDM.

The Wambiana paddock grazing treatments and approximate stocking rates with 1 AE = 1 animal equivalent or 450 kg steer (only steers are used in the trial) are:

- Medium stocking rate—relatively constant stocking at 8–10 ha/AE.
- Heavy stocking rate—relatively constant stocking at 4–5 ha/AE to May 2005; thereafter stocked at 6 ha/AE until May 2009, when stocking rates were returned to 4 ha/AE.
- Variable stocking—stocking rates adjusted upwards or downwards in May based on end of wet season feed availability (3–12 ha/AE).
- SOI (Southern Oscillation Index) variable stocking—stocking rates adjusted upwards or downwards in October based on feed availability and SOI forecasts for the next wet season (3–12 ha/AE).
- Rotational wet season spelling—spell a third of the paddock each wet season; relatively constant stocking at 7–8 ha/AE until November 2003 and at 8–10 ha/AE thereafter.

Pasture growth is strongly influenced by rainfall. The average long-term annual rainfall for the nearest climate station (17 km northwest of the site) is 643 mm, but annual rainfall is highly variable ranging from 207 to 1409 mm. The seasons related to this study were below average and are discussed in further detail below.

2.2. Optical Satellite Imagery and Products

Landsat data originating from the United States Geological Survey were utilised in this study. All available imagery for October/November 2014 and May/June 2015 were included (Table 1). The Landsat imagery were atmospherically corrected, and cloud-masked following the standardised pre-processing steps, as described in [30].

Table 1. Landsat image dates and sensors used in the observation period.

Dry Season Date	Sensor	Wet Season Date	Sensor
2 October 2014	OLI	6 May 2015	ETM+
10 October 2014	ETM+	14 May 2015	OLI
18 October 2014	OLI	22 May 2015	ETM+
3 November 2014	OLI	7 June 2015	ETM+
11 November 2014	ETM+	23 June 2015	ETM+
19 November 2014	OLI		
27 November 2014	ETM+		

Biophysically meaningful standardized data products were used here:

(a) Foliage Projective Cover (FPC)

Foliage projective cover is a metric describing the vertical projection of the foliated tree canopy in units of percent [25]. FPC is an important variable as the study site is located in anopen woodland, and thus are reflectance or backscatter signals influenced by the tree canopy. The authors of [31] have developed a state-wide FPC data product based on dry-season Landsat imagery at 30 m pixel size. This product is based on a multiple regression of Landsat imagery with field observation of stand basal area (RMSE < 10%), validated with independent FPC estimates from LiDAR aerial survey (RMSE 5.3%) across the major vegetation communities in Queensland.

FPC predictions from all available Landsat 7 ETM+ and Landsat 8 OLI imagery within the period of October to November 2014 and May to June 2015 were accessed (Table 2).

(b) Fractional Vegetation Cover (FVG)

Vegetative ground cover is a key piece of information in natural resource management and important for pasture biomass estimation [32]. A 30-m Landsat Fractional Vegetation Cover dataset (FVC) was developed by [33]. The data product contains fractional cover estimates of green vegetation, non-green vegetation, and bare ground summing to 100 percent plus model error. The authors incorporated 968 fractional vegetation cover field data points, collected at one hectare field sites [23] across the states of Queensland and New South Wales (Australia) with the closest Landsat image observation (no more than 60 days apart). These data were used to derive image-based endmember spectra of green vegetation, non-green vegetation, and bare soil, which were applied in a spectral unmixing to generate Landsat-based predictions for these fractions with an RMSE of 11.8%.

All available single-date Landsat 8 images within the time interval of October to November 2014 and April to May 2015 were processed to FVC cover (Table 1). The non-green vegetation component in the FVC product is a combined estimate of the senescent (non-green) vegetation and litter component, which will now be herein referred to as "dry vegetation".

2.3. X-Band SAR Imagery: TerrarSar-X (TSX)

Imagery from the TSX instrument in StripMap mode were acquired for the end of the dry season (October/November) 2014 and the end of the wet season (May) 2015 (Figure 1) with a 3.1 cm wavelength. Three overpasses in the dry season of 2014 (two with dual polarisation: HH/HV) and two dual-polarisation images (HH/HV) in the wet season of 2015 were obtained. Table 2 lists the most relevant metadata of the images used.

Table 2. TSX observation dates and metadata.

Date	26 October 2014	14 November 2014	17 November 2014	20 May 2015	23 May 2015
Polarisation	HH	HH/HV	HH/HV	HH/HV	HH/HV
Orbit	Descending	Ascending	Descending	Ascending	Descending
Incidence Angle	39.22°	33.73°	38.73°	32.93°	38.72°
Pixel resolution	3.25 m	3.75 m	3.75 m	4.0 m	3.75 m
Time (UTC)	19.50 h	08.39 h	19.50 h	08.39 h	19.50 h
Local time	5.50 h *	16.39 h	5.50 h *	16.39 h	5.50 h *

* +1 day.

The level 1b enhanced ellipsoid corrected imagery were calibrated and processed to terrain corrected γ_0 in decibels with the science toolbox exploitation platform (SNAP) provided by the European Space Agency [34], following:

$$\beta_0 = k \times DN$$

$$\sigma_0 = \beta_0 \times \sin(\theta_{loc}) \; ; assuming \; a \; flat \; terrain$$

$$\gamma_0 = \frac{\sigma_0}{\cos(\theta_{loc})} \; ; fully \; terrain \; corrected$$

where *DN* is the digital number, *k* is the calibration coefficient, and θ_{loc} is the local incidence angle.

2.4. Ancillary Data

Daily rainfall data were extracted from the "Data Drill" option of the SILO climate database (Scientific Information for Land Owners; [35]), which contains daily interpolated surfaces from available climate station data. The rainfall total of the displayed time interval (Figure 2) of the dry season 2014 and the wet season 2015 was 369.6 mm, which was well below the long-term average (643 mm). Soil volumetric water data were recorded with a data logger located in Paddock 8 with a time domain reflectometry probe (TDR).

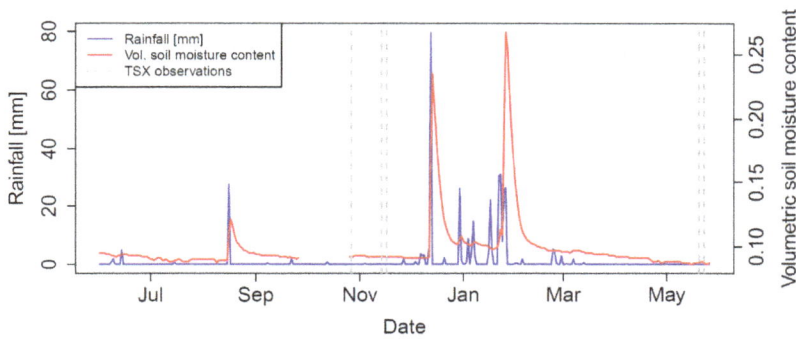

Figure 2. Time series of daily rainfall and soil volumetric water content in 0–30 cm (Paddock 8) time series for Wambiana station (June 2014 to May 2015). Vertical lines indicate the TSX observation dates.

The wet season rainfall for the site was uncharacteristically low, coinciding with a strong El Niño event, which has a strong link with rainfall and vegetation growth in northeast Queensland [36]. The soil surface was quite dry for all image acquisitions and slightly drier in the May 2015 data acquisitions. The average maximum day and minimum night air temperatures were 41.1 °C and 14.1 °C during the TSX observation period October 2014; and 32.4 °C and 11.0 °C in May 2015.

The vapour pressure deficit (VPD) was calculated from the SILO vapour pressure data following [37]. VPD is the difference (deficit) between the amount of moisture in the air and how much moisture the air can hold when it is saturated. VPD was calculated as the evaporative power of the atmosphere in the boundary layer above the canopy and has been shown to be correlated with overstorey FPC [25,31].

2.5. Spatial Data Analysis

The TSDM transect data were spatially averaged by paddock-and-land type parcels, resulting in 37 observations for October 2014 and 37 observations for May 2015. The parcel averages may not have represented the barest areas, which had consistently low TSDM; therefore, four additional polygons were digitised from high-resolution imagery for areas with low TSDM and included as additional data. These low-TSDM areas had FPC values ranging from 0% to 18%.

To ensure a sample from a high-TSDM region was represented, a visual field estimate of TSDM was made from an area exclosed from grazing, located in Paddock 8. The parcels formed the basis for all further analysis. All available raster data were spatially averaged to match these units (Figure 1).

The available single-date TSX data were used as well as temporal aggregations for a noise reduction: rasters of temporal minimum, maximum, and mean were calculated for the TSX HH and HV time series in 2014 and 2015, respectively. Mean rasters were calculated for the FPC and FVC data for the observation time intervals.

The Eureqa package [38] was used to conduct an extensive search for linear and nonlinear functions using a large set of predictor variables. A robust multiple regression analysis was performed in the R package [39] with the most prominent candidate variables: the single date and temporally aggregated datasets with the aim to generate TSDM maps for the two seasons in 2014 and 2015. The adjusted r^2 was used in reporting instead of the multiple r^2, as it adjusts for the number of terms in a model (i.e., it decreases when a predictor improves the model by less than expected by chance). The two season-specific models were compared with a combined model using all observations.

Correlations were done for several versions of model runs, starting with a one-variable model to models with an increased complexity of up to four variables (without interaction terms). All model combinations were reported, incorporating optical and SAR variables (FPC, FVC, and TSX variables).

An analysis of variance (ANOVA) was performed on the nested three variable models in comparison to a four-variable model, to test if there was a significant model improvement if a fourth variable was added; the Wald test uses a chi-square distribution to test for a model improvement.

3. Results

The TDSM paddock averages for 2014 and 2015 are displayed in Figure 3, categorised by paddock (grazing strategy) and land type in the Wambiana grazing trial. TSDM was generally higher at the end of the dry season on October 2014 than at the end of the wet season in May 2015 due to the impact of grazing and a failed wet season. The mean TSDM across all paddocks was 1190 kg/ha for the end of the 'dry' season and 612 kg/ha for the end of the 'wet' season.

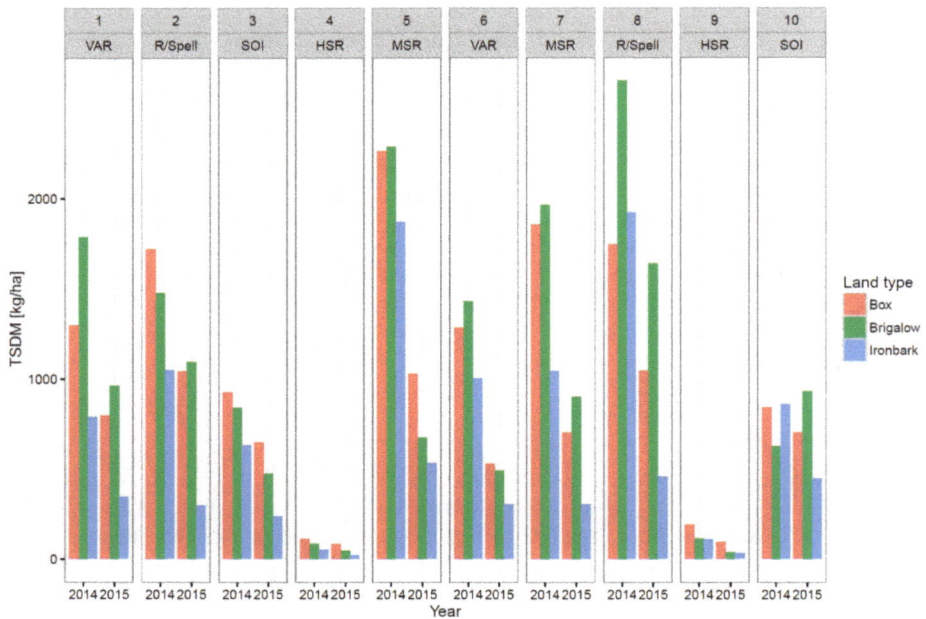

Figure 3. TSDM parcel averages recorded for October 2014 and May 2015, categorised by paddock (grazing strategy) and land type in the Wambiana grazing trial.

The HSR strategy (highest pasture utilisation) had the lowest TSDM, while the MSR strategy generally displayed the highest TSDM values. An exception to this was the Brigalow land type in Paddock 8 (R/Spell strategy), where high TSDM values were reported in 2014 as well as in 2015. In this paddock and land type there is an artificial topographic feature in the form of a levee bank, which may retain some excess moisture and could explain the higher TSDM values.

The increase in green cover from October 2014 to May 2015 (Figure 4) is likely due to an increase in overstorey greenness and not the pastures' greenness.

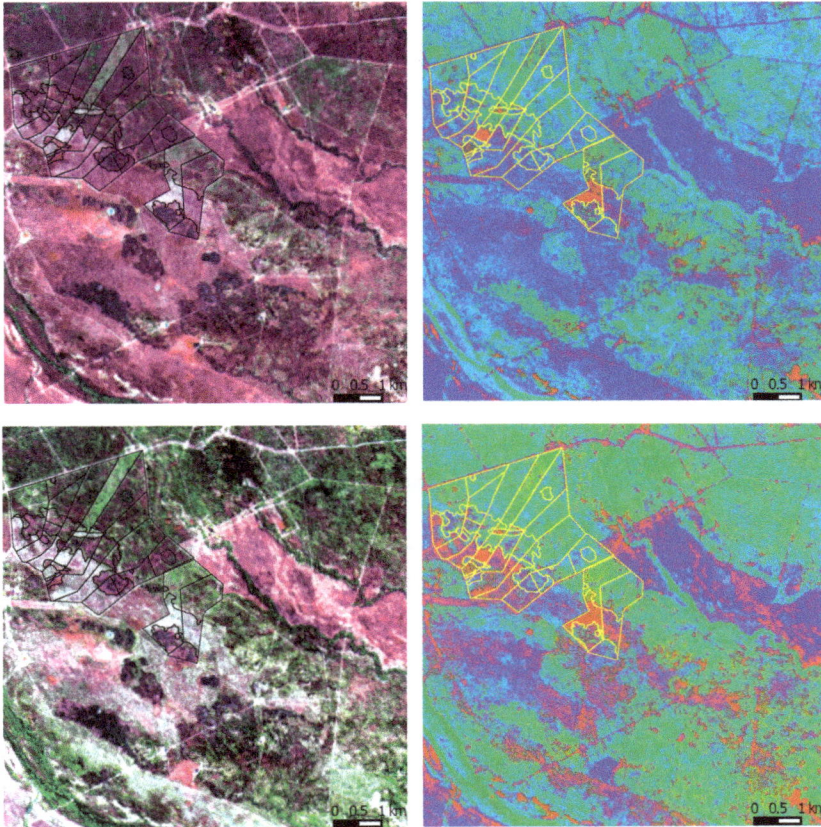

Figure 4. Landsat 8 R/G/B with spectral bands 6/5/4 for a dry season image from 19 November 2014 (**upper left panel**) and a wet season image from 14 May 2015 (**lower left panel**) at the same colour stretch. FVC images show a difference in vegetative cover between the 2014 dry (**upper right panel**) and the 2015 wet season (**lower right panel**). In this representation of FVC, R = bare ground, G = green vegetation, B = dry vegetation. The Wambiana grazing trial outline is superimposed in black on R/G/B imagery in the left panels and in yellow on FVC on the right.

The paddocks with different grazing treatments are visible in the R/G/B and the FVC imagery as well as the difference in greenness and vegetative cover between the two seasons.

In order to establish a TSDM model from the variables available, several regressions with differing variables were tested. The best-performing and simplest model revealed a robust multiple regression with four variables:

(1) FPC;
(2) dry fraction of the FVC (dryFVC);
(3) the maximum HH (HHmax); and
(4) the minimum HH (HHmin).

Without interaction terms: TSDM (dryFVC, FPC, HHmax, HHmin) as shown in Figure 5. All four variables were highly significant ($p < 0.001$).

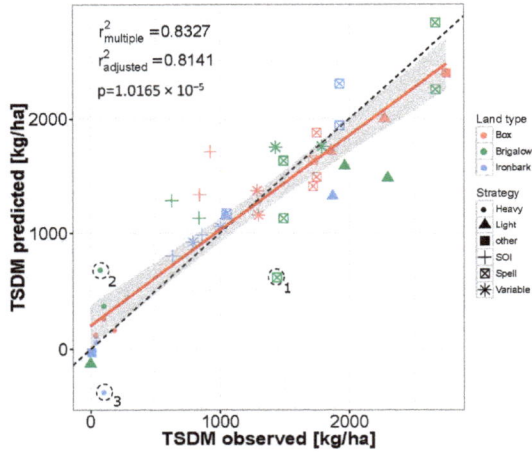

Figure 5. Regression for the dry season 2014 TSDM at paddock scale, with a mean standard error of 332 kg/ha. With TSDM = $-13{,}118.4 + 254.5 \times$ FPC $+ 85.6 \times$ dryFVC $- 4149.9 \times$ HHmax $+ 2559.0 \times$ HHmin.

The same variables were used in 2015 for a robust multiple regression for the wet season (Figure 6). All variables were highly significant ($p < 0.001$) with the exception of HHmin ($p < 0.1$).

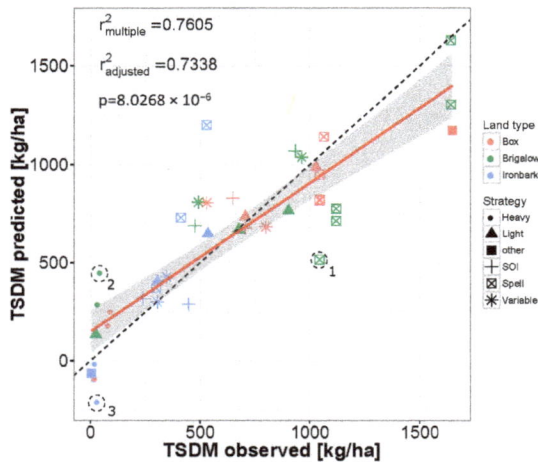

Figure 6. Regression for the dry season 2015 TSDM at paddock scale, with a mean standard error of 240 kg/ha. TSDM = $-10{,}406.4 + 73.5 \times$ FPC $+ 84.7 \times$ dryFVC $- 1048.3 \times$ HHmax $+ 357.3 \times$ HHmin.

Three data points that appear as outliers in both models are indicated with a dotted circle (Figures 5 and 6). Point 1 refers to a small scalded area, approximately 90 m in diameter on Brigalow land type in Paddock 2. Given the nature of the TSDM field data transects, it is conceivable that the observations are too high in both seasons—the transects may not have intersected in this small area and the surrounding high yield box land type area estimates were attributed instead. Point 2 has high predictions in both models, which may be attributable to a levee bank to retain water and the associated disturbed soil when it was established. This would most certainly alter the TSX backscatter signal. Point 3 in both seasons predicted negative TSDM values. It happens that this is the paddock with lowest cover at low FPC and therefore the data point is at the extreme end. The red square represents the grazing exclosure (shown as Ex in Figure 1) and the blue square is a bare area taken just outside the grazing trial area.

The four input variables were tested for their relevance in predicting TSDM for 2014 and 2015. All variable combinations were tested in multiple regression models. Table 3 summarises the variable importance and the respective predictive qualities for TSDM via the adjusted r^2 values and the mean residual standard error (MRSE). This was also performed for a model combining all data from both seasons.

Table 3. Adjusted r^2 values with regression variables and TSDM for 2014, 2015, and both seasons combined. The mean residual standard error (MRSE) is given in kg/ha. The number of variables included in the models increases down the table.

r^2 2014	MRSE 2014	r^2 2015	MRSE 2015	r^2 2014/15	MRSE 2014/15	FPC	dryFVC	HHmax	HHmin
0.06	792	0.03	430	0.17	666	x			
0.57	572	0.57	293	0.61	456		x		
0.09	842	0.13	443	0.23	642			x	
0.15	776	0.17	421	0.28	622				x
0.61	432	0.69	292	0.61	458	x	x		
0.07	827	0.14	437	0.24	640	x		x	
0.15	810	0.18	437	0.30	216	x			x
0.59	497	0.63	296	0.63	442		x	x	
0.60	447	0.63	296	0.63	450		x		x
0.63	554	0.22	421	0.45	544			x	x
0.67	425	0.73	235	0.68	411	x	x	x	
0.62	434	0.70	230	0.64	442	x	x		x
0.60	493	0.20	431	0.45	543	x		x	x
0.72	395	0.63	285	0.67	419		x	x	x
0.81	332	0.74	240	0.76	358	x	x	x	x

x represents the variables in the regression model to predict TSDM.

As a single variable, dryFVC exhibits the highest correlation with pasture TSDM. The optical data alone (FPC and dryFVC) have r^2 values of 0.61 and 0.69 in 2014 and 2015. The r^2 increases when adding the HHmax by 6% and 4%, respectively. The addition of HHmin resulted in an increase in r^2 of 20% in 2014 and 5% in 2015.

Despite the suspicion of collinearity for HHmin and HHmax reduced the inclusion the model variance for high and low TSDM. An ANOVA for the 2014 data indicated a highly significant model improvement by including either HHmin or HHmax compared to a three-variable model. In 2015 the inclusion of HHmin was significant at the 1% level with a lower p value: $p < 0.05$. The inclusion of all other variables was highly significant (Table 4).

Table 4. Significance code for the ANOVA-based Wald test for variable significance for 2014 and 2015 compared to the four-variable model, TSDM (dryFVC, FPC, HHmax, HHmin), when leaving either HHmax or HHmin out in a three-variable model (also without interaction terms).

Year	Model	Model
2014	TSDM (dryFVC, FPC, HHmax) p-value: 1.289×10^{-11} ***	TSDM (dryFVC, FPC, HHmin) p-value: 1.743×10^{-12} ***
2015	TSDM (dryFVC, FPC, HHmax) p-value: 0.04922 *	TSDM (dryFVC, FPC, HHmin) p-value: 2.2×10^{-16} ***

Significance codes: 0.0001 = ***, 0.01 = *.

The two models established are different for the dry and the wet season as the environmental conditions differ accordingly. Figure 7 shows the input data separated by season as scatterplots and histograms as a probability density function.

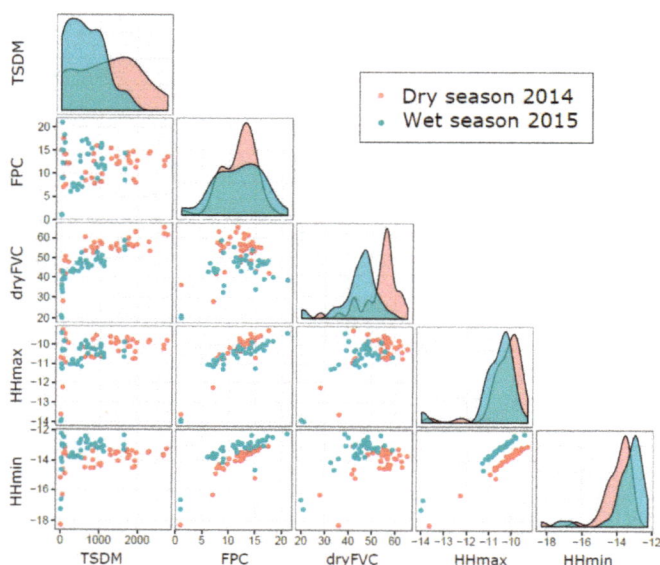

Figure 7. Variable density plot (diagonal) and scatterplot for the variables used in the model for 2014 and 2015.

The histogram of TSDM in Figure 7 reveals a clear difference in the data distribution, with lower TSDM in the 2015 wet season, as indicated in Figure 3. FPC appears with a different distribution at the high and low end of the distribution. More foliage may have been produced in the over- and midstorey during the wet season. The dryFVC (dry grass and tree litter) is clearly lower in 2015—also visible in Figure 4. HHmin and HHmax display distinct differences for the wet and dry season. As expected, the two backscatter variables with the same polarisation are correlated. However, the different overpasses and view angles generate sufficient additional information to improve the model performance (Table 4). The TSX data visually show a relation to FPC, which can be attributed to the interaction of X-band with the leaf canopy.

There appears to be an offset between the 2014 dry season and 2015 wet season data, most evident in the data pairs of FPC and TSX HHmin and also HHmax. The tree species in the study area are likely to exhibit an increase in litter fall and hence reduced FPC during a dry season. In another tropical savanna system, the woody species within crown overstorey FPC changed up to 20% between the wet and dry seasons for some evergreen species and up to 40% when averaged over 49 species including evergreen, partly, and fully deciduous [40].

This effect would also influence the total leaf water content held in a crown. The overall greenness (FVC green vegetation component) at paddock scale increased by 12.9% from the 2014 dry season to the 2015 wet season, while the FPC only changed 2.9% and the dryFVC declined by 17%. Table 5 summarises the single-date HH TSX observations and the relation to the respective seasonal FPC dataset. There appear to be a different slope and intercept in a simple regression with wet and dry seasons.

Table 5. Correlation of FPC (spatial median) with single-date (spatial median) HH polarised TSX data.

Polarisation: HH	Date	r^2 (Adjusted)	Slope	Intercept
Median FPC 2014	26 October 2014	0.84	4.8	73.7
	14 November 2014	0.85	5.4	79.5
	17 November 2014	0.85	4.9	76.7
Median FPC 2015	20 May 2015	0.74	7.0	97.5
	23 May 2015	0.74	7.3	108.3

One possible explanation for this disparity is the different levels of moisture in the system. Although the rainfall was very low for both the wet and the dry season, the water vapour deficit was substantially higher in the dry season (Figure 8). The higher overall greenness, as observed in Figure 2, also indicates a higher canopy greenness and FPC. It can be assumed that the vegetative water content in the overstorey is also higher, which may influence the HH backscatter signal. Figure 8 shows the vapour pressure deficit (VPD) in relation to the TSX observations. There is a large difference in the 2014 dry season observation between the descending and ascending overpasses (day, night). In addition to seasonal changes in tree canopy, it is possible that hydroscopic moisture from the atmosphere is influencing backscatter from the soil surface and dead grass.

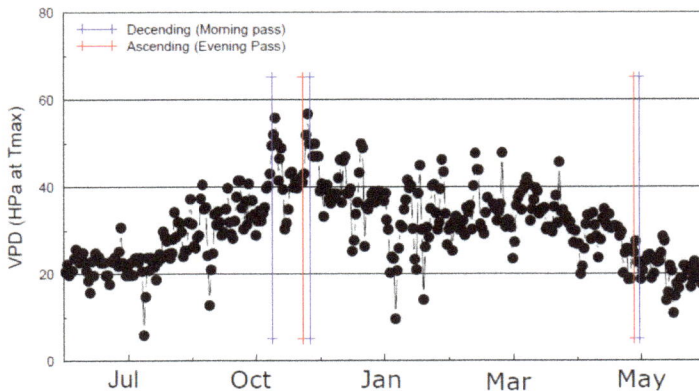

Figure 8. Estimated daily average vapour pressure deficit in relation to the TSX observation dates.

To accommodate the different conditions in the wet and dry season in a more robust manner, a combined model (wet and dry season) was established, where 'season' is a factor in the prediction (Figure 9).

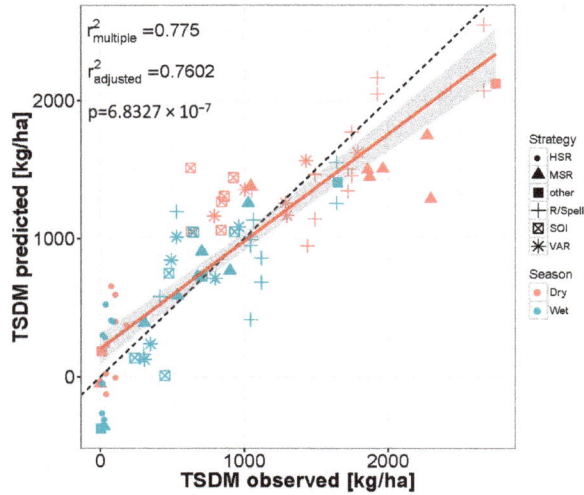

Figure 9. Combined model for pasture TSDM for wet and dry season observations, with a mean residual standard error of 358 kg/ha. Wet and dry season data points are colour coded differently, as well as the paddock averages and the bare areas. TSDM = −8517.1 + 123.6 × FPC + 82.2 × dryFVC −2531.5 × HHmax + 1579.1 × HHmin for the dry season and the wet season intercept of −10,355.1.

The spatial distribution of the TSDM model output for the grazing trial area is displayed in Figure 10. The model was restrained to the available data range such that negative predictions were set to zero and the upper values were capped at 3500 kg/ha TSDM.

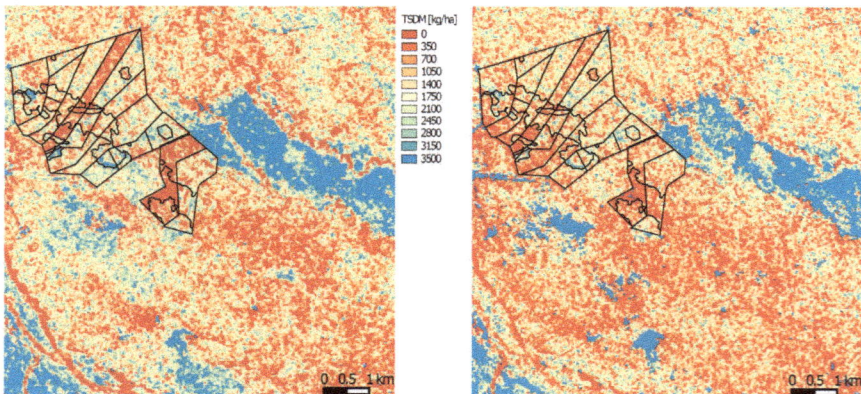

Figure 10. TSDM maps generated from the analyses for the trial and surrounding area, showing the dry season 2014 (**left**) and the wet season 2015 (**right**) using the seasonally adjusted function TSDM. The grazing trial outline is superimposed (as per Figure 1).

The TSDM maps for the grazing trial and the surrounding areas (Figure 10) show a general decrease in TSDM from 2014 to 2015, with some localised deviations. The high stocking rate Paddocks 4 and 9 stand out clearly and are low in comparison to the other grazing strategies, such as the MSR in Paddock 5 and 7, which had the highest TSDM. The Box land type in the HSR paddocks (4 and 9) show a relative increase in TSDM in 2015 compared to 2014, which is in contrast to the field observations

(Figure 3)—though the decrease in field measured TSDM is only 30 kg/ha (Paddock 4) and 98 kg/ha (Paddock 9). The lower TSDM in the other paddocks and land types for 2015, as indicated by the field data in Figure 3, is also reflected in the TSDM maps (Figure 10). The only area with higher field observations in TSDM in 2015 compared to 2014 (Figure 3) is in Paddock 10 on the Brigalow land type, which is also mapped in Figure 10 with increased TSDM.

4. Discussion

Pasture biomass measurements, in the form of TSDM, at the Wambiana grazing trial were spatially averaged to paddock scale (100 ha) and parcels of land type subdivisions. The spatial averaging of the satellite imagery (TSX, FPC, and FVC) to the paddock and parcel scale makes the satellite observations less spatially explicit, but more precise. This form of noise reduction may have contributed to establishing good relationships in this complex natural environment for a tree-grass ecosystem. A second form of noise reduction in the optical domain was the temporal averaging for the FPC and FVC product across several individual satellite images. In the case of FVC, only Landsat 8 OLI imagery were used, as there appeared to be a calibration difference in the FVC products between Landsat 7 and 8. This effect was not observed in the FPC product, therefore all available Landsat imagery was used in the FPC temporal averaging.

The dry component of FVC is the most important variable in the pasture TSDM estimation, with an r^2 of 0.57 alone, for both the dry season and wet season with TSDM (Table 3). The generally low vegetative ground cover in 2015 makes dryFVC a relatively good predictor in combination with FPC in low cover areas. However, the dryFVC component incorporates tree leaf litter components and it is unclear how robust the parameters will be in systems with different tree cover, as the cover-to-biomass relationship tends to saturate [5].

There was a noticeable difference in the overstorey FPC between 2014 and 2015 (Figure 4) within the study site. The parcel average FPC values did not change much (2.9% difference), but as the greenness in the system is almost exclusively attributable to the overstorey, it appears that with the surplus in greenness (12.9%) there might also be a surplus of vegetative moisture in the overstorey. This additional moisture may have altered the X-band backscatter, resulting in a changed relationship between FPC and HHmin or HHmax (Table 3, Figure 7). The variable HHmin appears to have an important influence in this context, so a series of images with differing viewing geometry seems to add to the information content. Only two HH overpasses in 2015 were available for the study region and therefore the difference between HHmax and HHmin was not as pronounced as with three observations in 2014, which may have contributed to a poorer model fit.

The robust multiple regression approach (without interaction terms) has represented the data well for the wet and dry season models as well as the combined model (Figures 5, 6 and 9). The largest scatter in the point clouds appear between 1000 and 2000 TSDM, while the predictions at the low (e.g., <500 TSDM) and high TSDM fit the data better. However, at very low TSDM values (<50 TSDM) some scatter and also negative predicted values appear. This could potentially be avoided by the application of an appropriate data transformation. These points were digitised and may be more error-prone than the TSDM observations.

Pasture utilisation rates resulting from the different stocking regimes in the grazing trial are clearly visible in the modelled TSDM maps for the dry season 2014 and the wet season 2015 (Figure 10). The relative increase of TSDM in the HSR Paddock 4 and 9 on box land type is in contrast to the field observations, which may be influenced by the relative significance of the FPC in the very low TSDM areas and the influence of the invasive shrub Carissa, which is highest in the box land type of the HSR treatment.

The MRSE was lower for 2015, perhaps because the total TSDM values were lower. A combined model for both seasons was established—with this approach there was some nominal model accuracy (r^2 and MRSE) sacrificed, with the benefit of a more robust model with more data points. The MRSE values for the calibrated dataset are encouraging; however, the MRSE as a percentage of the mean

TSDM for dry, wet, and combined seasons is 28%, 41%, and 40% respectively, which is still high relative to the planned pasture usage by cattle (about 20% of annual growth). A 40% error in TSDM estimation is, however, still an improvement over the current state as landholders often have only a vague idea about TSDM on their property as destructive measurements are expensive and impractical. Other forms of TSDM estimates vary greatly between methods, observers, and grass species [3].

The Wambiana grazing trial is a small study area with relatively good observed field data, with the whole trial site of 10 paddocks (about 1000 ha) being typically about the size of a single paddock in the grazing systems for the region. Upscaling of the findings for wider applications would be desirable, e.g., using TSX ScanSar mode (up to 18.5 m resolution) or Wide ScanSar (up to 40 m resolution) with swath width of approximately 100 km and 270 km, respectively. The HV polarisation did not add significantly to a TSDM model and was thus not included in further analysis. The investigation of quad polarisation may give more insight, or dual polarisation with VV as an additional variable. A time series of X-band imagery over the full duration of a wet and dry season would be ideal to evaluate the 'within-season' temporal behaviour. The characterisation of a more general TSDM model would also entice monitoring over longer time spans to cover a range of different seasons temporally, a range of different soil types spatially, and a range of different vegetation types structurally.

In order to apply a TSDM model across a wider area, more field data across space and time would be required. The dataset provided did not include a high proportion of green pasture biomass due to drought conditions and dry soils. The establishment of a pasture biomass data library would be desirable for larger scale studies. The c-band radar from Sentinel-1 may in future provide a useful time series in conjunction with X-band data and optical imagery of high temporal frequency for phenological analysis [41].

5. Conclusions

A statistical model estimating pasture biomass could be established using a combination of optical and SAR imagery. The inclusion of X-band TerrarSar-X data improved the model over optical imagery alone; and vice versa. The correlations with field data revealed an adjusted r^2 of 0.81 in the dry season and 0.74 in the wet season. The respective standard errors were 332 kg/ha and 240 kg/ha. The wet and dry season conditions differed largely due to the change in overstorey vegetation. A more generally applicable combined season model was established with an adjusted r^2 of 0.76 and a mean standard error of 358 kg/ha. A clear improvement in the model performance could be demonstrated when integrating the TSX HH imagery with optical satellite image products.

Acknowledgments: Our thanks go to Airbus Defense and Space for providing the TSX data for our analyses. Special thanks to Chris Holloway for advice on the area and the contribution of the initial soil and land type mapping. Thanks to Meat and Livestock Australia and the Lyons family for supporting Wambiana station. The authors would like to thank the internal reviewers (Matthew Pringle and Dan Tindall) and the anonymous reviewers who helped to improve this contribution. Thanks also to the DLR SAR-Edu program.

Author Contributions: M.S. performed the data analysis and prepared this manuscript. J.C. had vital input in the analysis and performed the variable comparison in Eureqa. G.S. helped with the data analysis, provided context, and helped in editing and improving this contribution. P.O. recorded the field data used in this study and provided local knowledge.

Conflicts of Interest: The authors declare no conflict of interest.

Appendix A

In this paper, we describe features, actions, and practices in an Australian vernacular that may not be apparent to all readers. To avoid any misinterpretation, we include some explanations to terminology used within the text:

- domestic livestock—cattle/cows, sheep
- AE (Adult Equivalent)—a standard animal: a steer or dry female (non-lactating or pregnant) weighing 450 kg

- stocking rates—instantaneous livestock density on a property (ha/AE)
- trial (grazing trial)—treatments imposed using a variation of stocking rate
- spelling—a management action to withhold livestock from areas of a property (or paddock) to allow pastures/grasses to recover or not be further degraded
- grazing—the act of livestock in grasslands/rangelands
- paddock—a subset of a property, fenced to contain livestock to a more manageable area
- rangeland—pasture lands grazed and not grazed
- pastures/grasses—predominantly C_4 perennial grass species that are consumed by domestic livestock, native, domestic, and feral herbivores—respectively kangaroos/wallabies; cattle/sheep; and wild goats/horses
- quadrat—a square or rectangular object to sample or measure groundcover or pasture grasses. In this study a 0.5 m × 0.5 m = 0.25 m^2 quadrat was used.
- TSDM (total standing dry matter)—the oven dry biomass (kg dry matter/ha)

References

1. Werner, A. *Savanna Ecology and Management, Australian Perspectives and International Comparisons*; Blackwell Scientific Publications: London, UK, 1991.
2. McKeon, G.M. *Pasture Degradation and Recovery in Australia's Rangelands: Learning from History*; Queensland Department of Natural Resources, Mines and Energy: Indooroopilly, Australia, 2004.
3. Harmoney, K.R.; Moore, K.J.; George, J.R.; Brummer, E.C.; Russell, J.R. Determination of pasture biomass using four indirect methods. *Agron. J.* **1997**, *89*, 665–672. [CrossRef]
4. Carter, J.O.; Hall, W.B.; Brook, K.D.; McKeon, G.M.; Day, K.A.; Paull, C.J. Aussie GRASS: Australian grassland and rangeland assessment by spatial simulation. In *Applications of Seasonal Climate Forecasting in Agricultural and Natural Ecosystems: The Australian Experience*; GL Hammer, N.N., Mitchell, C.D., Eds.; Kluwer Academic Publishers: Dordrecht, The Netherlands; Boston, MA, USA, 2000; p. 469.
5. Hobbs, T.J. The use of NOAA-AVHRR NDVI data to assess herbage production in the arid rangelands of central Australia. *Int. J. Remote Sens.* **1995**, *16*, 1289–1302. [CrossRef]
6. Barrachina, M.; Cristóbal, J.; Tulla, A.F. Estimating above-ground biomass on mountain meadows and pastures through remote sensing. *Int. J. Appl. Earth Obs. Geoinform.* **2015**, *38*, 184–192. [CrossRef]
7. Edirisinghe, A.; Hill, M.J.; Donald, G.E.; Hyder, M. Quantitative mapping of pasture biomass using satellite imagery. *Int. J. Remote Sens.* **2011**, *32*, 2699–2724. [CrossRef]
8. Dusseux, P.; Hubert-Moy, L.; Corpetti, T.; Vertès, F. Evaluation of SPOT imagery for the estimation of grassland biomass. *Int. J. Appl. Earth Obs. Geoinform.* **2015**, *38*, 72–77. [CrossRef]
9. Hanna, M.M.; Steyn-Ross, D.A.; Steyn-Ross, M. Estimating biomass for New Zealand pasture using optical remote sensing techniques. *Geocarto Int.* **1999**, *14*, 89–94. [CrossRef]
10. Kawamura, K.; Akiyama, T.; Yokota, H.; Tsutsumi, M.; Yasuda, T.; Watanabe, O.; Wang, S. Comparing MODIS vegetation indices with AVHRR NDVI for monitoring the forage quantity and quality in Inner Mongolia grassland, China. *Grassl. Sci.* **2005**, *51*, 33–40. [CrossRef]
11. Holm, A. The use of time-integrated NOAA NDVI data and rainfall to assess landscape degradation in the arid shrubland of Western Australia. *Remote Sens. Environ.* **2003**, *85*, 145–158. [CrossRef]
12. Milne, J.; Danaher, T.; Scarth, P.; Carter, J.; Armston, J.; Henry, B.; Cronin, N.; Hassett, R.; Stone, G.; Williams, P.; et al. *Evaluation of MODIS for Groundcover and Biomass/Feed Availability Estimates in Tropical Savannah Systems*; Meat & Livestock Australia: North Sydney, Australia, 2007.
13. Eisfelder, C.; Kuenzer, C.; Dech, S. Derivation of biomass information for semi-arid areas using remote-sensing data. *Int. J. Remote Sens.* **2012**, *33*, 2937–2984. [CrossRef]
14. Wang, X.; Ge, L.; Li, X. Pasture monitoring using SAR with COSMO-SkyMed, ENVISAT ASAR, and ALOS PALSAR in Otway, Australia. *Remote Sens.* **2013**, *5*, 3611–3636. [CrossRef]
15. Kumar, L.; Sinha, P.; Taylor, S.; Alqurashi, A.F. Review of the use of remote sensing for biomass estimation to support renewable energy generation. *J. Appl. Remote Sens.* **2015**, *9*, 97696. [CrossRef]
16. Cumming, I.G.; Wong, F.H. *Digital Processing of Synthetic Aperture Radar Data: Algorithms and Implementation*; Artech House: Boston, MA, USA, 2005.

17. Zhang, C.; Guo, X.; Wilmshurst, J.; Sissons, R. Application of RADARSAT imagery to grassland biophysical heterogeneity assessment. *Can. J. Remote Sens.* **2006**, *32*, 281–287. [CrossRef]

18. Hill, M.J.; Ticehurst, C.J.; Lee, J.-S.; Grunes, M.R.; Donald, G.E.; Henry, D. Integration of optical and radar classifications for mapping pasture type in Western Australia. *IEEE Trans. Geosci. Remote Sens.* **2005**, *43*, 1665–1681. [CrossRef]

19. Dusseux, P.; Corpetti, T.; Hubert-Moy, L.; Corgne, S. Combined use of multi-temporal optical and radar satellite images for grassland monitoring. *Remote Sens.* **2014**, *6*, 6163–6182. [CrossRef]

20. Hajj, M.; Baghdadi, N.; Belaud, G.; Zribi, M.; Cheviron, B.; Courault, D.; Hagolle, O.; Charron, F. Irrigated grassland monitoring using a time series of TerraSAR-X and COSMO-SkyMed X-band SAR data. *Remote Sens.* **2014**, *6*, 10002–10032. [CrossRef]

21. Dhar, T.; Menges, C.; Douglas, J.; Schmidt, M.; Armston, J. Estimation of pasture biomass and soil-moisture using dual-polarimetric X and L band SAR—accuracy assessment with field data. In Proceedings of the 2010 IEEE International Geoscience and Remote Sensing Symposium (IGARSS), Honolulu, HI, USA, 25–30 July 2010; pp. 1450–1453.

22. McNeill, S.J.; Pairman, D.; Belliss, S.E.; Dalley, D.; Dynes, R. Robust estimation of pasture biomass using dual-polarisation TerraSAR-X imagery. In Proceedings of the 2010 IEEE International Geoscience and Remote Sensing Symposium (IGARSS), Honolulu, HI, USA, 25–30 July 2010; pp. 3094–3097.

23. Muir, J.; Schmidt, M.; Tindall, D.; Trevithick, R.; Scarth, P.; Steward, J. *Field Measurement of Fractional Ground Cover: A Technical Handbook Supporting Ground Cover Monitoring for Australia*; Australian Bureau of Agricultural and Resource Economics and Sciences (ABARES): Canberra, Australia, 2011.

24. Tothill, J.C.; McDonald, C.K.; Jones, R.M.; Hargreaves, J.N.G. *Botanal*, 3rd ed.; Commonwealth Scientific and Industrial Research Organisation (CSIRO), Division of Tropical Crops and Pastures: Brisbane, Australia, 1992.

25. Specht, R.L. Foliage projective covers of overstorey and understorey strata of mature vegetation in Australia. *Austral Ecol.* **1983**, *8*, 433–439. [CrossRef]

26. Anderson, E. *Plants of Central Queensland: Their Identification and Uses*; Department of Primary Industries: Brisbane, Australia, 1993.

27. Boland, D.J.; Brooker, M.I.H.; Chippendale, G.M.; Hall, N.; Hyland, B.P.M.; Kleinig, D.A.; McDonald, M.W.; Turner, J.D. *Forest Trees of Australia*, 5th ed.; Commonwealth Scientific and Industrial Research Organisation (CSIRO) Publishing: Collingwood, Australia, 2006.

28. O'Regain, P.J.; Bushell, J.J. *The Wambiana Grazing Trial: Key Learnings for Sustainable and Profitable Management in a Variable Environment*; Department of Employment, Economic Development and Innovation: Brisbane, Australia, 2011.

29. O'Reagain, P.J.; Brodie, J.; Fraser, G.; Bushell, J.J.; Holloway, C.H.; Faithful, J.W.; Haynes, D. Nutrient loss and water quality under extensive grazing in the upper Burdekin river catchment, North Queensland. *Mar. Pollut. Bull.* **2004**, *51*, 37–50. [CrossRef] [PubMed]

30. Flood, N.; Danaher, T.; Gill, T.; Gillingham, S. An Operational Scheme for Deriving Standardised Surface Reflectance from Landsat TM/ETM+ and SPOT HRG Imagery for Eastern Australia. *Remote Sens.* **2013**, *5*, 83–109. [CrossRef]

31. Armston, J.D.; Denham, R.J.; Danaher, T.J.; Scarth, P.F.; Moffiet, T.N. Prediction and validation of foliage projective cover from Landsat-5 TM and Landsat-7 ETM+ imagery. *J. Appl. Remote Sens.* **2009**, *3*, 33540. [CrossRef]

32. Schmidt, M.; Scarth, P. Spectral Mixture Analysis for Ground-Cover Mapping. In *Innovations in Remote Sensing and Photogrammetry*; Jones, S., Reinke, K., Eds.; Springer: Berlin/Heidelberg, Germany, 2009; pp. 349–359.

33. Scarth, P.; Röder, A.; Schmidt, M.; Denham, R. Tracking grazing pressure and climate interaction—The role of Landsat fractional cover in time series analysis. In Proceedings of the 15th Australasian Remote Sensing and Photogrammetry Conference, Alice Springs, Australia, 13–17 September 2010.

34. *Sentinel Application Platform (SNAP)—ESA*, version 2.0.2; European Space Agency: Frascati, Italy, 2016.

35. Jeffrey, S.J.; Carter, J.O.; Moodie, K.B.; Beswick, A.R. Using spatial interpolation to construct a comprehensive archive of Australian climate data. *Environ. Model. Softw.* **2001**, *16*, 309–330. [CrossRef]

36. Schmidt, M.; Raupach, M.; Briggs, P. Use of lagged time series correlations to relate climate drivers and vegetation response. In Proceedings of the 15th Australasian Remote Sensing and Photogrammetry Conference, Alice Springs, Australia, 13–17 September 2010.

Remote Sens. **2016**, *8*, 989

37. Tanner, C.B.; Sinclair, T.R. Efficient water use in crop production: research or re-search? In *Limitation to Efficient Water Use in Crop Production*; American Society of Agronomy: Madison, WI, USA, 1983; pp. 1–27.

38. *Eureqa*, version 1.24.0; Nutonian: Boston, MA, USA, 2013.

39. R Development Core Team. *R: A Language and Environment for Statistical Computing*; R Foundation for Statistical Computing: Vienna, Austria, 2008.

40. Williams, R.J.; Myers, B.A.; Muller, W.J.; Duff, G.A.; Eamus, D. Leaf phenology of woody species in a North Australian tropical savanna. *Ecology* **1997**, *78*, 2542–2558. [CrossRef]

41. Schmidt, M.; Udelhoven, T.; Gill, T.; Röder, A. Long term data fusion for a dense time series analysis with MODIS and Landsat imagery in an Australian Savanna. *J. Appl. Remote Sens.* **2012**, *6*. [CrossRef]

remote sensing

MDPI

Article

Potential of ALOS2 and NDVI to Estimate Forest Above-Ground Biomass, and Comparison with Lidar-Derived Estimates

Gaia Vaglio Laurin [1,*], Francesco Pirotti [2], Mattia Callegari [3], Qi Chen [4], Giovanni Cuozzo [3], Emanuele Lingua [2], Claudia Notarnicola [3] and Dario Papale [1]

[1] Department for Innovation in Biological, Agro-Food and Forest Systems (DIBAF),
     University of Tuscia, 01100 Viterbo, Italy; darpap@unitus.it
[2] Dipartimento Territorio e Sistemi Agro-Forestali (TESAF)/Interdepartmental Research Center
     of Geomatics (CIRGEO), University of Padova, 35020 Legnaro, Italy; francesco.pirotti@unipd.it (F.P.);
     emanuele.lingua@unipd.it (E.L.)
[3] European Academy of Bozen EURAC-Institute for Applied Remote Sensing, Viale Druso 1, 39100 Bolzano,
     Italy; mattia.callegari@eurac.edu (M.C.); giovanni.cuozzo@eurac.edu (G.C.);
     claudia.notarnicola@eurac.edu (C.N.)
[4] Department of Geography, University of Hawai'i at Manoa, 422 Saunders Hall, 2424 Maile Way, Honolulu,
     HI 96822, USA; qichen@hawaii.edu
*   Correspondence: gaia.vl@unitus.it; Tel.: +39-392-3952-067

Academic Editors: Lalit Kumar, Onisimo Mutanga, Xiaofeng Li and Prasad S. Thenkabail
Received: 7 September 2016; Accepted: 21 December 2016; Published: 29 December 2016

Abstract: Remote sensing supports carbon estimation, allowing the upscaling of field measurements to large extents. Lidar is considered the premier instrument to estimate above ground biomass, but data are expensive and collected on-demand, with limited spatial and temporal coverage. The previous JERS and ALOS SAR satellites data were extensively employed to model forest biomass, with literature suggesting signal saturation at low-moderate biomass values, and an influence of plot size on estimates accuracy. The ALOS2 continuity mission since May 2014 produces data with improved features with respect to the former ALOS, such as increased spatial resolution and reduced revisit time. We used ALOS2 backscatter data, testing also the integration with additional features (SAR textures and NDVI from Landsat 8 data) together with ground truth, to model and map above ground biomass in two mixed forest sites: Tahoe (California) and Asiago (Alps). While texture was useful to improve the model performance, the best model was obtained using joined SAR and NDVI (R^2 equal to 0.66). In this model, only a slight saturation was observed, at higher levels than what usually reported in literature for SAR; the trend requires further investigation but the model confirmed the complementarity of optical and SAR datatypes. For comparison purposes, we also generated a biomass map for Asiago using lidar data, and considered a previous lidar-based study for Tahoe; in these areas, the observed R^2 were 0.92 for Tahoe and 0.75 for Asiago, respectively. The quantitative comparison of the carbon stocks obtained with the two methods allows discussion of sensor suitability. The range of local variation captured by lidar is higher than those by SAR and NDVI, with the latter showing overestimation. However, this overestimation is very limited for one of the study areas, suggesting that when the purpose is the overall quantification of the stored carbon, especially in areas with high carbon density, satellite data with lower cost and broad coverage can be as effective as lidar.

Keywords: ALOS2; mixed forest; biomass; lidar; NDVI

1. Introduction

Forests have a major role in the exchange of carbon between the land surface and the atmosphere, and can represent both carbon sinks and sources through forest growth and by means of deforestation and degradation [1]. Forests are also rich in biodiversity, with higher diversity levels often associated with higher above ground biomass (AGB) [2]. Forest carbon stock estimates are required for reducing uncertainty in the global carbon budget, as well as for biodiversity conservation purposes and forested area management and planning.

Satellite-based remote sensing is fundamental in forest biomass monitoring, as it can support the extrapolation of local ground forest measurements to large extents [3]. The collection of ground data is time and resource demanding, usually covers only limited areas and accessible locations, and is organized in plots of relatively small areas.

Remote sensing of forest carbon stock can be considered a challenging task as the information recorded by a remote sensing instrument is only indirectly related to carbon. Passive optical instruments do not penetrate the dense forest canopy and do not sense the forest compartment where most of the carbon is stored; usually they saturate at low biomass values, with their use being strongly limited by cloud presence; however examples of successful optical-based AGB estimations are reported, especially using high spatial resolution data [4].

Among active sensors, lidar (light detection and ranging) usually generates highly accurate biomass estimates, thanks to its ability to provide detailed vertical forest structure information and to consequently link the strong relationship between forest height and biomass. Lidar is at present considered the premier instrument to quantify carbon stocks [5,6]. A common and successful method to use lidar data for AGB estimation consists in the extraction of forest height measures from the lidar point cloud, use these metrics with field data to build and validate a regression model, and apply the model to the whole area covered by lidar data [7–10]. However, lidar data are presently collected only through on-demand airborne surveys, and thus available only with limited spatial and temporal coverage.

The relationship between radar backscattering and biomass has been illustrated more than two decades ago [11]. Since then, various studies focused on the retrieval of forest structural features from synthetic aperture radar (SAR) due to several advantages, including: the availability of satellites equipped with different SAR sensors, high spatial and temporal resolution of the datasets, extended and often global coverage, and radar insensitivity to cloud cover [12]. The SAR system frequency strongly influences the backscattering to biomass relationship, with the P-band characterized by major sensitivity due to its greater penetration [13]. While waiting for the future launch of the P-band European Space Agency BIOMASS satellite mission, specifically designed to monitor forests [14], AGB could be estimated using SAR at different frequencies, including L-band SAR data.

Data from the Japanese Earth Resources Satellite (JERS) and the Advanced Land Observing Satellite (ALOS) Phased Array type L-band Synthetic Aperture Radar (PALSAR), the former operational until 1998 and the latter until 2011, were extensively employed to estimate AGB in different forest ecosystems [15–21]. Previous literature results suggests that there is a saturation level above which there is a loss of sensitivity between AGB and backscattered signal [22]. This saturation point is influenced by forest type and structure, environmental conditions, as well as sensor characteristics, with the commonly observed saturation points usually occurring between 30 and 100 Mg/ha [13,23–25].

There is a recognized influence of plot size on AGB estimates, with larger size often resulting in better accuracy of estimates. Using plots of reduced size and ALOS PALSAR data, AGB estimates with moderate or even limited accuracy are usually obtained: a positive correlation of backscatter at all polarizations with tropical AGB at 0.25, 0.5, and 1 ha scales was found by [17], but they noted that the spatial variability of forest structure and speckle noise in SAR data contributed equally to degrading the sensitivity of radar to AGB at scales less than 1.0 ha. In a Chinese forest, He et al. [26] observed a poor relationship between SAR backscatter and AGB at plot level, while at stand level a logarithm equation could be used to describe the relationship in different biomass ranges. A strong monotonical statistical dependence between ALOS PALSAR and AGB was found by [27] in savanna pine woodlands using field data collected at 0.25, 0.5, and 1 ha, and a weak dependence using data at 0.1 ha. However, due to the high amount of resources needed to set up and monitor forest plots, it is difficult to establish large sampling areas and obtain field datasets based on large plots [28]; thus, small plots are much more used than large ones in biomass research and mapping activities.

Texture features extracted from SAR data have proved useful in some studies to improve AGB estimates: using RADARSAT-2 C-band dual-polarization data in a complex subtropical forest, Sarker et al. [29] found the Grey Level Co-occurrence Measure (GLCM; [30]) texture features more effective than the original bands; in [31], they found that the addition of GLCM textures improved a joined Landsat-ALOS PALSAR model for AGB estimation in Iranian forests; and Champion et al. [32] found high correlation between GLCM textures extracted from airborne P-band cross polarization data and AGB in a French Guyana forests characterized by high carbon density.

Another option to improve AGB estimates based on SAR could be the addition of multispectral optical data, to exploit the response from different regions of the electromagnetic spectrum. In this respect, Deng et al. [33] found beneficial the addition of WorldView-2 to ALOS PALSAR in mountain Chinese forests; both Basuki et al. [34] and Fedrigo et al. [35] improved the ALOS PALSAR-based estimation by integrating Landsat 7 ETM+ data in tropical forests.

The ALOS2 continuity mission was launched on May 2014 and currently produces dual and full polarization data. Even if full polarization data are available only in selected regions, the availability of polarizations ratio from dual-pol data, such as in this study, is recognized as an advantage when using SAR for biomass estimation [33]. ALOS2 improvements with respect to the former ALOS satellite include: the exploitation of a dual receiving antenna allowing to broaden the imaged swath, improved spatial resolution with spotlight (from 1 to 3 m) and strip-map mode (from 3 to 10 m); a reduced revisit time, from 46 to 14 days; and the possibility to provide both left and right looking for fast coverage in emergency cases [36,37]. To the best of our knowledge, only one study has evaluated the performance of this new sensor for AGB estimation [38]. Specifically, the improved spatial resolution could represent an advantage to build regression models using small field plots.

In our study, we aimed at using small plots to develop regression models between remote sensing derived data—SAR and NDVI (Normalized Difference Vegetation Index)—and ground measured variables. We also used a lidar-based regression model, and developed a new one, for results comparison purposes. Limited research quantitatively compared AGB estimates obtained using SAR plus NDVI and lidar data types, which have quite different data acquisition costs. Through this effort, the present research aims at providing useful insights for forest monitoring, focusing on two different study sites that were chosen for the availability of accurate ground and lidar data, and for representing two different types of mixed conifer forest.

2. Materials and Methods

2.1. Field Data

The Asiago study site (Figure 1) is part of the Asiago plateau (Province of Vicenza—Italy), and is located in a karstic plain area on the esalpic range of the north-eastern Alps. This site is divided into two areas: the Boscon southern area has an extent of about 32 km^2, while the Verena northern area covers approximately 23 km^2. Slopes are quite mild and elevation ranges come from about 1100 to 1300 m a.s.l. and from about 1300 to 1750 m a.s.l. in the northern and southern part respectively. Vegetation is mixed conifer forest, composed mostly by spruce stands (*Picea abies*), with presence of silver fir (*Abies alba*), beech (*Fagus sylvatica*), and larch (*Larix decidua*). In 2012, in the framework of a national research project, 33 circular plots of 19.95 m radius (0.1256 ha) were set up, in which height and diameter at breast height (DBH) were measured for each tree with DBH greater than 5 cm; AGB was calculated according to species-specific allometric models [39,40]. Plots were set up according to a stratified random sampling, with stratification based on height classes.

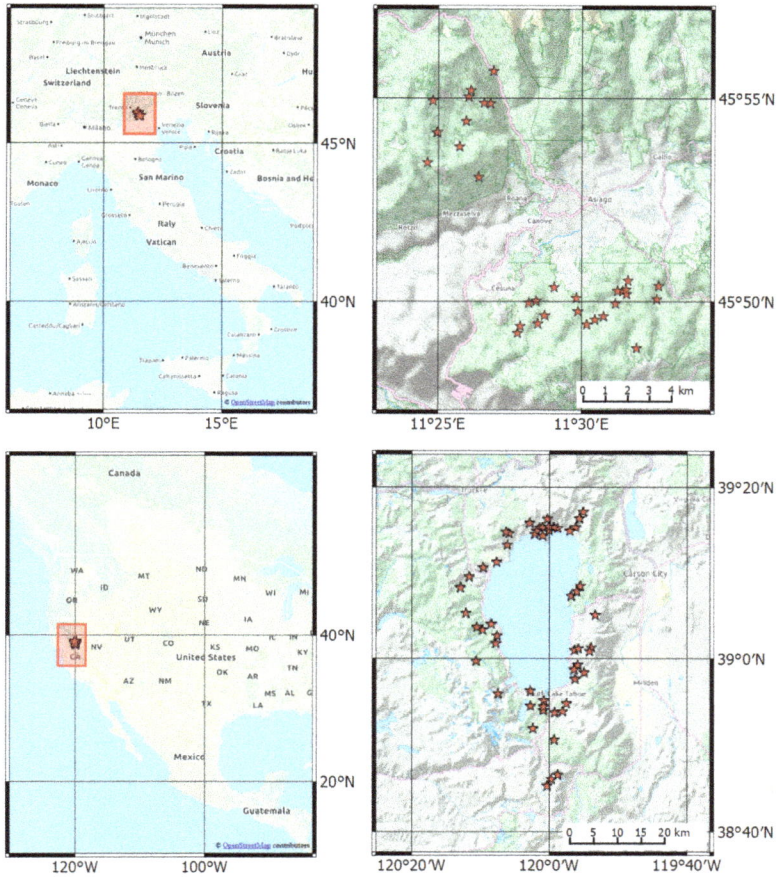

Figure 1. The Asiago (**top**) and Tahoe (**bottom**) study sites. On the left side, the geographic location of the areas are illustrated; on the right side a zoom over the area is shown, with red star symbols indicating field plots.

The Tahoe study site is located on the eastern slope of the Sierra Nevada mountain range and is named as the United States Department of Agriculture Forest Service (USDA-FS) Lake Tahoe Basin Management Unit. The Tahoe site (Figure 1) covers about 936 km^2, but in this analysis we considered an area of 786.4 km^2 after removal of all water bodies and a small southern area affected by cloud presence in optical data. The elevation is between 1900 to 2500 m a.s.l. and slopes are usually mild, although a stronger variability is present, especially in the eastern part of the area. The major vegetation type in Tahoe is mixed conifer forest including: Jeffrey pine (*Pinus jeffreyi*), white fir (*Abies concolor*), California red fir (*Abies magnifica*), lodgepole pine (*Pinus contorta*), incense cedar (*Calocedrus decurrens*), quaking aspen (*Populus tremuloides*), western white pine (*Pinus monticola*), sugar pine (*Pinus lambertiana*), western juniper (*Juniperus occidentalis*), and mountain hemlock (*Tsuga mertensiana*). At Lake Tahoe, over 1000 trees were mapped in 2012 for 56 circular plots of 17.6 m radius (0.0973 ha) using a Nikon DTM-322 total station. These plots were initially established through the Multi-Species Inventory and Monitoring project and the Lake Tahoe Urban Biodiversity project. Plot locations were selected using a combination of systematic/grid sampling and stratified random sampling. At each plot, all trees greater than 2 cm in DBH were measured. Tree measurements include species, DBH, tree height, height to live crown, and tree status (live, dead, unhealthy, or sick) [9,41,42]. Tree AGB was calculated according to the Component Ratio Method, adopted by the USDA-FS since 2012 [43,44].

The 89 plots were screened against vegetation changes that occurred between the 2012 field data collection and the 2014 remote sensing datasets acquisition; additionally, plots in areas that were affected by distortion effects (layover, foreshortening, shadowing) in SAR-processed images were excluded. Vegetation changes were detected using very high resolution orthophotos (<1 m) and additional Google Earth imagery. After the screening procedure, we reduced the ground dataset to 75 plots, 52 from Tahoe and 23 from Asiago. The two areas significantly differ in terms of AGB content: Tahoe's AGB ranged between 21 and 305 Mg/ha, with an average of 146 Mg/ha and standard deviation of 70 Mg/ha, while Asiago had between 210 and 530 Mg/ha, an average of 358 Mg/ha and standard deviation of 83 Mg/ha.

2.2. Remote Sensing Data

We used four ALOS2 dual-pol stripmap SAR scenes, one for Asiago dated 8 October 2014, and three for the Tahoe site, from 4 September and 2 October (two images) 2014. These scenes were selected for being as temporally close as possible to field data collection, during days without precipitations. Conversion from digital number (DN) to the backscattering coefficient (σ^0 sigma-naught, in decibel or dB) was performed according to the method described in [45]:

$$\sigma^0 = 10 \, \log_{10} Q^2 + I^2 + CF_1 - A \tag{1}$$

where *CF* and *A* are constants, respectively −83 dB and 32 dB and *Q* and *I* are the real and imaginary part of the digital number.

The HH and HV scenes, having a resolution about 7 m in range and 3.5 m in azimuth, were multi-looked (one look in range and two in azimuth) and filtered with a Lee Filter (7 × 7 window size) in order to reduce the speckle noise. The images were then geocoded and radiometrically calibrated by using the 30 m SRTM digital elevation model (DEM) for the SAR scenes acquired in Italy and a 3 m lidar-derived DEM for the scenes acquired in California. The radiometric calibration, which is the correction of the σ^0 coefficient obtained considering a flat terrain assumption with the local incidence angle θ_i, was computed using the DEM and orbital data. In particular, the terrain calibrated normalized radar cross-section σ_c^0 is computed as:

$$\sigma_c^0 = \sigma^0 \frac{\sin \theta_i}{\sin \theta_F} \tag{2}$$

where θ_F is the local incidence angle under the flat terrain assumption. After geocoding, the final spatial resolution was set to 0.000083 degrees (approximately 9 m). The remote sensing processing was conducted using the SARscape module in ENVI5.0 software (Exelis Visual Information Solutions, Boulder, CO, USA).

Two atmospherically corrected and calibrated Landsat 8 images from September 2014 were downloaded from the Google Earth Engine facility to cover the two study areas. The Normalized Difference Vegetation Index (NDVI) was then computed for each image.

2.3. Lidar Data and Derived AGB Maps

For the Tahoe area, lidar data were acquired surrounding Lake Tahoe from 11 August to 24 August 2010 using two Leica ALS50 Phase II laser systems mounted in a Cessna Caravan 208B. The Leica systems were set to a pulse frequency of 83–105.9 kHz, flight height of 900–1300 m, and scan angle of ±14°. The resulting point density is eight pulses per square meter. The airborne lidar data were processed using the Toolbox for Lidar Data Filtering and Forest Studies (Tiffs) [46] to extract the lidar metrics within each plot. A biomass prediction map was then calculated for the area, using the lidar quadratic mean height (QMH) selected with stepwise procedure as input in a power model (AGB = a × H^b_{qm}; where a and b are coefficients and H_{qm} is the QMH of all returns), trained and validated (five-fold cross validation) with the original 56 field plots, at a 31 m spatial resolution in agreement with the plot size; the datasets and models are also documented by [41].

For the Asiago area, lidar data were collected on 5 June 2012, using a helicopter and an Optech ALTM 3100 sensor with a scan angle ±29° and a scan frequency of 100 kHz. With a relative flight height of 200–725 m, the resulting average point density was 10–12 pulses per square meter [47]. A biomass prediction map at 31 m resolution was calculated, fitting the same model form used in Tahoe with the QMH lidar metric, selected by stepwise procedure, and trained and validated (leave-one-out cross validation) with the original 33 field plots; the full set of available plots were used because they were collected at the time of lidar survey. We used LOO validation for the Asiago site, with respect to the five-fold cross validation used in Tahoe, due to the fewer plots available in Asiago that would have resulted in a limited number of folds and, consequently, less effective validation. However, unreported tests indicated non-significant difference in results when using five-fold for Asiago.

2.4. Data Analysis

The SAR data for Tahoe, acquired in two different dates, were first masked to exclude water pixels and then normalized. For Asiago and Tahoe areas, all the pixels having >70% of their area inside the plot, thus the majority, were extracted from the ALOS2 scenes and weighted according to the percentage of area included into the plot. A set of basic statistics per plot were computed: mean, standard deviation, minimum and maximum backscattering per HH and HV polarizations, and their difference and sum. Two Tahoe plots were covered by overlapping scenes: values from both scenes were extracted and used to generate the statistics.

Similarly to the SAR data, Landsat 8 values from pixels inside the plot for >70% were extracted by weighted average. Preliminary tests suggested that NDVI is not inferior to single bands for forest biomass estimation, and thus the index was retained for the estimation phase. The following eight GLCM texture features were generated for each NDVI, HH, and HV scenes: mean, variance, homogeneity, contrast, dissimilarity, entropy, second moment, and correlation [30]. A 64-bit quantization level and three different window sizes—3 × 3, 5 × 5, and 7 × 7—were used. The values from different offsets (0,−1; −1,−1; 1,0; 0,1) were averaged assuming non-directional effects; then pixels included >70% inside plot area and their weighted average was extracted.

We tested a set of progressively more complex inputs, performing stepwise regression for feature selection and linear regression (or multiple linear regression when multiple inputs were selected) for each set of inputs. The results, validated with leave-one-out (LOO) and 10-fold cross validation, applied at plot level and with plots from both areas, indicate the selected inputs and the accuracy of the model based on those inputs. The inputs were:

1. the SAR statistics (minimum, maximum, mean, standard deviation per HH and HV; HH and HV sum and difference; total 10 inputs);
2. the SAR statistics plus SAR-GLCM textures per HH and HV (total 26 inputs);
3. the SAR statistics plus NDVI and NDVI-GLCM textures (total 19 inputs);
4. SAR HH + HV selected by Test 1 plus the SAR-GLCM texture type selected by Test 2 and the NDVI feature type selected by Test 3 (totaling three inputs).

Applying the best model derived from the aforementioned tests, after resampling the NDVI texture to the same spatial resolution of the SAR data, two AGB prediction maps were generated for the Tahoe and Asiago area. The AGB maps were masked to ensure full overlapping in area extent with the corresponding lidar-derived AGB maps, and resampled to meet their spatial resolution. Finally, the AGB maps derived from multispectral and SAR inputs were compared to those generated using lidar data. Considering that for one of the study areas the AGB model was previously developed using the stepwise approach [40], for comparison purposes we preferred to maintain this method over more complex statistical ones. All the data analyses were conducted using MATLAB [48] and R [49] software.

3. Results

With the aim of comparing the different tested datasets, we first introduce the results obtained by lidar: the AGB map for Tahoe (Figure 2, left), based on a power regression model using QMH as input, was characterized by a R^2 of 0.92 and RMSE of 23.6 Mg/ha, obtained using field values and five-fold cross validation. The Asiago AGB map, realized with the same model form and input feature, was characterized by a R^2 of 0.75 and RMSE of 62.3 Mg/ha (Figure 2, right), obtained using field values and leave-one-out validation.

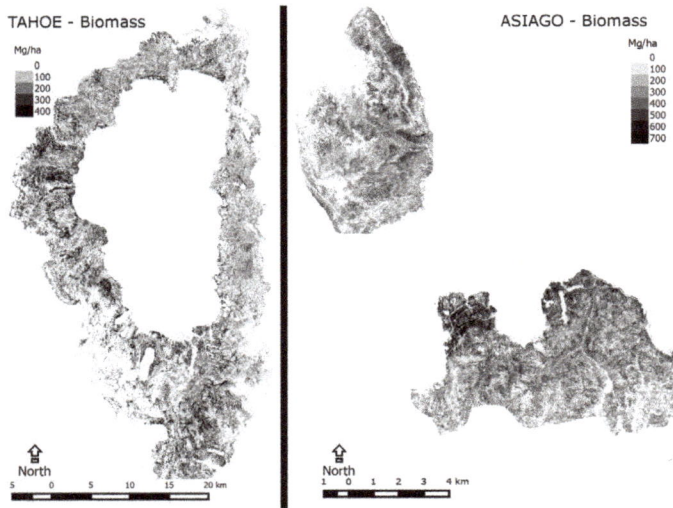

Figure 2. Lidar-derived AGB map for Tahoe area (**left**) and Asiago area (**right**).

For the other SAR and NDVI combined datasets, the inputs for the different tests were selected via stepwise procedure. Cross-validated accuracies for each test allowed selection of the best model for AGB estimation; results are presented in Table 1. In the first (a) test based on SAR data, we used as stepwise regression inputs the statistics derived from HH and HV channels averaged at plot level; the stepwise procedure selected the sum of HH and HV backscattering values, and the linear regression R^2 resulted equal to 0.59 with RMSE equal to 78.76 Mg/ha with LOO validation, and 0.59 and 78.33 Mg/ha with 10-fold validation.

Table 1. Results of the tests conducted with different SAR and optical inputs, using stepwise selection and multiple linear regression, validated with ground truth. RMSE is expressed both in Mg/ha and as percentage.

Inputs for Tests	Selected via Stepwise Criteria	R^2 LOO	RMSE LOO (Mg/ha)	R^2 10-Fold	RMSE 10-Fold (Mg/ha)
(a). SAR HH and HV various statistics	HH + HV	0.59	78.76 (0.15%)	0.59	78.33 (0.15%)
(b). SAR HH and HV various statistics + GLCM HH and HV textures	HH + HV 5 × 5 HHmean	0.65	71.95 (0.14%)	0.65	72.09 (0.14%)
(c). SAR HH and HV various statistics + NDVI + GLCM NDVI textures	HH + HV 5 × 5 NDVImean	0.66	71.62 (0.14%)	0.66	71.59 (0.14%)
(d). SAR HH and HV various statistics + 5 × 5 HHmean + 5 × 5 NDVImean	HH + HV 5 × 5 NDVImean	0.66	71.62 (0.14%)	0.66	71.59 (0.14%)

In the second (b) test, we used as input the SAR statistics and the different SAR GLCM features, using textures calculated with a different window size each time. The stepwise procedure always selected the HH + HV sum and the mean texture derived from HH SAR polarization, but the model using the 5 × 5 mean SAR texture performed better than those using textures calculates with the other window sizes; the multiple linear validated regression showed a R^2 of 0.65 (both with LOO and 10-fold) and a RMSE of 71.95 (LOO) and 72.09 (10-fold) Mg/ha.

For the third (c) test, we used as input the SAR statistics, the NDVI and the GLCM NDVI textures, using each time textures calculated with different window size. The NDVI in the two areas showed different averaged values. Namely, 0.52 for Tahoe and 0.69 for Asiago. The stepwise procedure always selected the HH + HV backscattering and the mean NDVI texture, but the model using the 5 × 5 mean NDVI texture performed better than those using NDVI textures calculated with the other window sizes; R^2 for multiple linear regression was equal to 0.66 for both LOO and 10-fold validation, and RMSE to 71.62 (LOO) and 71.59 (10-fold) Mg/ha.

In the fourth (d) test, we used as input the 5 × 5 window mean texture from HH SAR data selected in (b), the 5 × 5 window mean NDVI from Landsat 8 data selected in (c), and the HH + HV backscattering value selected in (a); the stepwise procedure selected only HH + HV and the 5 × 5 window mean NDVI texture, with multiple linear regression having a R^2 (coefficient of determination) and a RMSE (root mean square error) equal to those obtained with test (c).

Among test results, we considered as the best the one produced by the fourth (d) test, having the same results of (c) but being based on much reduced number of input data; selected inputs were HH + HV (dB sum) and the 5 × 5 mean NDVI texture, with the related equation and scatterplot of the predicted vs. observed AGB values presented in Figure 3.

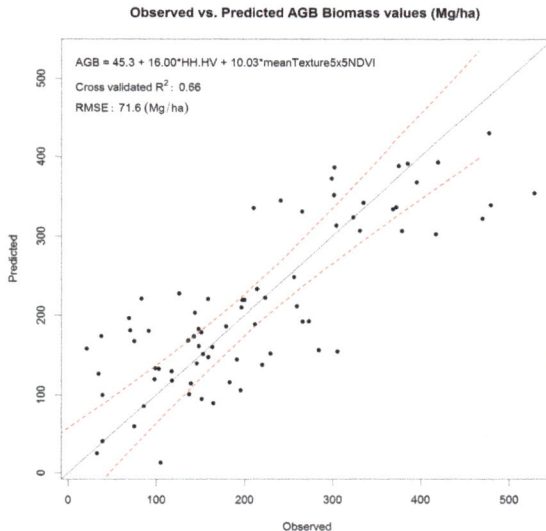

Figure 3. Scatterplot of the observed and predicted AGB values using a multiple linear regression model with HH + HV backscattering and 5 × 5 mean NDVI GLCM texture as inputs.

To visualize the dynamic range of the SAR input, the scatterplot AGB and HH + HV backscattering values are presented in Figure 4. The backscatter range is included between approximately −11 and −30 dB, with saturation occurring approximately over 350 Mg/ha.

The statistical comparison of the AGB maps produced with different datasets, namely the multispectral and SAR dataset and the lidar one, are summarized in Table 2. The maps generated applying the model that uses SAR and NDVI as inputs are presented in Figure 5; Figure 6 illustrates the difference in the AGB distributions of the two maps, and Figure 7 shows the difference of the SAR + NDVI AGB estimate with respect to the lidar map AGB intervals. Based on Table 2, we calculated the difference of total AGB stored in Asiago according to the SAR plus NDVI and lidar map, which is equal to 1.00×10^5 and represents 6.4% of the AGB amount estimated by lidar. Similarly, for Tahoe this difference is equal to 2.3×10^6 and corresponds to 23.7% of the lidar-based estimated AGB. The number of 31×31 m pixels in the maps are 54,299 for Asiago and 818,315 for Tahoe.

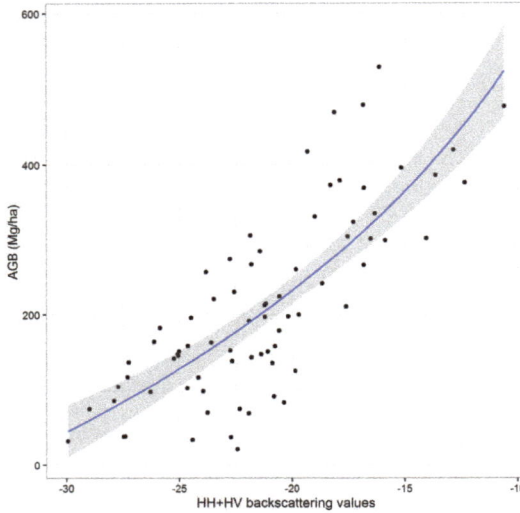

Figure 4. Scatterplot of AGB and HH + HV backscattering values.

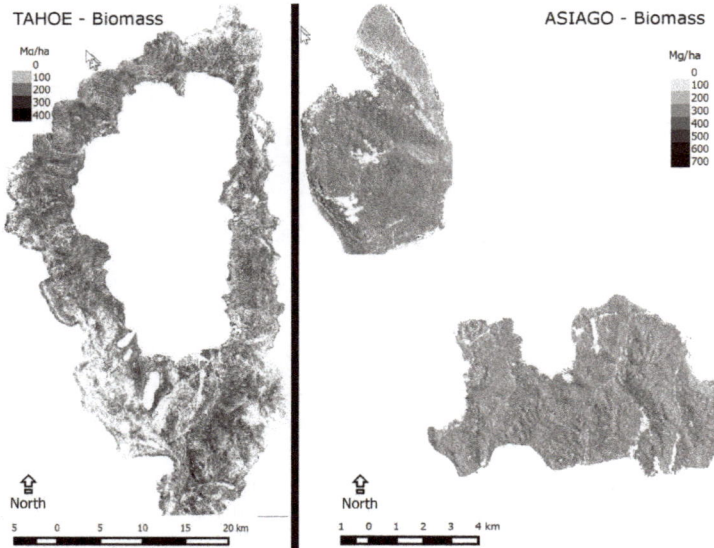

Figure 5. SAR + NDVI based map for Tahoe area (**left**) and Asiago area (**right**).

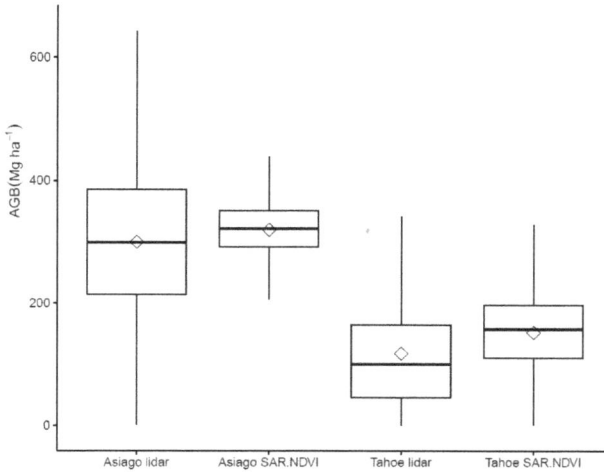

Figure 6. Distribution of AGB values in SAR plus NDVI and lidar-based maps. Boxes represent upper (75%) and lower (25%) quartiles; vertical segments out of boxes are minimum and maximum values; boxes are divided by the median; diamonds represent the mean.

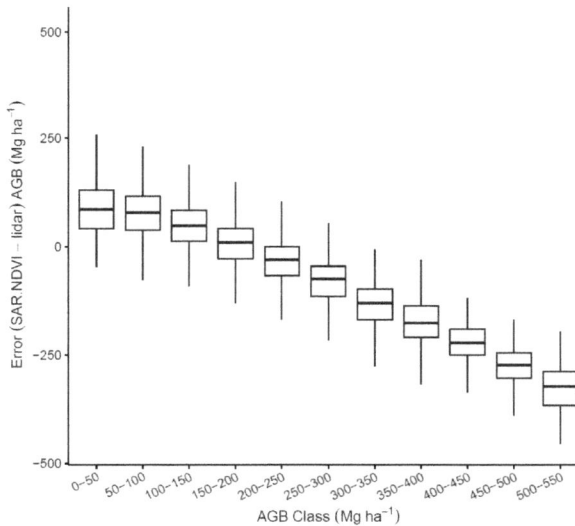

Figure 7. Difference per AGB intervals of the SAR + NDVI map with respect to lidar map. Boxes represent upper (75%) and lower (25%) quartiles; vertical segments out of boxes are minimum and maximum values; boxes are divided by the median.

Table 2. Comparison of the lidar-derived and the SAR + NDVI derived AGB maps for the two study areas.

	Asiago SAR + NDVI	Asiago Lidar	Tahoe SAR + NDVI	Tahoe Lidar
Mean AGB (Mg/ha)	321	301	153	117
Standard deviation of AGB (Mg/ha)	49	117	64	94
Total AGB (Mg)	1.67×10^6	1.57×10^6	1.20×10^7	0.97×10^7

4. Discussion

The lidar-derived AGB maps are characterized by high (Asiago) to very high (Tahoe) accuracies, with differences expected for operating distinct instruments over different sites. For Tahoe, the used linear model was developed and published by Chen [41]. For Asiago, the same model form was adopted as preliminary unreported tests indicated it as the most accurate option; as in Tahoe, the QMH resulted to be the most useful regression input. Both models explain the majority of the AGB variability in the areas, but they are characterized by differences in accuracy. High variability in the accuracy of lidar-based estimates obtained by different studies has been previously observed [50]. It is known that different sources of error, generated while propagating from the ground tree level to the landscape estimation level, may affect lidar-based AGB predictions [51], and it is reasonable to consider that the impact of these errors differs according to the study site. Even if at both sites mixed forests are present, the sites and the collected data are characterized by relevant differences, such as: the lidar systems used, the aerial and field survey characteristics, the species communities, the tree cover density, the biomass density—with significantly larger values of AGB found in Asiago with respect to Tahoe—and the allometric relationships adopted. All these factors are reported as possible sources of error in AGB estimations [51] and may explain the observed difference in accuracy. Overall, the accuracies are in a high range, adding evidence to the value that lidar has for biomass studies.

With respect to results from SAR tests, the sum of HH and HV backscattering values was always selected as input in stepwise procedure, even if this input performed only slightly better in regression than the single channel cross-polarized backscattering value usually used for AGB estimations. Previous research showed that L-band cross-polarized backscatter is more sensitive to biomass variations, whereas the co-polarized signal better captures differences in forest cover fraction [52]. High resolution imagery used in the initial screening phase, as well the difference in NDVI average values found in the two areas, confirms a difference in forest cover, with Asiago having a higher density of trees. The cover information brought by HH polarization, together with the fact that the sum of the two channels reduces the extreme backscattering values and acts as an additional filter effect, may be the reason for the selection in our models of the HH and HV sum.

The use of SAR GLCM features in test (b) consistently improved the accuracy of the model: the stepwise selected texture was the GLCM mean, which is a statistical feature useful to describe homogeneous regions and that further suppress the noise in the data. The NDVI information added to the SAR data in the third (c) test also improved the accuracy of SAR data alone, slightly more than that done by the SAR mean texture. This indicates a complementarity of the SAR and optical datatypes, as already shown by other researches [31,33–35,53], and underlines the importance of forest cover information especially when considering simultaneously different vegetation communities, as in this study. When—in test (d)—both mean NDVI and mean HH SAR textures are added to the SAR backscattering statistics, only the first input is selected by the stepwise procedure, possibly due to the redundant cover information present in both HH and NDVI datasets. However, the result from using SAR data with texture and SAR + NDVI are not so different; considering the broad and free availability of NDVI data from multiple sensors, as well as the extra resources needed to compute texture (with large increase of features in input) we support the use of NDVI as a way of improving accuracy of SAR estimates.

The AGB variation explained by SAR inputs only (test (a)) is in the range of the results obtained by other studies based on ALOS PALSAR data [53–55]. The best model, having HH + HV and mean NDVI texture as input features, shows an accuracy higher than some results obtained by studies using small field plots [17,27,56]. These results support the view that ALOS2 is a valuable tool for AGB estimation, and that its potential can be boosted by the integration of an optical feature such as NDVI.

The scatterplot of the observed vs. predicted values obtained with the most accurate model (Figure 3) reveals that slight saturation effects are present at the high Asiago biomass values. However our results, also according to Figure 4, do not clearly show the usual 100–150 Mg/ha saturation limit reported in literature [1,57,58]; and it is difficult to understand if at higher AGB levels there is a scarce saturation effect or increased error in model. These results from a single research might be caused also by environmental characteristics of the study areas (e.g., soil and forest structure). Several sensor improvements suggest that ALOS2 is a valuable tool, able to characterize biomass in dense forests and offering continuity to ALOS. Some of the improved features to consider, with respect to the previous ALOS, are: higher geolocation accuracy, generating a better correspondence between remote sensing and ground data; the higher spatial resolution, allowing the collection of multiple backscatter information, and reducing noise, even from plots of reduced size; and the improved radiometric accuracy, resulting in finer backscattering response at different AGB levels. For full sensor details see the information available online: https://directory.eoportal.org/web/eoportal/satellite-missions/a/alos-2. Even if additional study is needed to evaluate the use of this sensor for AGB research, and its response in different forests, our research indicates the suitability of the new ALOS2 for AGB estimation, even using small field plots as ground truth.

The availability in our study areas of lidar-derived AGB prediction maps allowed the comparison of the estimates obtained using this on-demand data or satellite data. The evaluation of results obtained with different instruments, which acquire data at very different costs, can be of high relevance for forest and natural resources management, especially in areas where repeated carbon density monitoring is requested.

In terms of total AGB stored in each area, SAR and NDVI data overestimated the total AGB of 6.4% for Asiago and 23.7% for Tahoe respectively, compared to lidar data. The analysis of the errors illustrated in Figure 7 shows that SAR and NDVI moderately overestimates at lower biomass ranges, approximately until 200 Mg/ha, while it underestimates at higher AGB density. The larger overall overestimation found in Tahoe compared to Asiago is possibly due to the sparser tree cover characterizing this site, in which the SAR backscattering is therefore more influenced by the ground signal component. Viergever [59] also reported overestimation modeling biomass with SAR data in sparse savanna woodlands, while [60] suggested that pantropical carbon maps may overestimate AGB in savanna areas. However, underestimation of tree heights has also been reported for lidar-derived tree height models, and attributed to the laser beams missing the tree tops, especially at low laser point densities [61–63], even if there are several examples of AGB models developed with low point density lidar [64,65]. It is possible that the lidar-based AGB maps are slightly influenced by this effect, especially for Tahoe where the sparse tree density may have exacerbated the problem. The distribution of the AGB values in maps produced with the two different systems (Table 2) show that the central measures are not so dissimilar; however, the range of local variation captured by lidar-based maps is higher than in SAR plus NDVI maps, which tend to reduce the AGB variation over the sites, especially at the high AGB Asiago range. This suggests that when the purpose is the overall quantification of the stored carbon in a given area, satellite data which has lower cost and is widely available can be effective, while when precise information on AGB spatial distribution is needed, lidar is a better choice. For Asiago, considering that accuracy of the AGB lidar-based estimate is moderate ($R^2 = 0.75$) and that the overestimation of SAR plus NDVI is quite limited (6.3%), the cost-effectiveness of using lidar for biomass monitoring has to be carefully evaluated.

5. Conclusions

Airborne lidar remains, when possible, the first choice for local studies and fine scale information on biomass spatial distribution. The present research shows the continuity from ALOS to ALOS2 in providing reasonably accurate AGB estimates at a coarse scale, in this case also using small forest plots at considerable biomass densities. Given the reduced number of input features used in modeling, and the large availability of NDVI information vs. the extra resources needed to compute and select SAR textures features, the use of NDVI as integration is advisable. The results also underline the importance of considering the degree of forest cover, similarly to what has been observed in other research [66], which in our study influenced both lidar and SAR + NDVI results.

Considering the larger availability of small plots with respect to larger ones from forest research and inventory efforts, and the increasing number or SAR missions with low-cost data expected in forthcoming years, to map AGB with SAR might be a cost-effective choice not only for large global or country-level analysis but also at smaller scales. When both data are available, lidar-based estimation can represent an accurate baseline, as in this study, with satellite data offering repeated monitoring in time.

Acknowledgments: Gaia Vaglio Laurin and Dario Papale thank the EU for supporting the BACI project funded by the EU's Horizon 2020 Research and Innovation Program under grant agreement 640176. Francesco Pirotti and Emanuele Lingua thank the NEWFOR "NEW technologies for a better mountain FORest timber mobilization"—contract number 2-3-2-FR, Alpine Space Program Interreg IIIB 2007/2013. Gaia Vaglio Laurin thanks Japan Aerospace Exploration Agency (JAXA) for providing ALOS2 scenes (P.I. 1012) and Google Earth Engine for providing Landsat 8 imagery.

Author Contributions: Gaia Vaglio Laurin and Francesco Pirotti conceived and designed the experiments; Gaia Vaglio Laurin, Emanuele Lingua, Qi Chen, and Dario Papale performed the experiments; Claudia Notarnicola, Mattia Callegari, Francesco Pirotti, and Gaia Vaglio Laurin analyzed the data; Qi Chen contributed reagents/materials/analysis tools; Gaia Vaglio Laurin wrote the paper.

Conflicts of Interest: The authors declare no conflict of interest.

References

1. Dixon, R.; Brown, S.; Houghton, R.E.A.; Solomon, A.M.; Trexler, M.C.; Wisniewski, J. Carbon pools and flux of global forest ecosystems. *Science* **1994**, *263*, 185–189. [CrossRef] [PubMed]
2. Strassburg, B.B.N.; Kelly, A.; Balmford, A.; Davies, R.G.; Gibbs, H.K.; Lovett, A.; Miles, L.; Orme, C.D.L.; Price, J.; Turner, R.K.; et al. Global congruence of carbon storage and biodiversity in terrestrial ecosystems. *Conserv. Lett.* **2010**, *3*, 98–105. [CrossRef]
3. Turner, W.; Spector, S.; Gardiner, N.; Fladeland, M.; Sterling, E.; Steininger, M. Remote sensing for biodiversity science and conservation. *Trends Ecol. Evol.* **2003**, *18*, 306–314. [CrossRef]
4. Lu, D.; Chen, Q.; Wang, G.; Liu, L.; Li, G.; Moran, E. A survey of remote sensing-based aboveground biomass estimation methods in forest ecosystems. *Int. J. Digit. Earth* **2016**, *9*, 63–105. [CrossRef]
5. Clark, M.L.; Roberts, D.A.; Ewel, J.J.; Clark, D.B. Estimation of tropical rain forest aboveground biomass with small-footprint lidar and hyperspectral sensors. *Remote Sens. Environ.* **2011**, *115*, 2931–2942. [CrossRef]
6. Chen, Q. Lidar remote sensing of vegetation biomass. *Remote Sens. Nat. Resour.* **2013**, *399*, 399–420.
7. Chen, Q.; Laurin, G.V.; Battles, J.J.; Saah, D. Integration of airborne lidar and vegetation types derived from aerial photography for mapping aboveground live biomass. *Remote Sens. Environ.* **2012**, *121*, 108–117. [CrossRef]
8. Naesset, E. Estimating timber volume of forest stands using airborne laser scanner data. *Remote Sens. Environ.* **1997**, *61*, 246–253. [CrossRef]
9. Lu, D.; Chen, Q.; Wang, G.; Moran, E.; Batistella, M.; Zhang, M.; Vaglio Laurin, G.; Saah, D. Aboveground forest biomass estimation with Landsat and LiDAR data and uncertainty analysis of the estimates. *Int. J. For. Res.* **2012**, *2012*, 1–16. [CrossRef]
10. Patenaude, G.; Hill, R.A.; Milne, R.; Gaveau, D.L.A.; Briggs, B.B.J.; Dawson, T.P. Quantifying forest above ground carbon content using LiDAR remote sensing. *Remote Sens. Environ.* **2004**, *93*, 368–380. [CrossRef]
11. Riom, J.; Le Toan, T. Relation entre des types de forets de pin maritime et la retrodiffusion radar en bande L. en polarisation HH.[teledetection]. *Colloq. l'INRA (France)* **1981**, *5*.
12. Tanase, M.A.; Panciera, R.; Lowell, K.; Tian, S.; Garcia-Martin, A.; Walker, J.P. Sensitivity of L-band radar backscatter to forest biomass in semiarid environments: A comparative analysis of parametric and nonparametric models. *IEEE Trans. Geosci. Remote Sens.* **2014**, *52*, 4671–4685. [CrossRef]
13. Dobson, M.C.; Ulaby, F.T.; LeToan, T.; Beaudoin, A.; Kasischke, E.S.; Christensen, N. Dependence of radar backscatter on coniferous forest biomass. *IEEE Trans. Geosci. Remote Sens.* **1992**, *30*, 412–415. [CrossRef]
14. Le Toan, T.; Quegan, S.; Davidson, M.W.J.; Balzter, H.; Paillou, P.; Papathanassiou, K.; Plummer, S.; Rocca, F.; Saatchi, S.; Shugart, H.; et al. The BIOMASS mission: Mapping global forest biomass to better understand the terrestrial carbon cycle. *Remote Sens. Environ.* **2011**, *115*, 2850–2860. [CrossRef]

15. Cartus, O.; Santoro, M.; Kellndorfer, J. Mapping forest aboveground biomass in the Northeastern United States with ALOS PALSAR dual-polarization L-band. *Remote Sens. Environ.* **2012**, *124*, 466–478. [CrossRef]
16. Harrell, P.A.; Bourgeau-Chavez, L.L.; Kasischke, E.S.; French, N.H.F.; Christensen, N.L. Sensitivity of ERS-1 and JERS-1 radar data to biomass and stand structure in Alaskan boreal forest. *Remote Sens. Environ.* **1995**, *54*, 247–260. [CrossRef]
17. Saatchi, S.; Marlier, M.; Chazdon, R.L.; Clark, D.B.; Russell, A.E. Impact of spatial variability of tropical forest structure on radar estimation of aboveground biomass. *Remote Sens. Environ.* **2011**, *115*, 2836–2849. [CrossRef]
18. Santoro, M.; Eriksson, L.; Askne, J.; Schmullius, C. Assessment of stand-wise stem volume retrieval in boreal forest from JERS-1 L-band SAR backscatter. *Int. J. Remote Sens.* **2006**, *27*, 3425–3454. [CrossRef]
19. Santos, J.R.; Lacruz, M.S.P.; Araujo, L.S.; Keil, M. Savanna and tropical rainforest biomass estimation and spatialization using JERS-1 data. *Int. J. Remote Sens.* **2002**, *23*, 1217–1229. [CrossRef]
20. Lucas, R.; Armston, J.; Fairfax, R.; Fensham, R.; Accad, A.; Carreiras, J.; Kelley, J.; Bunting, P.; Clewley, D.; Bray, S.; et al. An evaluation of the ALOS PALSAR L-band backscatter—Above ground biomass relationship Queensland, Australia: Impacts of surface moisture condition and vegetation structure. *IEEE J. Sel. Top. Appl. Earth Obs. Remote Sens.* **2010**, *3*, 576–593. [CrossRef]
21. Carreiras, J.; Melo, J.B.; Vasconcelos, M.J. Estimating the above-ground biomass in miombo savanna woodlands (Mozambique, East Africa) using L-band synthetic aperture radar data. *Remote Sens.* **2013**, *5*, 1524–1548. [CrossRef]
22. Imhoff, M.L. A theoretical analysis of the effect of forest structure on synthetic aperture radar backscatter and the remote sensing of biomass. *IEEE Trans. Geosci. Remote Sens.* **1995**, *33*, 341–352. [CrossRef]
23. Le Toan, T.; Quegan, S.; Woodward, I.; Lomas, M.; Delbart, N.; Picard, G. Relating radar remote sensing of biomass to modelling of forest carbon budgets. *Clim. Chang.* **2004**, *67*, 379–402. [CrossRef]
24. Saatchi, S.; Halligan, K.; Despain, D.G.; Crabtree, R.L. Estimation of forest fuel load from radar remote sensing. *IEEE Trans. Geosci. Remote Sens.* **2007**, *45*, 1726–1740. [CrossRef]
25. Sandberg, G.; Ulander, L.M.H.; Fransson, J.E.S.; Holmgren, J.; Le Toan, T. L-and P-band backscatter intensity for biomass retrieval in hemiboreal forest. *Remote Sens. Environ.* **2011**, *115*, 2874–2886. [CrossRef]
26. He, Q.-S.; Cao, C.-X.; Chen, E.-X.; Sun, G.-Q.; Ling, F.-L.; Pang, Y.; Zhang, H.; Ni, W.-J.; Xu, M.; Li, Z.-Y.; et al. Forest stand biomass estimation using ALOS PALSAR data based on LiDAR-derived prior knowledge in the Qilian Mountain, western China. *Int. J. Remote Sens.* **2012**, *33*, 710–729. [CrossRef]
27. Michelakis, D.G.; Stuart, N.; Woodhouse, I.H.; Lopez, G.; Linares, V. Establishing the sensitivity of ALOS PALSAR to above ground woody biomass: A case study in the pine savannas of Belize, Central America. In Proceedings of the 2013 IEEE International Geoscience and Remote Sensing Symposium (IGARSS), Melbourne, Australia, 21–26 July 2013; pp. 953–956.
28. Hansen, E.H.; Gobakken, T.; Bollandsås, O.M.; Zahabu, E.; Næsset, E. Modeling aboveground biomass in dense tropical submontane rainforest using airborne laser scanner data. *Remote Sens.* **2015**, *7*, 788–807. [CrossRef]
29. Sarker, L.R.; Nichol, J.; Iz, H.B.; Ahmad, B.B.; Rahman, A.A. Forest biomass estimation using texture measurements of high-resolution dual-polarization C-band SAR data. *IEEE Trans. Geosci. Remote Sens.* **2013**, *51*, 3371–3384. [CrossRef]
30. Haralick, R.M. Statistical and structural approaches to texture. *Proc. IEEE* **1979**, *67*, 786–804. [CrossRef]
31. Attarchi, S.; Gloaguen, R. Improving the estimation of above ground biomass using dual polarimetric PALSAR and ETM+ data in the Hyrcanian Mountain Forest (Iran). *Remote Sens.* **2014**, *6*, 3693–3715. [CrossRef]
32. Champion, I.; Da Costa, J.P.; Godineau, A.; Villard, L.; Dubois-Fernandez, P.; Le Toan, T. Canopy structure effect on SAR image texture versus forest biomass relationships. *EARSeL eProc.* **2013**, *12*. [CrossRef]
33. Deng, S.; Katoh, M.; Guan, Q.; Yin, N.; Li, M. Estimating forest aboveground biomass by combining ALOS PALSAR and WorldView-2 data: A case study at Purple Mountain National Park, Nanjing, China. *Remote Sens.* **2014**, *6*, 7878–7910. [CrossRef]
34. Basuki, T.M.; Skidmore, A.K.; Hussin, Y.A.; Van Duren, I. Estimating tropical forest biomass more accurately by integrating ALOS PALSAR and Landsat-7 ETM+ data. *Int. J. Remote Sens.* **2013**, *34*, 4871–4888. [CrossRef]
35. Fedrigo, M.; Meir, P.; Sheil, D.; Van Heist, M.; Woodhouse, I.H.; Mitchard, E.T.A. Fusing radar and optical remote sensing for biomass prediction in mountainous tropical forests. In Proceedings of the 2013 IEEE International Geoscience and Remote Sensing Symposium (IGARSS), Melbourne, Australia, 21–26 July 2013; pp. 975–978.

36. Azcueta, M.; d'Alessandro, M.M.; Zajc, T.; Grunfeld, N.; Thibeault, M. ALOS-2 preliminary calibration assessment. In Proceedings of the 2015 IEEE International Geoscience and Remote Sensing Symposium (IGARSS), Milan, Italy, 26–31 July 2015; pp. 4117–4120.

37. Rosenqvist, A.; Shimada, M.; Suzuki, S.; Ohgushi, F.; Tadono, T.; Watanabe, M.; Tsuzuku, K.; Watanabe, T.; Kamijo, S.; Aoki, E. Operational performance of the ALOS global systematic acquisition strategy and observation plans for ALOS-2 PALSAR-2. *Remote Sens. Environ.* **2014**, *155*, 3–12. [CrossRef]

38. Nguyen, L.V.; Tateishi, R.; Nguyen, H.T.; Sharma, R.C.; To, T.T.; Le, S.M. Estimation of tropical forest structural characteristics using ALOS-2 SAR data. *Adv. Remote Sens.* **2016**, *5*, 131–144. [CrossRef]

39. Tabacchi, G.; Di Cosmo, L.; Gasparini, P. Aboveground tree volume and phytomass prediction equations for forest species in Italy. *Eur. J. For. Res.* **2011**, *130*, 911–934. [CrossRef]

40. Tabacchi, G.; Di Cosmo, L.; Gasparini, P.; Morelli, S. Stima del volume e della fitomassa delle principali specie forestali italiane. Equazioni di previsione, tavole del volume e tavole della fitomassa arborea epigea. Consiglio per la Ricerca e la sperimentazione in Agricoltura, Unit{à} di Ricerca per il Moni. In *Trento: Consiglio per la Ricerca e la Sperimentazione in Agricoltura, Unita di Ricerca per il Monitoraggio e la Pianificazione Forestale*; Consiglio per la Ricerca e la sperimentazione in Agricoltura (CRA): Roma, Italy, 2011. (In Italian)

41. Chen, Q. Modeling aboveground tree woody biomass using national-scale allometric methods and airborne lidar. *ISPRS J. Photogramm. Remote Sens.* **2015**, *106*, 95–106. [CrossRef]

42. White, A.; Manley, P. Wildlife Habitat Occurrence Models for Project and Landscape Evaluations in the Lake Tahoe Basin. Final Report to the U.S. Department of Interior, Bureau of Land Management. 2012. Available online: https://www.fs.fed.us/psw/partnerships/tahoescience/documents/p050_FinalReportWildlifeHabitat.pdf (accessed on 23 December 2016).

43. Heath, L.S.; Hansen, M.; Smith, J.E.; Miles, P.D.; Smith, B.W. Investigation into calculating tree biomass and carbon in the FIADB using a biomass expansion factor approach. In Proceedings of the Forest Inventory and Analysis (FIA) Symposium 2008, Park City, UT, USA, 21–23 October 2008; U.S. Department of Agriculture, Forest Service, Rocky Mountain Research Station: Fort Collins, CO, USA, 2009.

44. Woodall, C.W.; Heath, L.S.; Domke, G.M.; Nichols, M.C. Methods and equations for estimating aboveground volume, biomass, and carbon for trees in the US forest inventory, 2010. In *General Technical Report NRS-88*; U.S. Department of Agriculture, Forest Service, Northern Research Station: Newtown Square, PA, USA, 2011.

45. ALOS-2/Calibration Result of JAXA Standard Products. Available online: http://www.eorc.jaxa.jp/ALOS-2/en/calval/calval_index.htm (accessed on 6 December 2015).

46. Chen, Q. Airborne lidar data processing and information extraction. *Photogramm. Eng. Remote Sens.* **2007**, *73*, 109–112.

47. Pirotti, F.; Guarnieri, A.; Vettore, A. Analysis of correlation between full-waveform metrics, scan geometry and land-cover: An application over forests. *ISPRS Ann. Photogramm. Remote Sens. Spat. Inf. Sci.* **2013**, *II-5/W2*, 235–240. [CrossRef]

48. *Matlab*, version 2010b; MathWorks Inc.: Natick, MA, USA, 2012.

49. RC Team. *R: A Language and Environment for Statistical Computing*, R Foundation for Statistical Computing: Vienna, Austria, 2013.

50. Zolkos, S.G.; Goetz, S.J.; Dubayah, R. A meta-analysis of terrestrial aboveground biomass estimation using lidar remote sensing. *Remote Sens. Environ.* **2013**, *128*, 289–298. [CrossRef]

51. Chen, Q.; Laurin, G.V.; Valentini, R. Uncertainty of remotely sensed aboveground biomass over an African tropical forest: Propagating errors from trees to plots to pixels. *Remote Sens. Environ.* **2015**, *160*, 134–143. [CrossRef]

52. Tanase, M.A.; Panciera, R.; Lowell, K.; Aponte, C.; Hacker, J.M.; Walker, J.P. Forest biomass estimation at high spatial resolution: Radar versus lidar sensors. *IEEE Geosci. Remote Sens. Lett.* **2014**, *11*, 711–715. [CrossRef]

53. Goh, J.; Miettinen, J.; Chia, A.S.; Chew, P.T.; Liew, S.C. Biomass estimation in humid tropical forest using a combination of ALOS PALSAR and SPOT 5 satellite imagery. *Asian J. Geoinform.* **2014**, *13*.

54. Bharadwaj, P.S.; Kumar, S.; Kushwaha, S.P.S.; Bijker, W. Polarimetric scattering model for estimation of above ground biomass of multilayer vegetation using ALOS-PALSAR quad-pol data. *Phys. Chem. Earth Parts A/B/C* **2015**, *83*, 187–195. [CrossRef]

55. Morel, A.C.; Saatchi, S.S.; Malhi, Y.; Berry, N.J.; Banin, L.; Burslem, D.; Nilus, R.; Ong, R.C. Estimating aboveground biomass in forest and oil palm plantation in Sabah, Malaysian Borneo using ALOS PALSAR data. *For. Ecol. Manag.* **2011**, *262*, 1786–1798. [CrossRef]

56.   Peregon, A.; Yamagata, Y. The use of ALOS/PALSAR backscatter to estimate above-ground forest biomass: A case study in Western Siberia. *Remote Sens. Environ.* **2013**, *137*, 139–146. [CrossRef]

57.   Thumaty, K.C.; Fararoda, R.; Middinti, S.; Gopalakrishnan, R.; Jha, C.S.; Dadhwal, V.K. Estimation of Above Ground Biomass for Central Indian Deciduous Forests Using ALOS PALSAR L-Band Data. *J. Indian Soc. Remote Sens.* **2016**, *44*, 31–39. [CrossRef]

58.   Hamdan, O.; Aziz, H.K.; Hasmadi, I.M. L-band ALOS PALSAR for biomass estimation of Matang Mangroves, Malaysia. *Remote Sens. Environ.* **2014**, *155*, 69–78. [CrossRef]

59.   Viergever, K.M. Establishing the Sensitivity of Synthetic Aperture Radar to above-Ground Biomass in Wooded Savannas. 2008. Available online: https://www.era.lib.ed.ac.uk/handle/1842/3180 (accessed on 23 December 2016).

60.   Michelakis, D.; Stuart, N.; Lopez, G.; Linares, V.; Woodhouse, I.H. Local-scale mapping of biomass in tropical lowland pine savannas using ALOS PALSAR. *Forests* **2014**, *5*, 2377–2399. [CrossRef]

61.   Evans, J.S.; Hudak, A.T.; Faux, R.; Smith, A. Discrete return lidar in natural resources: Recommendations for project planning, data processing, and deliverables. *Remote Sens.* **2009**, *1*, 776–794. [CrossRef]

62.   Morsdorf, F.; Meier, E.; Kötz, B.; Itten, K.I.; Dobbertin, M.; Allgöwer, B. LiDAR-based geometric reconstruction of boreal type forest stands at single tree level for forest and wildland fire management. *Remote Sens. Environ.* **2004**, *92*, 353–362. [CrossRef]

63.   Shendryk, I.; Hellström, M.; Klemedtsson, L.; Kljun, N. Low-density LiDAR and optical imagery for biomass estimation over boreal forest in Sweden. *Forests* **2014**, *5*, 992–1010. [CrossRef]

64.   Jakubowski, M.K.; Guo, Q.; Kelly, M. Tradeoffs between LiDAR pulse density and forest measurement accuracy. *Remote Sens. Environ.* **2013**, *130*, 245–253. [CrossRef]

65.   Magnussen, S.; Næsset, E.; Gobakken, T. Reliability of LiDAR derived predictors of forest inventory attributes: A case study with Norway spruce. *Remote Sens. Environ.* **2010**, *114*, 700–712. [CrossRef]

66.   Montesano, P.M.; Nelson, R.F.; Dubayah, R.O.; Sun, G.; Cook, B.D.; Ranson, K.J.R.; Kharuk, V. The uncertainty of biomass estimates from LiDAR and SAR across a boreal forest structure gradient. *Remote Sens. Environ.* **2014**, *154*, 398–407. [CrossRef]

remote sensing

MDPI

Article

Estimating Biomass of Native Grass Grown under Complex Management Treatments Using WorldView-3 Spectral Derivatives

Mbulisi Sibanda [1,*], Onisimo Mutanga [1], Mathieu Rouget [1] and Lalit Kumar [2]

[1] School of Agriculture, Earth and Environmental Science, University of KwaZulu-Natal, P. Bag X01, Scottsville, Pietermaritzburg 3209, South Africa; mutangao@ukzn.ac.za (O.M.); rouget@ukzn.ac.za (M.R.)
[2] School of Environmental & Rural Science, University of New England, Armidale NSW 2351, Australia; lkumar@une.edu.au
* Correspondence: sibandambulisi@gmail.com; Tel.: +27-33-260-5779

Academic Editors: Lenio Soares Galvao, Xiaofeng Li and Prasad S. Thenkabail
Received: 20 September 2016; Accepted: 4 January 2017; Published: 11 January 2017

Abstract: The ability of texture models and red-edge to facilitate the detection of subtle structural vegetation traits could aid in discriminating and mapping grass quantity, a challenge that has been longstanding in the management of grasslands in southern Africa. Subsequently, this work sought to explore the robustness of integrating texture metrics and red-edge in predicting the above-ground biomass of grass growing under different levels of mowing and burning in grassland management treatments. Based on the sparse partial least squares regression algorithm, the results of this study showed that red-edge vegetation indices improved above-ground grass biomass from a root mean square error of perdition (RMSEP) of 0.83 kg/m^2 to an RMSEP of 0.55 kg/m^2. Texture models further improved the accuracy of grass biomass estimation to an RMSEP of 0.35 kg/m^2. The combination of texture models and red-edge derivatives (red-edge-derived vegetation indices) resulted in an optimal prediction accuracy of RMSEP 0.2 kg/m^2 across all grassland management treatments. These results illustrate the prospect of combining texture metrics with the red-edge in predicting grass biomass across complex grassland management treatments. This offers the detailed spatial information required for grassland policy-making and sustainable grassland management in data-scarce regions such as southern Africa.

Keywords: grass biomass; SPLSR; vegetation indices; estimation accuracy

1. Introduction

Understanding above-ground grass biomass variations at various scales has become increasingly critical among stakeholders, such as farmers, ecologists and scientists, amongst others. Grasslands are significant carbon sinks, accounting for 18% of the global terrestrial carbon sinks [1]. Furthermore, grasslands are one of the biodiversity hot spots harbouring a wide variety of plants and animals [2], while facilitating soil formation and preservation. From an agricultural perspective, native grasses are the cheapest source of stock feed available. Moreover, grasslands are also a significant source of livelihood, especially to rural communities in southern Africa, where natural disasters and socio-economic hardships are frequent. Collectively, these factors drive the growing interest of accurately monitoring grassland biomass variations for developing optimal management regimes.

A total of 7.5% of the world's grasslands have been degraded, while about 16% are currently being degraded [3]. Tropical grasslands, specifically, are often at risk of degradation because of increasing pressure from human activities due to population increase [4]. For instance, infrastructural development, crop farming and overgrazing have been cited as the major causes of tropical grassland degradation [3]. Livestock farming has been considered as the fastest growing agricultural sector due to the demand for meat and milk products. Consequently, overstocking and overgrazing have been reported as drivers of grassland degradation. To optimise productivity, while preserving native grasses, numerous grass management practices have been introduced [5]. These include burning, mowing, fertiliser application, as well as controlled grazing [5]. However, insights on the effectiveness of these grass management treatments on grass productivity are limited. This is because there are no cost-effective monitoring systems that have hitherto been developed. Furthermore, the use of existing methods has not been comprehensively evaluated across space and time to the extent that is sufficient for meaningful decision-making and management in data-scarce regions, such as southern Africa.

To acquire comprehensive quantitative information on grass biomass, the utility of earth observation (EO) data has recently become more popular and feasible with an increase, as well as advances, in the available sensors [6]. EO data have been renowned for facilitating rapid, repeated and ongoing biomass observations over various spatial and temporal scales. This is because EO enables comparatively convenient data acquisition dating back over several years, while offering satisfactory ranges of accuracy on above-ground biomass estimation over larger spatial scales. Despite the fact that numerous EO methodologies have been evaluated in quantifying above-ground biomass, no study has hither to illustrate an operational technique that is consistent, precise and repeatable for estimating biomass at local to continental scales. This is caused by the variations in the biophysical, environmental and topographic traits of vegetation in space and time [7,8].

A growing body of literature illustrates that the common approach for estimating biomass, based on EO data, has been to examine the possible association between the ground measured biomass and the EO data, since biomass quantities cannot be directly derived from remotely sensed data [9,10]. Landsat data is the most widely used EO data in vegetation above-ground biomass estimation studies due to its limited costs. However, the majority of the studies have used Landsat for forest inventories [11,12]. The few studies that have been conducted on grass productivity have focused only on a limited number of grass management treatments [13,14].

Furthermore, primary vegetation indices (VIs), such as the normalised difference vegetation index (NDVI), have been widely used for estimating above-ground grass biomass [13,14]. VIs have been widely used because they tend to supersede the influences of the soil background, atmospheric impurities and the viewing and zenith angle effects, while magnifying the signature of vegetation [15,16]. However, these have attained only moderate success in the tropical and subtropical regions [17,18] characterised by complex management treatments, with high spatial heterogeneity. This is due to the lack of strategically located wavebands [19,20], such as the red-edge (i.e., in the Landsat data series). Furthermore, these indices are affected by saturation, soil background and the coarse spatial resolutions for application in grass grown across different grassland management treatments, which still remains a challenge [17,21,22]. This is aggravated by the lack of a clear criterion on the appropriateness of specific EO sensors, proxies, as well as repeatable operational techniques that could provide accurate biomass information from a variety of grass management treatments.

Red-edge (680–740 nm) and texture models seem to offer better proxies, which suppress the soil-background effect, saturation issues [17] and high spatial heterogeneity. Literature shows that the red-edge is sensitive to chlorophyll, as well as leaf structure reflection (i.e., leaf area index, leaf angle distribution), thereby providing more information for the characterization of vegetation [23,24]. More specifically, when the concentration of foliar chlorophyll increases, it results in the bulging of the optical chlorophyll absorption feature, shifting away from the long wavelength margin, and thereby shifting the red-edge to longer wavelengths [25]. Meanwhile, the concentration of leaves of a certain vegetation canopy, as well as the angular nature of those leaves, directly affects the spectral reflectance of that vegetation, especially in the red-edge portion of the electromagnetic spectrum [26]. Subsequently, the biomass of vegetation with a high chlorophyll concentration or leaf area index can then be detected from that with less concentration, based on these shifts. In this regard, it is perceived that the red-edge waveband and its derivatives can better estimate above-ground biomass, when compared to primary bands and vegetation indices [17].

On the other hand, literature indicates that grey level co-occurrence optical texture models also relate better with field measured above-ground vegetation biomass when compared with vegetation indices [7,27]. For instance, work by Cutler et al. [28] indicated that integrating texture metrics data improved biomass estimation from R^2 of 0.05, 0.23 and 0.16 to 0.79, 0.79 and 0.84 in Thailand, Malaysia and Brazil, respectively, when compared with multispectral data. Furthermore, texture models offer information that could characterize the subtle structural characteristics of the vegetation canopy, such as those induced by different grassland management treatments. Texture metrics i.e., the grey level co-occurrence matrix, distinguishes minute, but critical, vegetation details, based on a local spectral variation in the image [6]. This is due to the fact that texture models can also suppress the influence of atmospheric effects, the sensor view-angle and the sun view angle, which improve the vegetation spectral signature required for the accurate estimation of above-ground grass biomass [7,29,30]. It is, therefore, important to note that texture variables can optimize the discrimination of vegetation spatial information independently from the tone, while spectral features, i.e., the red-edge, provides detailed vegetation tonal variations that are paramount for accurate vegetation mapping. Based on the above premise, the combination of optimal texture models and red-edge wavebands has a high potential for improving above-ground biomass estimation across different grassland management treatments, superseding the saturation effect of spectral data. To the best of our knowledge, very few studies, if any, have been conducted, based on texture models, to predict above-ground grass biomass.

The majority of the studies that utilised texture metrics were focused on forest above-ground biomass [6,10,30–33]. In addition, most of these studies utilised the moderate resolution Landsat data, which does not capture the minute variations that could be induced by different grass treatments in a grassland landscape that is characterised by high spatial heterogeneity [1]. Considering the lack of suitable specific proxies for accurate biomass information in southern African grasslands, due to limited resources and data scarcity [30], there is a need to evaluate the performance of possible sources of spatial information, such as texture models and red-edge wavebands. The advent of a new generation of multispectral sensors, such as the newly launched Sentinel-2 multispectral imager and WorldView-3, offers an opportunity to improve the accuracy of above-ground grass biomass estimation in southern Africa. This is because of their spectral regions—such as red-edge, which are crucial for vegetation mapping, as well as their optimal spatial resolution—could offer the critical spatial information that is required in well-informed grassland management practices.

Despite the relatively high costs associated with high spatial resolution EO data, these data sources offer abundant texture information, which could better characterize the spatial distribution of different grassland management treatments [29]. For example, the new WorldView-3 (WV-3) sensor, characterized by a fine spatial resolution of 2 m, as well as the strategically positioned red-edge waveband, offers better spatial information, when compared to other sensors, such as Landsat, which has a moderate spatial resolution and lacks the red-edge waveband. In that regard, WorldView-3 texture models, combined with red-edge band derivatives, could have better spectral responses to grass above-ground biomass estimation with complex grass management treatments [7].

The aim of this study, therefore, is to test whether combining WV-3 optical texture models with red-edge can improve the accuracies of predicting above-ground biomass of native grass grown under different levels of mowing, burning and fertilizer treatments using the sparse partial least squares regression algorithm. To achieve the above aim we tested the strength of (i) WV-3 wavebands with that of broadband Vis; (ii) WV-3 standard wavebands combined with broadband VIs compared with that of red-edge-derived Vis; (iii) WV-3 wavebands, broadband and red-edge VIs combined compared to single-band texture models; (iv) all variables combined compared to that of all texture models in estimating above-ground biomass of grass grown under different grassland treatments.

2. Methods and Materials

2.1. Study Area Description

This study was undertaken at the Ukulinga Research Farm in Pietermaritzburg, KwaZulu-Natal, South Africa (29°24′E, 30°24′S) (Figure 1). The weather at Pietermaritzburg is characterised by cold winters and hot summers, with a minimum mean monthly temperature of 6 °C, as well as a maximum mean monthly temperature of ±27 °C. Ukulinga is a 228 ha farm that is situated on a plateau, hence it is characterized by a generally flat terrain with an altitude ranging between 838 and 847 m above sea level [34]. The major grass species at the grassland trials on the University farm are *Themeda triandra*, *Heteropogon contortus*, *Eragrostis plana*, *Panicum maximum*, *Setaria nigrirostrosis* and *Tristachya leucothrix*. The mean height of these grasses was about 40 cm. The soils at the research farm are generally infertile, acidic and of the Westleigh type [34]. The experimental site at Ukulinga was established by JD Scott in 1950 [35], with the aim of understanding the influence of different management practices on grass quantity and quality. In general, these grasslands in South Africa have a total economic value of R 9.7 billion, which includes a consumptive value of R 1.59 million as well as an indirect value of about R 8 million [36].

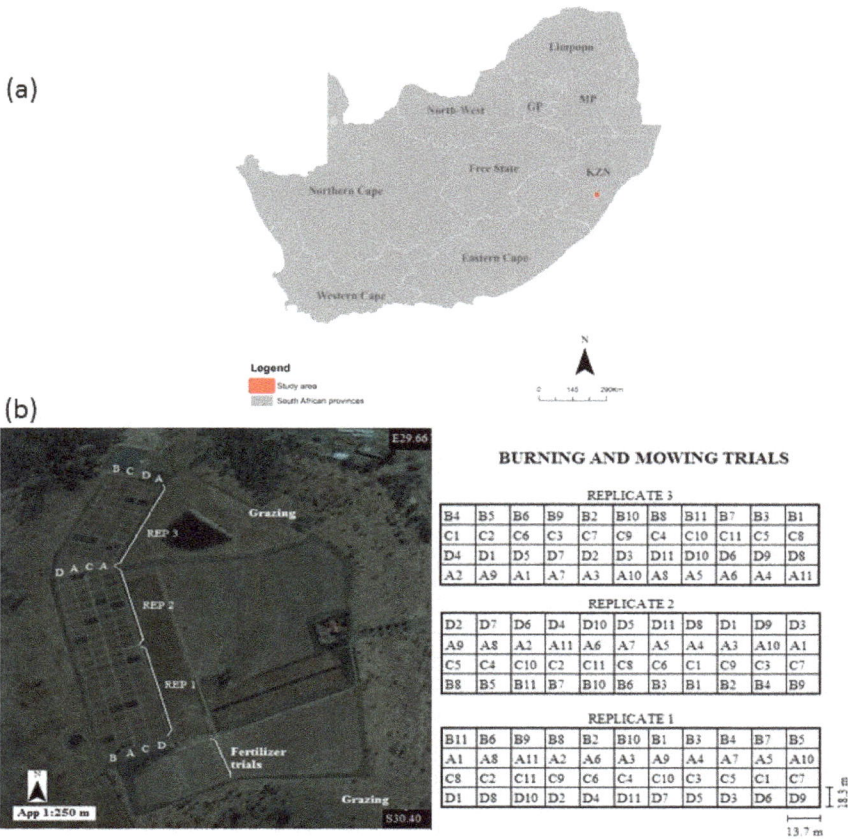

(a)

(b)

BURNING AND MOWING TRIALS

REPLICATE 3

B4	B5	B6	B9	B2	B10	B8	B11	B7	B3	B1
C1	C2	C6	C3	C7	C9	C4	C10	C11	C5	C8
D4	D1	D5	D7	D2	D3	D11	D10	D6	D9	D8
A2	A9	A1	A7	A3	A10	A8	A5	A6	A4	A11

REPLICATE 2

D2	D7	D6	D4	D10	D5	D11	D8	D1	D9	D3
A9	A8	A2	A11	A6	A7	A5	A4	A3	A10	A1
C5	C4	C10	C2	C11	C8	C6	C1	C9	C3	C7
B8	B5	B11	B7	B10	B6	B3	B1	B2	B4	B9

REPLICATE 1

B11	B6	B9	B8	B2	B10	B1	B3	B4	B7	B5
A1	A8	A11	A2	A6	A3	A9	A4	A7	A5	A10
C8	C2	C11	C9	C6	C4	C10	C3	C5	C1	C7
D1	D8	D10	D2	D4	D11	D7	D5	D3	D6	D9

Figure 1. (**a**) Location of the grassland sites at Ukulinga University of KwaZulu-Natal experimental Farm, Pietermaritzburg, South Africa; (**b**) shows the experimental setup and design at Ukulinga research farm (Image source: Google Earth).

2.2. Experimental Design

The experiment consisted of grass burning, mowing and fertilisation treatments at timely intervals. A total of 54 plots measuring 13.7 m × 18.3 m, with native grass growing under mowing and burning, were utilised in this study (Table 1). Burning treatments were undertaken at three levels, namely: (i) annually; (ii) biennially (after two years); and (iii) triennially (after three years). Mowing was also implemented at three levels. At Level 1, there was no mowing, at Level 2 grass was mown once in August, and at Level 3, grass was mown twice in August and after the first Spring rains.

Table 1. Reflectance samples measured on each rangeland management treatment.

Treatment Level	Treatment	Samples	Plots
C1	Control	60	3
C2	Annual burn (in August)	60	3
C3	Annual burn (after Spring rain)	60	3
C4	Biennial burn (in August)	60	3
C5	Biennial burn (after Spring rain)	60	3
C7	Triennial burn (in August)	60	3
C8	Triennial burn (after Spring rain)	60	3
C10	Mowing (in August)	60	3
C11	Mowing (after Spring rain)	60	3
D1	Control	60	3
D2	Annual burn (in August)	60	3
D3	Annual burn (after Spring rain)	60	3
D4	Biennial burn (in August)	60	3
D5	Biennial burn (after Spring rain)	60	3
D7	Triennial burn (in August)	60	3
D8	Triennial burn (after Spring rain)	60	3
D10	Mowing (in August)	60	3
D11	Mowing (after Spring rain)	60	3
Total		1080	54

Note: Grass on C treatments are removed end of February, while those in D are removed twice in February and December.

2.3. Field Campaign

To extract spectra from each plot, 20 points were randomly generated in a Geographic Information System (GIS) environment. Ultimately, 1080 points were derived from 54 plots and used to extract all WV-3 variables, using an overlay function in a GIS (Table 1). To test the capability of the combined red-edge and texture models in estimating above-ground grass biomass, we conducted a field survey on the 10 February 2016. During the field campaigns, plots with native grasses grown under mowing, burning, as well as no-treatment, were surveyed and the grass biomass clipped. The wet biomass of grass from each level of treatment was derived after cutting during the field survey. The samples were then taken to the laboratory, where moisture content was determined and dry grass biomass, hereafter referred to as above-ground grass biomass, was derived.

2.4. Remotely Sensed Data

A WorldView-3 image, acquired on a cloudless day on 16 February 2016, was used in this study to evaluate the strength of red-edge, combined with texture models, in predicting above-ground biomass. The WV-3 image has eight multispectral bands, i.e., coastal blue at 400–450 nm, blue at 450–510 nm, green at 510–589 nm, yellow at 585–625 nm, red at 630–690 nm, red-edge at 705–895 nm and two near-infrared bands, which overlap, at 770–895 and 860–1040 nm, respectively. The spatial resolution of all wavebands was 2 m. The image was first pre-processed to correct for the influence of atmospheric effects, using the Fast Line of Sight Atmospheric Analysis of Spectral Hypercubes (FLAASH), based on the parameters that were provided with the image. The FLAASH analysis was conducted after converting the image into radiance in Envi 5.2. Subsequently, the WorldView-3 image was geometrically corrected, based on ten locations measured using a handheld Trimble GeoXH 6000 global positioning system with a sub-meter accuracy. The image was then to resample using the first order polynomial transformation and nearest-neighbor resampling technique as in Sibanda et al. [37]. As mentioned earlier, the atmospherically corrected image was used in an overlay analysis, in conjunction with the point map, in order to derive spectral signatures of grass growing under different levels of grassland management treatments.

2.5. Modelling Above-Ground Grass Biomass

Single wavebands, broadband and red-edge vegetation indices, as well as grey level co-occurrence single-band and band-ratio texture models, were derived in Envi 4.3 from the pre-processed WV-3 image. The vegetation indices used in this study were chosen based on their optimal performance in literature [17,22]. Formulae for computing vegetation indices are detailed in Schumacher et al. [38]. The window sizes for deriving the grey-level co-occurrence texture models used in this study were 3×3, 5×5 and 7×7 pixels [39,40]. These window sizes were selected because their area was not bigger than that of a single plot of grass used in this study. The co-occurrence shifts considered in this study were 0:1, 1:1, 1:0, -1:1, 1:-1 which were chosen based on literature [30,41] and a quantization level of 64 was used in this study. The texture models computed in this study were mean, variance, homogeneity, contrast, dissimilarity, entropy, second moment and correlation. More details about the formulae for computing these texture models are summarised in Dube and Mutanga [30], as well as Schumacher et al. [38]. All the variables used in this study, and the formulae used to compute them, are detailed in Table 2. The derived spectral signatures were saved in a table format and exported to Microsoft Excel as comma separated values. These were then imported into Statistica Version 7 and R statistical software for statistical modelling.

Table 2. Variable categories used in this study.

Phase	Analysis	Variable	Description	Reference
1	Bands	WV-3 B2-B8	Single-bands—reflectance values	
	vs.			
	Broadband VIs	*Broadband VIs*		
		Chlorophyll Index Green	$CGM = \dfrac{NIR}{G} - 1$	Kang et al. [42], Gitelson et al. [43]
		Green normalised difference VI	$GNDVI = \dfrac{NIR - G}{NIR + G}$	Fernández-Manso et al. [44]
		Green blue normalised difference VI	$GBNDVI = \dfrac{NIR - (G + B)}{NIR + (G + B)}$	Santoso et al. [45]
		Normalised difference VI	$NDVI = \dfrac{NIR - R}{NIR + R}$	Tucker [46]
		Soil adjusted vegetation index	$SAVI = \dfrac{NIR - R}{NIR + R + 0.5} \times (1 + 0.5)$	Huete [47]
		Enhanced vegetation index	$EVI = \dfrac{2.5 \times NIR - R}{NIR + 6 \times R - 7.5 \times B + 1}$	Cabezas et al. [48]
2	Broadband VIs + bands	*Red-Edge Indices*		
	vs.	Browning reflectance index	$BRI = \dfrac{\frac{1}{G} - \frac{1}{RE}}{NIR}$	Merzlyak et al. [49]
	Red-Edge Vis	Canopy chlorophyll content index	$CCCI = \dfrac{\frac{NIR - RE}{NIR + RE}}{\frac{NIR - R}{NIR + R}}$	El-Shikha et al. [50]
		Normalised difference near-infrared red-edge index	$NDNRE = \dfrac{NIR - RE}{NIR - RE}$	
		Normalised difference red-edge index	$NDRE = \dfrac{RE - R}{RE + R}$	Fitzgerald et al. [51]
		Tasseled cap: Soil brightness Index	$TCSBI = 0.332 \times G + 0.603 \times R + 0.675 \times RE - 0.262 \times NIR$	Cabezas et al. [48]
		Anthocyanin reflectance Index		Gitelson et al. [52]

Table 2. *Cont.*

Phase	Analysis	Variable	Description	Reference
3	All VI + Bands	*Single Band Textures, windows (3 and 5)*		
	vs.	*Texture type:*		
	Single-band textures	Mean	$Mn = \sum_{i,j=0}^{N-1}(P_{i,j})$	Wallis [31] Kelsey et al. [53] Schumacher et al. [38] Ouma et al. [54] Salas et al. [33] Zhao et al. [21]
		Variance	$Var = \sum_{i,j=0}^{N-1} Pi, j(i - ME)^2$	
		Homogeneity	$Hom = \sum_{i,j=0}^{N-1} \dfrac{Pi, j}{1 + (i, j)^2}$	Wallis [31] Kelsey et al. [53] Schumacher et al. [38] Ouma et al. [54] Salas et al. [33] Zhao et al. [21]
		Contrast	$Con = \sum_{i,j=0}^{N-1} Pi, j(i - j)^2$	
		Dissimilarity	$Dis = \sum_{i,j=0}^{N-1} Pi, j|i - j|$	
		Entropy	$Ent = \sum_{i,j=0}^{N-1} Pi, j\left(-, \ln P_{i,j}\right)$	Wallis [31] Kelsey et al. [53] Schumacher et al. [38] Ouma et al. [54] Salas et al. [33] Zhao et al. [21]
		Second moment	$Sec = \sum_{i,j=0}^{N-1} P^2i, j$	
		Correlation	$Cor = \sum_{i,j=0}^{N-1} Pi, j\left[\dfrac{(I - ME)(j - ME)}{\sqrt{VA_I}VA_J}\right]$	
4	Band texture variables	*Band-ratios* texture	B2/B3, B2/B5, B2/B7, B2/B8, B3/B5, B3/B7, B3/B8, B5/B7, B5/B8, B2/B6, B3/B6, B6/B7, B6/B8, B6/B8, B8/B7,	
	vs.			
	All combined data			

Note: $Pi, j = \sum_{I,J=0}^{N-1} Vi, j$ where Vµ is the value in cell i, j and N is the number of rows or columns.

2.5.1. Statistical Modelling of Above-Ground Grass Biomass

The initial step was to conduct exploratory analysis and to derive descriptive statistics in Statistica Version 7. Under the exploratory data analysis procedure, we tested whether above-ground grass biomass data measured in the field significantly deviated ($\alpha = 0.05$) from the normal distribution, based on the Lilliefors test. We then tested whether there was significant difference in the amount of above-ground biomass of grass grown under different levels of mowing and burning treatments based on analysis of variance and Tukey's honest significant difference post hoc test.

2.5.2. Regression Modelling

In this study, we used Chun and Keleş's [55] sparse partial least regression (SPLSR) algorithm. The SPLSR algorithm converts the variables into new orthogonal factors to circumvent multicollinearity and overfitting issues, considering the large number of variables used in this study. In converting the variables into orthogonal factors, SPLSR imparts sparsity into the models and then selects the optimal variables that correlate better to grass above-ground biomass. Because of these capabilities, SPLSR is appropriate for application on data with multicollinearity issues, such as the texture models of this study, relative to other algorithms (i.e., partial least squares regression (PLSR)) [55,56]. In this study, the aim was to test whether combining WV-3 optical texture models with red-edge derivatives improves accuracies. Therefore, SPLSR was chosen and utilised because of its ability to select optimal variables.

2.5.3. Assessing the Accuracy of Above-Ground Grass Biomass Models

To evaluate the accuracy of above-ground grass biomass models in this study, a leave-one-out cross-validation (LOOCV) procedure was followed, as detailed in Ritcher et al. [18]. In implementing the LOOCV procedure, 1080 samples, derived from 54 grassplots, were eliminated one by one and above-ground grass biomass estimation errors for each latent variable were derived. The latent variables that exhibited the least root mean square errors were considered as the optimal models for estimating above-ground grass biomass across different levels of grassland management treatments. We computed the coefficient of determination (R^2), root mean square error (RMSEP) as well as the relative root mean square error (RMSEP_rel), as in Frazer et al. [57], to evaluate the models derived using band indices, as well as texture models. Models that exhibited small RMSEs and a high R^2 were considered to be best in estimating above-ground biomass. Considering that SPLSR has the capability of identifying selecting optimal variables, we then used the variable importance (VIP) scores allocated for each of the selected variables by SPLSR, to distinguish the most influential ones from the best models [56].

Finally, an analysis of variance was used to test whether there were significant differences between the accuracies (RMSEP) of: (i) WV-3 wavebands; (ii) broadband Vis; (iii) Wavebands combined with broadband VIs; (iv) red-edge VIs; (v) combination of all VIs and wavebands; (vi) single-band texture models; (vii) combination of single-band and band-ratio texture models; and (viii) all variables combined in predicting above-ground biomass. These combinations were derived from literature [30,38]. Analysis of variance (ANOVA) was used after the normality test and it indicated that the data did not significantly deviate from the normal distribution.

2.5.4. Phases of Estimating Above-Ground Grass Biomass

Table 2 summarises the four phases that were followed. In phase one, the strength of WV-3 wavebands was compared with that of broadband vegetation indices. In the second phase, wavebands were combined with broadband vegetation indices and then compared with the performance of red-edge vegetation indices. In the third phase, the wavebands, broadband and red-edge vegetation indices were combined and compared to the performance of single-band texture models. Lastly, the combination of all variables were then compared with the performance of all texture models. The optimal bands, indices and texture models that are derived using the variable selection capability of SPLSR were then used to estimate above-ground biomass across all grassland management treatments in this study. Figure 2 conceptually illustrates the phases followed.

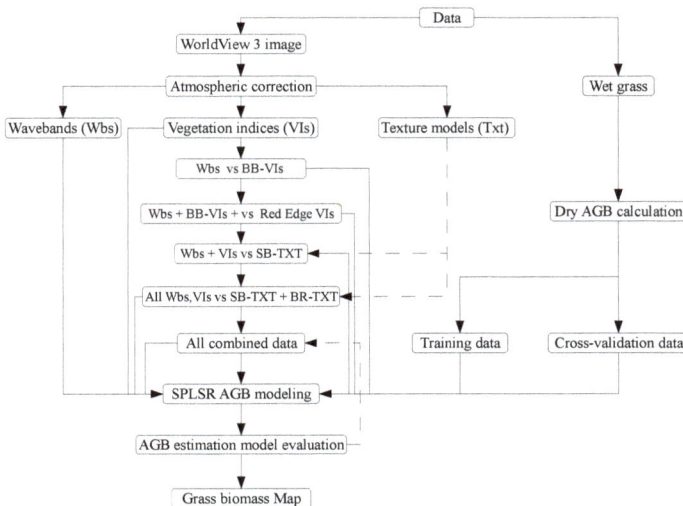

Figure 2. Flowchart illustrating stages in estimating above-ground (ABG) grass biomass in this study. Wbs represents WV-3 wavebands, VIs are vegetation indices, BB-VIs are broadband vegetation indices, SB-TXT represents single band texture models and BR-TXT represents band ratio texture models.

3. Results

3.1. Descriptive Statistical Analysis and ANOVA Tests

Normality test results based on the Lilliefors test, showed that above-ground grass biomass did not significantly deviate from the normal distribution ($\alpha = 0.05$), as illustrated in Figure 3a. Consequently, ANOVA and SPLSR were then conducted. Figure 3a illustrates other descriptive statistics of grass above-ground biomass. The mean of 3.158 kg and a median of 3.149 kg were derived from the field-measured above-ground biomass of grass growing under different levels of burning and mowing treatments. Significant differences in the amount of above-ground biomass were observed amongst grasses growing under different grassland treatments (Figure 3b). Furthermore, Tukey's HSD post hoc test showed that there were significant differences in the quantity of grass biomass between different pairs of burning and mowing grass treatments, as illustrated in Table 3 (p-value < 0.05).

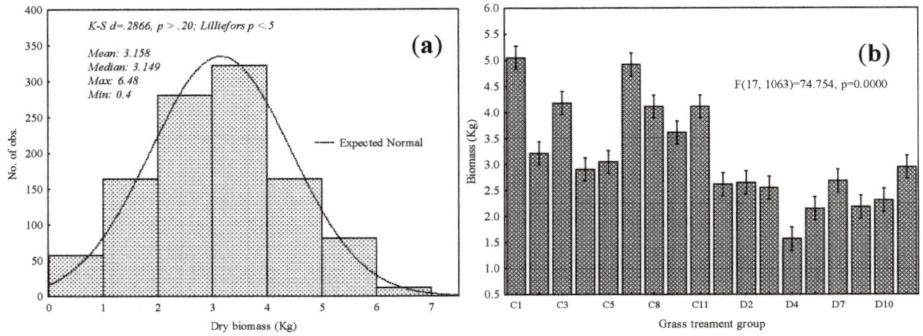

Figure 3. (**a**) Descriptive statistics of measured grass above-ground biomass; (**b**) significant difference amongst different levels of mowing and burning grassland management treatments based on analysis of variance test. Bars represent mean biomass of each management treatment level while whiskers represent confidence intervals of means at 95%.

Table 3. Significant differences between different pairs of grass above-ground biomass grown under different levels of mowing and burning treatments, based on the Tukey's HSD test.

	C1	C2	C3	C4	C5	C7	C8	C10	C11	D1	D2	D3	D4	D5	D7	D8	D10
C2	0.00																
C3	0.00	0.00															
C4	0.00	0.89	0.00														
C5	0.00	1.00	0.00	1.00													
C7	1.00	0.00	0.00	0.00	0.00												
C8	0.00	0.00	1.00	0.00	0.00	0.00											
C10	0.00	0.53	0.04	0.00	0.04	0.00	0.14										
C11	0.00	0.00	1.00	0.00	0.00	0.00	1.00	0.14									
D1	0.00	0.02	0.00	0.94	0.37	0.00	0.00	0.00	0.00								
D2	0.00	0.04	0.00	0.98	0.53	0.00	0.00	0.00	0.00	1.00							
D3	0.00	0.00	0.00	0.73	0.14	0.00	0.00	0.00	0.00	1.00	1.00						
D4	0.00	0.00	0.00	0.00	0.00	0.00	0.00	0.00	0.00	0.00	0.00	0.00					
D5	0.00	0.00	0.00	0.00	0.00	0.00	0.00	0.00	0.00	0.24	0.14	0.53	0.03				
D7	0.00	0.08	0.00	1.00	0.69	0.00	0.00	0.00	0.00	1.00	1.00	1.00	0.00	0.08			
D8	0.00	0.00	0.00	0.00	0.00	0.00	0.00	0.00	0.00	0.37	0.24	0.69	0.01	1.00	0.14		
D10	0.00	0.00	0.00	0.03	0.00	0.00	0.00	0.00	0.00	0.92	0.83	0.99	0.00	1.00	0.69	1.00	
D11	0.00	0.97	0.00	1.00	1.00	0.00	0.00	0.00	0.00	0.83	0.92	0.53	0.00	0.00	0.97	0.00	0.01
Treatment	C1	C2	C3	C4	C5	C7	C8	C10	C11	D1	D2	D3	D4	D5	D7	D8	D10

Legend: 0.00 Significant ($\alpha = 0.05$); 1.00 Non-Significant

Note: light grey cells illustrate significant differences between pairs of treatments, while dark grey cells represent non-significant differences ($\alpha = 0.05$). D1 to D11 and C1 to C11 represent the different levels of burning and mowing treatments illustrated in Table 1.

3.2. Comparing the Performance of WorldView-3 Wavebands Combined with Broadband Vegetation Indices (Vis) and Red-Edge VIs in Estimating Above-Ground Grass Biomass

Exploring the possibility that WV-3 wavebands could better estimate above-ground biomass in relation to broadband VIs resulted in very small and very high RMSEP indicating poor model fitting. In that regard, those results were not presented. It can be observed from Figure 4a,b that the red-edge-derived vegetation indices performed better than broadband vegetation indices combined with band reflectance values. Red-edge-derived VIs resulted in higher accuracies (lower RMSEP), when compared with combined broadband VIs and band reflectance values. Specifically, triennial burning treatment D7 (R^2 = 0.45, RMSEP = 0.26 kg/m^2, RMSEPrel = 12.83) exhibited the lowest prediction error, when red-edge-derived vegetation indices were used. Meanwhile, the highest prediction errors obtained based on the red-edge vegetation indices were observed in C5 (R^2 = 0.62, RMSEP = 0.87 kg/m^2, RMSEPrel = 28.49). Red-edge-derived vegetation indices improved the accuracies of above-ground grass biomass estimation. However, relatively high prediction errors were observed from the triennial burn treatment D7 (R^2 = 0.2, RMSEP = 0.34 kg/m^2, RMSEPrel = 13) and C5 (R^2 = 0.04, RMSEP = 1.81 kg/m^2, RMSEPrel = 92.21), when WV-3 bands were combined with broadband vegetation indices in estimating above-ground grass biomass. The optimal red-edge indices that were selected were the normalized difference near-infrared red-edge index, the normalized difference red-edge index, the canopy chlorophyll content index, the tasseled cap: soil brightness index, and the anthocyanin reflectance index, in order of influence.

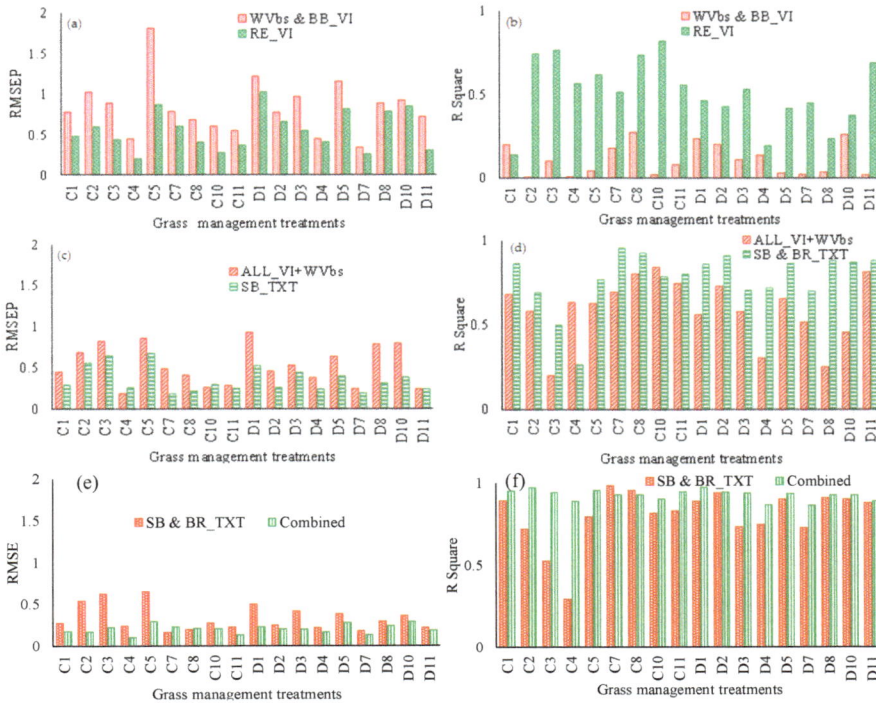

Figure 4. A comparison of estimation accuracies derived using different WV-3 satellite data and its derivatives. Root mean square error of prediction (RMSEP) and R squares obtained in comparing (**a,b**) WV-3 combined BB_VIs and red-edge vegetation (RE_VIs) (**c,d**), all VIs combined with WVbs and single-band texture models (SB_TXT) and (**e,f**) SB_TXT) and all data combined. C1–11 and D1–11 are illustrated in Table 1.

3.3. Comparing the Performance of Single-Band Texture Models with All WV-3 VIs and Band Reflectance Values in Estimating Above-Ground Grass Biomass

The results of this study showed that the single-band texture models derived using the SPLSR algorithm predicted above-ground grass biomass better than all vegetation indices and wavebands combined. Figure 4c,d shows accuracies derived from using single-band texture models, as well as combined vegetation indices and wavebands. Based on single-band texture models, triennial burn treatments C7 ($R^2 = 0.51$, RMSEP = 0.18 kg/m^2, RMSEPrel = 5.56) had the least prediction errors. The single-band texture predictions had relatively lower estimation errors, when compared with all vegetation indices, combined with wavebands (C7 $R^2 = 0.18$, RMSEP = 0.48 kg/m^2, RMSEPrel = 9.83). When single-band texture models were used, the optimal window sizes were 3×3 and 5×5 at [0:1] and [1:1] offsets. The mean, dissimilarity, homogeneity entropy, correlation, variance and second moment texture model types were frequently selected as optimal variables at this stage, based on the SPLSR algorithm. In this study, the single-band texture and band-ratio texture models did not perform significantly differently, hence those results were not included in this study.

3.4. Comparing the Performance of Combined Single-Band and Band-Ratio Texture Models with the Combination of All WV-3 VIs, Band Reflectance Values and Single-Band Texture Models in Estimating Above-Ground Grass Biomass

Results of this work also showed that all data combined (texture indices, vegetation indices a nd spectral wavebands), outperformed the texture models (i.e., single-band and band-ratio texture). Texture models individually exhibited slightly higher prediction errors when compared to the combination of single-band texture models' vegetation indices and wavebands. Based on all variables combined, biennial burn treatments C4 ($R^2 = 0.89$, RMSEP = 0.1 kg/m^2, RMSEPrel = 3.45) had the lowest estimation errors. The combination of texture models resulted in comparatively lower accuracies with higher errors (C4: $R^2 = 0.29$, RMSEP = 0.22 kg/m^2, RMSEPrel = 5.61) (see Figure 4e,f).

3.5. Estimating Above-Ground Grass Biomass across Different Levels of Grassland Management Treatments Using WV-3-Derived Texture Models Combined with Optimal Vegetation Indices Selected by the SPLSR Algorithm

When all data were combined and all treatments pooled, a comparatively lower prediction error was obtained, as illustrated in Figure 5. Further analysis (Figure 5b) illustrated that the stray points on Figure 5a were induced by those variables which exhibited low correlation coefficients such as B6, B6/B7 and NDRRE. However, the overall influence of stray points on error was minimal as indicated by an observed R^2 of 0.90 and RMSEP of 1.67 kg/m^2. It was also observed that the red-edge-derived texture and vegetation indices were the most influential variables that produced relatively lower accuracies (Figure 6). From the selected variables, the 5×5 second moment and variance simple band-ratio texture models derived from Bands 6 and 7 exhibited the highest scores in this study.

Figure 7 illustrates the spatial distribution of above-ground biomass (ABGB) across different levels of mowing and burning treatments. It can be observed that the triennial (C8) and biennial C5) treatments accumulate more biomass, compared to the annual burn (D3). On the other hand, the mowing treatments (C10) show less ABGB accumulation, due to the high removal of grass.

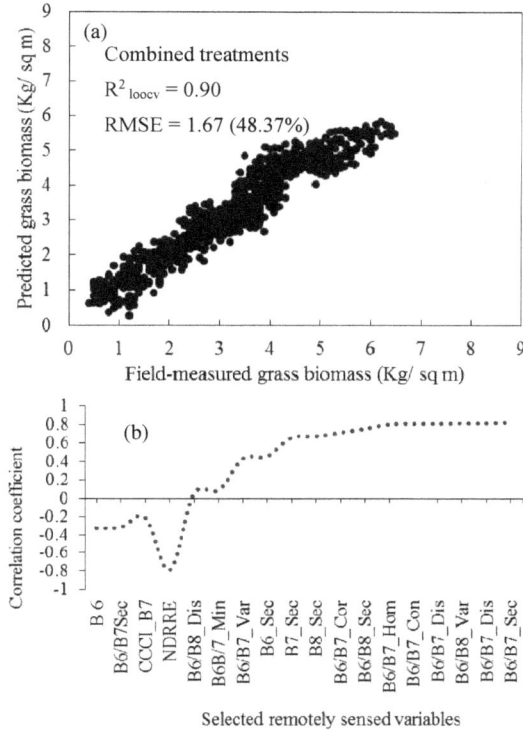

Figure 5. (**a**) Relationship between the field-measured and estimated grass above-ground biomass across all grass management treatments for validating sparse partial least regression (SPLSR) models, based on the leave-one-out cross-validation procedure. Note that the relative root mean square error is presented as a percentage; (**b**) illustrates the relationship between all the optimal variables and grass biomass across all treatments.

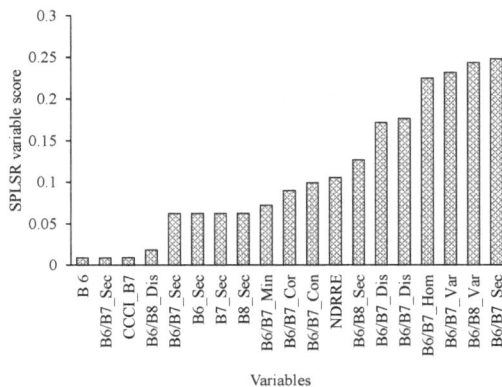

Figure 6. Best variables selected using SPLSR, in estimating above-ground grass biomass across different grassland management treatments. Note that on 'B6/B7' represents the ratio of WV-3 Bands 6 and 7 and NDRE is the normalized difference red-edge index.

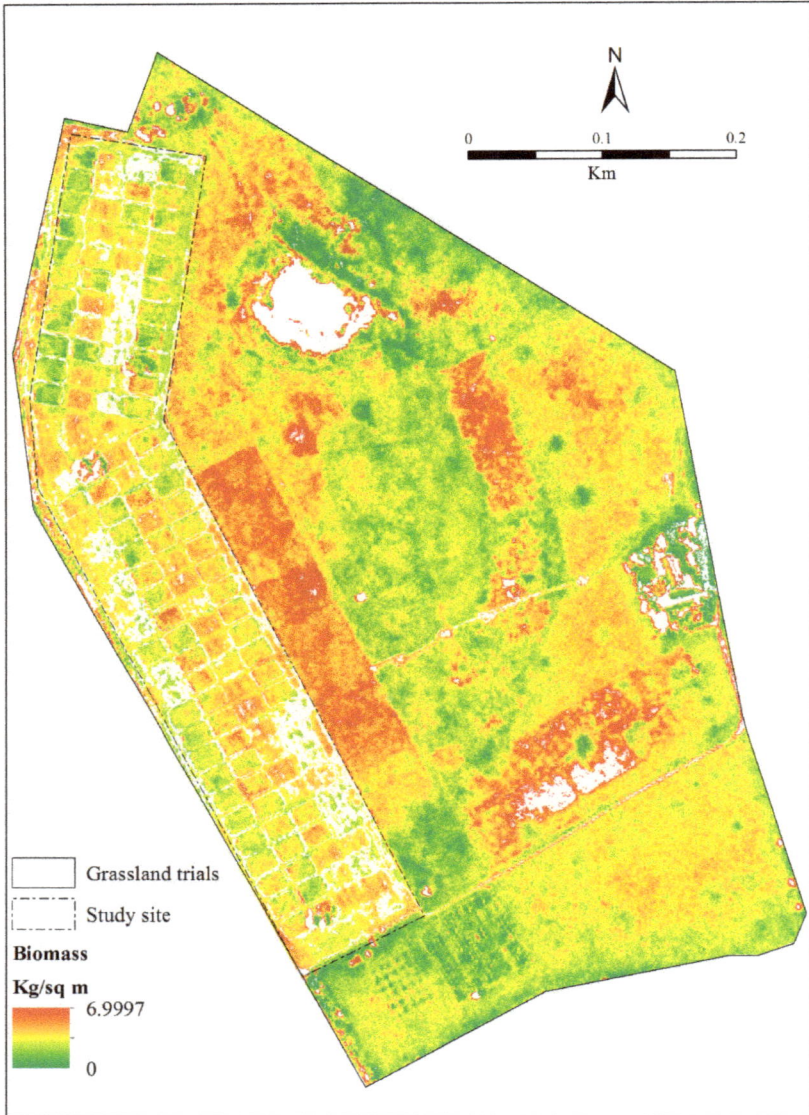

Figure 7. Spatial distribution of biomass across different grassland management treatments.

Figure 8 summarises the accuracies obtained, using single wavebands, broadband vegetation indices, red-edge vegetation indices, single-band and band-ratio texture models, in predicting ABGB across different levels of mowing and burning treatments. When single wavebands were used in estimating above-ground grass biomass, an average RMSEP of 1.02 kg/m^2 was obtained. These variables had the highest RMSEP and were the least accurate predictors for estimating grass ABGB in this study. The accuracy of estimating ABGB slightly improved to an average RMSEP of 0.83, 1.02 kg/m^2, when broadband vegetation indices were used. However, combining the broadband vegetation indices did not significantly improve the accuracy of ABGB estimation, as illustrated in

Figure 8. The red-edge vegetation indices significantly improved the accuracy of ABGB estimation to average RMSEP: 0.55 kg/m^2. The combination of red-edge vegetation indices with broadband vegetation indices, as well as single wavebands, did not significantly improve the accuracy of estimating grass ABGB in this study. When single-band grey level co-occurrence texture matrices were used the ABGB prediction accuracy significantly improved (average RMSEP: 0.35 kg/m^2). In comparison, the combination of single-band and band-ratio texture models did not significantly improve the accuracy of estimating ABGB. When all variables were combined (red-edge and texture models), optimal accuracies (average RMSEP: 0.2 kg/m^2) were obtained in this study.

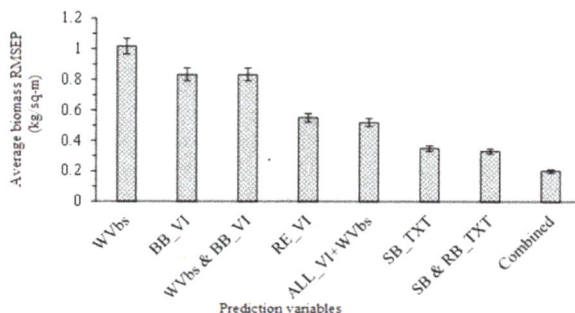

Figure 8. Average RMSEPs derived in predicting above-ground biomass, using WV-3 wavebands (WVbs) broadband (BB_VI), red-edge (RE_VI), single-band (SB_TXT), band-ratio texture (BR_TXT) indices and all combined data across different rangeland management treatments. Whiskers represent the upper and lower confidence intervals of the mean.

4. Discussion

This study tested the robustness of combining texture models with red-edge in estimating the ABGB across different rangeland management treatments, based on the recently launched WorldView-3 EO data. This study specifically sought to find out whether the integration of the red-edge with grey level co-occurrence texture models, extracted at different window sizes and offsets, could improve the accuracy of models for predicting grass above-ground biomass across different levels of mowing and burning treatments in the context of southern African grasslands.

4.1. Combining Texture Models with Red-Edge in Predicting above-Ground Grass Biomass

The findings of this study suggest that combining texture metrics and red-edge-derived vegetation indices has relatively higher prospects of improving the estimation accuracy of ABGB growing across different levels of grassland management treatments, when compared to the performance of texture metrics as stand-alone data.

This could be attributed to the sensitivity of the red-edge section of the electromagnetic spectrum to the variations in LAI and LAD changes [58,59], as well as foliar chlorophyll variability caused by different levels of mowing, and the influx of post-fire nutrients [60]. During the mowing process, grass twigs and leaves are reduced, according to different mowing treatment levels. This results in the alteration of the grass LAI as well as LAD across different levels of mowing. Accordingly, the spectral reflectance from these mowing different levels is better detected by the red-edge section of the electromagnetic spectrum, augmenting the performance of texture models. Furthermore, the red-edge is also sensitive to the variability in chlorophyll content, which accumulates after the burning treatment of grass. This also facilitates an improvement in the accuracy of the estimation of grass biomass, when the red-edge is combined with texture models.

Meanwhile, the textural variables are sensitive to the geographical distribution of minute, but crucial, tonal grass variations in the image induced by the reflectance of different levels of grassland management treatments on certain spectral bands, such as the red-edge and its derived band ratios [61]. This boosts the robustness of texture models and red-edge variables in estimating ABGB. Furthermore, texture is also sensitive to the variations in LAI and LAD induced by mowing, as well as the high chlorophyll content from post-fire nutrients in those grasses grown under different levels of burning and mowing treatments. Subsequently, high estimation accuracies of above-ground grass biomass are realised when texture models are combined with the red-edge derivatives. In addition texture optimises the characterisation of spatial information independently of the tone, while increasing the range of biomass to optimal levels [8]. This facilitates robustness and a plausible performance, when texture metrics are combined with red-edge waveband derivatives, optimising the accurate estimation of ABGB across complex grassland management treatments in this study. Our results are consistent with those of Zhang et al. [62], who noted that the models derived from a combination of spectrum and texture models of the Chinese high-resolution remote sensing satellite Gaofen-1, increased the estimation accuracies of Populus euphratica forest when compared with the performance of reflectance or texture models. In another similar study, Takayama and Iwasaki [63] showed that the combination of the spatial and spectral information from spectral responses and texture models optimally improved the estimation accuracies of tropical vegetation biomass from a RMSE of 66.16 t/ha to a RMSE of 62.62 t/ha in Hampangen, Central Kalimantan, Indonesia, based on WV-3 satellite data. Kelsey and Neff [53] also demonstrated that texture models improved the estimation of vegetation biomass at the San Juan National Forest in southwest Colorado, USA, from a RMSE of 56.4 to a RMSE of 45.6, based on Landsat data.

Results of this study also indicated that the single-band texture metrics improved the accuracy of ABGB estimation relative to red-edge and broadband VI, combined with single wave bands. This is because texture metrics are renowned for accurately capturing the heterogeneity of vegetation structural traits when compared to vegetation indices as a stand-alone dataset [29,32]. The local variance within pixels at a defined neighbourhood, induced by different levels of mowing and burning treatments in this study, is better distinguished by the texture variables when compared with their spectral signature variations at various WorldView-3 wavelengths. Specifically, the spectral responses of vegetation are computed on a pixel basis, while texture is computed from a desired neighbourhood of pixels that is adjustable, increasing the prospects of texture in credibly predicting biomass better than broadband and spectral reflectance [53].

Furthermore, the optimal performance of texture variables, in relation to red-edge and other wavebands and indices in this study, could be explained by the fact that the saturation levels of texture metrics in estimating biomass are considerably higher when compared to those of vegetation indices, such as NDVI, which saturate at lower levels of biomass [64,65]. This results in the underestimation of ABGB. In addition, the distinctive performance of texture models could also be attributed to the fact that the band-ratio textures are an amalgamation of strengths derived from different spectral wavebands, combined with image tone variations. This increases the sensitivity of texture and red-edge models to the spatial characteristics of different grass canopies, hence facilitating a comparatively higher estimation accuracy of ABGB, a mammoth challenge when using vegetation indices.

Our results are consistent with those of a growing body of literature that attests the optimal performance of grey level texture models, when compared to all vegetation indices [32,64,66]. For example, Zhang et al. [62] noted that when texture models from a high spatial resolution (2 and 16 m) GoaFen-1 optical EO data were integrated, the accuracy of above-ground of the *Populus euphratica* forest. In a related study, Sarker and Nichol [7] concluded that the spectral reflectance and traditional vegetation indices have low prospects for estimating biomass, when compared with texture models. Specifically, Sarker and Nichol [7] noted that texture models derived from ALSO AVNIR-2 improved the vegetation biomass estimation from a RMSE of 64 t/ha, based on traditional vegetation indices and spectral reflectance to a RMSE of 46 t/ha, as noted in this study. However, Sarker and Nichol's [46] results showed that band ratios further improved the accuracy of estimating biomass to a RMSE of 32 t/ha. Their results are contrary to those of this study, which indicated that band-ratio and single-band texture models did not perform significantly differently when predicting ABGB across different levels of grassland management treatments.

Furthermore, results of this study showed that the red-edge waveband derivatives improved the accuracy of the models for predicting ABGB at different grassland management treatments, when compared to broadband vegetation indices combined with wavebands. Based on the results of this study, the red-edge bands outperformed the broadband vegetation indices, combined with raw wavebands. These results were somewhat expected, as this has been noted in literature. This can be explained by the fact that the red-edge portion of the electromagnetic spectrum is highly sensitive to changes in the grass chlorophyll [67], induced by disturbances such as mowing and burning. Post-fire foliar nutrients, which are rich in nitrogen and phosphorus, induce high chlorophyll concentrations in the grass, which is then detected by the red-edge waveband derivatives in this study.

Meanwhile, the decreases in the leaf area distribution and LAI, due to mowing activities, induces a variation in the signature of grass, which is then detected better by the red-edge derivatives, when compared with the single wavebands and broadband vegetation indices. Our results are consistent with those found in a growing body of contemporary literature [68–71]. For instance, Fernández-Manso et al. [44] noted that red-edge derivatives detected the fire activities better and with higher accuracies (Modified Simple Ratio red-edge narrow R^2: 0.69), when compared to single wave bands and broadband vegetation indices (Red band R^2: 0.093, NIR R^2: 0.63, and NDVI R^2: 0.43) in Sierra de Gata (central-western Spain), based on Sentinel data. Gara, et al. [70] also noted that the inclusion of red-edge derivatives also improved the estimation of carbon stocks from an explained variance of 63%, based on NDVI, to 70% in the savanna dry forest of Zimbabwe.

4.2. Biological Behavior of Grasses at Ukulinga Research Farm Based on Literature Review

As highlighted earlier, mowing through defoliation reduces grass LAI as well as LAD. This markedly reduces the relative abundance of the dominant *Themeda triandra* (which is a highly palatable grass species), overall grass basal cover as well as the biomass [72]. The changes in grass species composition and dominance then could explain the spatial variability of grass biomass noted in this study. Furthermore, mowing at Ukulinga increased sward productivity in the season following the removal treatment when compared to burning which promotes growth of grasses with higher protein content [72]. This is illustrated by high estimates of biomass in some mowing (C10 and 11 as well as D10 and 11) treatments in relation to other burning treatments (C2 and 3 as well D2 and 3) in the results of this study through high biomass. Treatments with frequent fire administration would yield a variety of short grasses dominated by a *Themeda triandra*, *Hyparrhenia hirta* and *Tristachya leucothrix as shown by* Kirkman et al. [73] which could also explain some of the variabilities observed in treatments such as C1 and 2 with annual burning relative to other treatments. Kirkman et al. [73] reported that there is a high replacement rate of the dominant grass species between annually burned and unburned treatments at Ukulinga. These findings by Kirkman et al. [73] are in agreement with the results of this study which indicate a variability in the estimated ABGB between annually burned and the control treatments. Furthermore research shows that biennially burnt treatments tend to produce more biomass, on average, than treatments burnt less frequently or mown annually in winter [74]. Above all, the effects of mowing and burning, as well as their interaction on native grasses still requires further studies [75] especially from a remote sensing context.

5. Conclusions

The aim of this study was to assess the accuracy of combining red-edge derivatives with texture models in predicting the above-ground biomass of grass growing under different levels of grassland management treatments. Based on the findings of this study, we conclude that:

- combining texture models with red-edge derivatives provides a more accurate approach in estimating the above-ground biomass of grass grown under complex grassland management treatments. To the best of our knowledge, this is the first study to evaluate the utility of texture models and red-edge in estimating above-ground grass biomass, across a multitude of grassland management treatment levels,

- the best predictor in estimating above-ground biomass (ABGB) grown under complex grassland management treatments was derived using all data combined,
- texture models perform better than the red-edge vegetation indices in estimating grass above-ground biomass, and
- as expected, the red-edge spectrum-derived vegetation indices outperformed the broadband indices.

In testing specific objectives, our results suggests that (i) broadband vegetation indices such as normalised difference vegetation index (NDVI), enhanced vegetation index (EVI) and soil-adjusted vegetation index (SAVI) are comparatively better predictors of ABGB WorldView-3 (WV-3) standard wavebands; (ii) red-edge-derived vegetation indices are better predictors than standard wave bands combined with broadband vegetation indices; (iii) texture models are better predictors of ABGB in relation to red-edge, broadband vegetation indices (Vis) combined with all WV-3 bands; (iv) band texture ratios are better predictors of ABGB across different treatments when compared to all variables combined. Ultimately, when all variables were combined, red-edge VI texture and band-ratio texture exhibited optimal ABGB predictions in this study. The results of this work give insights into the estimation of grass biomass in complex grassland management treatments of arid tropical region grasses. The bulk of the studies that have demonstrated the utility of texture variables in above-ground biomass estimation have focused on the forests and crops of America and Europe. Therefore, to the best of our knowledge, the results of this study demonstrate, for the first time, the utility of texture models combined with red-edge waveband derivatives in estimating above-ground grass biomass across the complex grassland management treatments of the arid tropics, characterised by a high soil background effect. These results are an important footstool upon which critical spatial information required for grassland policy-making and sustainable grassland management in southern Africa could be derived.

Acknowledgments: The authors are grateful to the University of KwaZulu-Natal/National Research Fund and KwaZulu-Natal Sandstone Sourveld (KZNSS) forum, in conjunction with the eThekwini Municipality, also known as the DRAP, for funding this research. The authors would also like to thank K. Kirkman, Alison Young, Deepa Mangesh, T. Dube, Terence D Mushore, Thulile Vundla and Reneilwe Maake, for their assistance with field work, data collection and analysis as well as proofreading the manuscript. Finally, the authors extend their gratitude to the reviewers for their constructive criticism.

Author Contributions: Mbulisi Sibanda conceived and performed the experiment, collected data, analysed the data and wrote the paper under the supervision of Onisimo Mutanga, Mathieu Rouget and Lalit Kumar. Onisimo Mutanga, Mathieu Rouget and Lalit Kumar also edited the manuscript. Furthermore, Mathieu Rouget provided the funds for conducting this research. The research funds were provided by eThekwini Municipality through the Durban Research Action Partnership: KwaZulu-Natal Sandstone Sourveld Programme and the South African Research Chairs Initiative of the Department of Science and Technology and the National Research Foundation of South Africa (grant No. 84157).

Conflicts of Interest: The authors declare no conflict of interest.

References

1. Kumar, L.; Sinha, P.; Taylor, S.; Alqurashi, A.F. Review of the use of remote sensing for biomass estimation to support renewable energy generation. *J. Appl. Remote Sens.* **2015**, *9*, 097696. [CrossRef]
2. Wilson, J.B.; Peet, R.K.; Dengler, J.; Pärtel, M. Plant species richness: The world records. *J. Veg. Sci.* **2012**, *23*, 796–802. [CrossRef]
3. O'Mara, F.P. The role of grasslands in food security and climate change. *Ann. Bot.* **2012**, *110*, 1263–1270. [CrossRef] [PubMed]
4. Andrade, B.O.; Koch, C.; Boldrini, I.I.; Vélez-Martin, E.; Hasenack, H.; Hermann, J.-M.; Kollmann, J.; Pillar, V.D.; Overbeck, G.E. Grassland degradation and restoration: A conceptual framework of stages and thresholds illustrated by southern Brazilian grasslands. *Nat. Conserv.* **2015**, *13*, 95–104. [CrossRef]

5.   Conant, R.T.; Paustian, K.; Elliott, E.T. Grassland management and conversion into grassland: Effects on soil carbon. *Ecol. Appl.* **2001**, *11*, 343–355. [CrossRef]
6.   Bastin, J.-F.; Barbier, N.; Couteron, P.; Adams, B.; Shapiro, A.; Bogaert, J.; De Cannière, C. Aboveground biomass mapping of African forest mosaics using canopy texture analysis: Toward a regional approach. *Ecol. Appl.* **2014**, *24*, 1984–2001. [CrossRef]
7.   Sarker, L.R.; Nichol, J.E. Improved forest biomass estimates using ALOS AVNIR-2 texture indices. *Remote Sens. Environ.* **2011**, *115*, 968–977. [CrossRef]
8.   Rosenqvist, Å.; Milne, A.; Lucas, R.; Imhoff, M.; Dobson, C. A review of remote sensing technology in support of the Kyoto Protocol. *Environ. Sci. Policy* **2003**, *6*, 441–455. [CrossRef]
9.   Lu, D. The potential and challenge of remote sensing-based biomass estimation. *Int. J. Remote Sens.* **2006**, *27*, 1297–1328. [CrossRef]
10. Meng, S.; Pang, Y.; Zhang, Z.; Jia, W.; Li, Z. Mapping Aboveground Biomass using Texture Indices from Aerial Photos in a Temperate Forest of Northeastern China. *Remote Sens.* **2016**, *8*, 230. [CrossRef]
11. Timothy, D.; Onisimo, M.; Riyad, I. Quantifying aboveground biomass in African environments: A review of the trade-offs between sensor estimation accuracy and costs. *Trop. Ecol.* **2016**, *57*, 393–405.
12. Schino, G.; Borfecchia, F.; De Cecco, L.; Dibari, C.; Iannetta, M.; Martini, S.; Pedrotti, F. Satellite estimate of grass biomass in a mountainous range in central Italy. *Agrofor. Syst.* **2003**, *59*, 157–162. [CrossRef]
13. Griffith, J.A.; Price, K.P.; Martinko, E.A. A multivariate analysis of biophysical parameters of tallgrass prairie among land management practices and years. *Environ. Monit. Assess.* **2001**, *68*, 249–271. [CrossRef] [PubMed]
14. Xie, Y.; Sha, Z.; Yu, M.; Bai, Y.; Zhang, L. A comparison of two models with Landsat data for estimating above ground grassland biomass in Inner Mongolia, China. *Ecol. Model.* **2009**, *220*, 1810–1818. [CrossRef]
15. Huete, A.R. Separation of soil-plant spectral mixtures by factor analysis. *Remote Sens. Environ.* **1986**, *19*, 237–251. [CrossRef]
16. Bannari, A.; Morin, D.; Bonn, F.; Huete, A. A review of vegetation indices. *Remote Sens. Rev.* **1995**, *13*, 95–120. [CrossRef]
17. Mutanga, O.; Skidmore, A.K. Narrow band vegetation indices overcome the saturation problem in biomass estimation. *Int. J. Remote Sens.* **2004**, *25*, 3999–4014. [CrossRef]
18. Nichol, J.E.; Sarker, M.L.R. Improved biomass estimation using the texture parameters of two high-resolution optical sensors. *IEEE Trans. Geosci. Remote Sens.* **2011**, *49*, 930–948. [CrossRef]
19. Ngubane, Z.; Odindi, J.; Mutanga, O.; Slotow, R. Assessment of the Contribution of WorldView-2 Strategically Positioned Bands in Bracken fern (*Pteridium aquilinum* (L.) Kuhn) Mapping. *S. Afr. J. Geomat.* **2014**, *3*, 210–223. [CrossRef]
20. Ramoelo, A.; Skidmore, A.K.; Cho, M.A.; Schlerf, M.; Mathieu, R.; Heitkönig, I.M. Regional estimation of savanna grass nitrogen using the red-edge band of the spaceborne RapidEye sensor. *Int. J. Appl. Earth Obs. Geoinf.* **2012**, *19*, 151–162. [CrossRef]
21. Zhao, P.; Lu, D.; Wang, G.; Wu, C.; Huang, Y.; Yu, S. Examining Spectral Reflectance Saturation in Landsat Imagery and Corresponding Solutions to Improve Forest Aboveground Biomass Estimation. *Remote Sens.* **2016**, *8*, 469. [CrossRef]
22. Broge, N.H.; Leblanc, E. Comparing prediction power and stability of broadband and hyperspectral vegetation indices for estimation of green leaf area index and canopy chlorophyll density. *Remote Sens. Environ.* **2001**, *76*, 156–172. [CrossRef]
23. Pu, R.; Gong, P.; Biging, G.S.; Larrieu, M.R. Extraction of red edge optical parameters from Hyperion data for estimation of forest leaf area index. *IEEE Trans. Geosci. Remote Sens.* **2003**, *41*, 916–921.
24. Delegido, J.; Verrelst, J.; Rivera, J.P.; Ruiz-Verdú, A.; Moreno, J. Brown and green LAI mapping through spectral indices. *Int. J. Appl. Earth Obs. Geoinf.* **2015**, *35*, 350–358. [CrossRef]
25. Curran, P.J.; Dungan, J.L.; Gholz, H.L. Exploring the relationship between reflectance red edge and chlorophyll content in slash pine. *Tree Physiol.* **1990**, *7*, 33–48. [CrossRef] [PubMed]
26. Asner, G.P. Biophysical and biochemical sources of variability in canopy reflectance. *Remote Sens. Environ.* **1998**, *64*, 234–253. [CrossRef]
27. Lu, D. Aboveground biomass estimation using Landsat TM data in the Brazilian Amazon. *Int. J. Remote Sens.* **2005**, *26*, 2509–2525. [CrossRef]

28. Cutler, M.; Boyd, D.; Foody, G.; Vetrivel, A. Estimating tropical forest biomass with a combination of SAR image texture and Landsat TM data: An assessment of predictions between regions. *ISPRS J. Photogramm. Remote Sens.* **2012**, *70*, 66–77. [CrossRef]

29. Eckert, S. Improved forest biomass and carbon estimations using texture measures from WorldView-2 satellite data. *Remote Sens.* **2012**, *4*, 810–829. [CrossRef]

30. Dube, T.; Mutanga, O. Investigating the robustness of the new Landsat-8 Operational Land Imager derived texture metrics in estimating plantation forest aboveground biomass in resource constrained areas. *ISPRS J. Photogramm. Remote Sens.* **2015**, *108*, 12–32. [CrossRef]

31. Wallis, C.I.; Paulsch, D.; Zeilinger, J.; Silva, B.; Fernández, G.F.C.; Brandl, R.; Farwig, N.; Bendix, J. Contrasting performance of Lidar and optical texture models in predicting avian diversity in a tropical mountain forest. *Remote Sens. Environ.* **2016**, *174*, 223–232. [CrossRef]

32. Ozdemir, I.; Karnieli, A. Predicting forest structural parameters using the image texture derived from WorldView-2 multispectral imagery in a dryland forest, Israel. *Int. J. Appl. Earth Obs. Geoinf.* **2011**, *13*, 701–710. [CrossRef]

33. Salas, E.A.L.; Boykin, K.G.; Valdez, R. Multispectral and Texture Feature Application in Image-Object Analysis of Summer Vegetation in Eastern Tajikistan Pamirs. *Remote Sens.* **2016**, *8*, 78. [CrossRef]

34. Fynn, R.W.; O'Connor, T.G. Determinants of community organization of a South African mesic grassland. *J. Veg. Sci.* **2005**, *16*, 93–102. [CrossRef]

35. Morris, C.; Fynn, R. The Ukulinga long-term grassland trials: Reaping the fruits of meticulous, patient research. *Bull. Grassl. Soc. South. Afr.* **2001**, *11*, 7–22.

36. De Wit, M.; Blignaut, J.; Nazare, F. Monetary Valuation of the Grasslands in South Africa. 2006. Available online: http://biodiversityadvisor.sanbi.org/wp-content/uploads/2014/07/2006deWit_Background-InfoRep5_Strategic-Monetary-valuation.pdf (accessed on 6 January 2017).

37. Sibanda, M.; Mutanga, O.; Rouget, M. Testing the capabilities of the new WorldView-3 spaceborne sensor's red-edge spectral band in discriminating and mapping complex grassland management treatments. *Int. J. Remote Sens.* **2017**, *38*, 1–22. [CrossRef]

38. Schumacher, P.; Mislimshoeva, B.; Brenning, A.; Zandler, H.; Brandt, M.; Samimi, C.; Koellner, T. Do Red Edge and Texture Attributes from High-Resolution Satellite Data Improve Wood Volume Estimation in a Semi-Arid Mountainous Region? *Remote Sens.* **2016**, *8*, 540. [CrossRef]

39. Chica-Olmo, M.; Abarca-Hernandez, F. Computing geostatistical image texture for remotely sensed data classification. *Comput. Geosci.* **2000**, *26*, 373–383. [CrossRef]

40. Wang, L.; Sousa, W.P.; Gong, P.; Biging, G.S. Comparison of IKONOS and QuickBird images for mapping mangrove species on the Caribbean coast of Panama. *Remote Sens. Environ.* **2004**, *91*, 432–440. [CrossRef]

41. Safari, A.; Sohrabi, H. Ability of Landsat-8 OLI derived texture metrics in estimating aboveground carbon stocks of coppice Oak Forests. *ISPRS Int. Arch. Photogramm. Remote Sens. Spat. Inf. Sci.* **2016**, 751–754. [CrossRef]

42. Kang, Y.; Özdoğan, M.; Zipper, S.C.; Román, M.O.; Walker, J.; Hong, S.Y.; Marshall, M.; Magliulo, V.; Moreno, J.; Alonso, L. How Universal Is the Relationship between Remotely Sensed Vegetation Indices and Crop Leaf Area Index? A Global Assessment. *Remote Sens.* **2016**, *8*, 597. [CrossRef]

43. Gitelson, A.A.; Merzlyak, M.N. Remote estimation of chlorophyll content in higher plant leaves. *Int. J. Remote Sens.* **1997**, *18*, 2691–2697. [CrossRef]

44. Fernández-Manso, A.; Fernández-Manso, O.; Quintano, C. SENTINEL-2A red-edge spectral indices suitability for discriminating burn severity. *Int. J. Appl. Earth Obs. Geoinf.* **2016**, *50*, 170–175. [CrossRef]

45. Santoso, H.; Gunawan, T.; Jatmiko, R.H.; Darmosarkoro, W.; Minasny, B. Mapping and identifying basal stem rot disease in oil palms in North Sumatra with QuickBird imagery. *Precis. Agric.* **2011**, *12*, 233–248. [CrossRef]

46. Tucker, C.J. A critical review of remote sensing and other methods for non-destructive estimation of standing crop biomass. *Grass Forage Sci.* **1980**, *35*, 177–182. [CrossRef]

47. Huete, A.R. A soil-adjusted vegetation index (SAVI). *Remote Sens. Environ.* **1988**, *25*, 295–309. [CrossRef]

48. Cabezas, J.; Galleguillos, M.; Perez-Quezada, J.F. Predicting Vascular Plant Richness in a Heterogeneous Wetland Using Spectral and Textural Features and a Random Forest Algorithm. *IEEE Geosci. Remote Sens. Lett.* **2016**, *13*, 646–650. [CrossRef]

49. Merzlyak, M.; Gitelson, A.A.; Chivkunova, O.; Solovchenko, A.; Pogosyan, S. Application of reflectance spectroscopy for analysis of higher plant pigments. *Russ. J. Plant Physiol.* **2003**, *50*, 704–710. [CrossRef]

50. El-Shikha, D.M.; Barnes, E.M.; Clarke, T.R.; Hunsaker, D.J.; Haberland, J.A.; Pinter, P., Jr.; Waller, P.M.; Thompson, T.L. Remote sensing of cotton nitrogen status using the Canopy Chlorophyll Content Index (CCCI). *Trans. ASABE* **2008**, *51*, 73–82. [CrossRef]

51. Fitzgerald, G.; Rodriguez, D.; O'Leary, G. Measuring and predicting canopy nitrogen nutrition in wheat using a spectral index—The canopy chlorophyll content index (CCCI). *Field Crop. Res.* **2010**, *116*, 318–324. [CrossRef]

52. Gitelson, A.A.; Merzlyak, M.; Zur, Y.; Stark, R.; Gritz, U. Non-destructive and remote sensing techniques for estimation of vegetation status. *Pap. Nat. Resour.* **2001**, *273*, 205–210.

53. Kelsey, K.C.; Neff, J.C. Estimates of aboveground biomass from texture analysis of Landsat imagery. *Remote Sens.* **2014**, *6*, 6407–6422. [CrossRef]

54. Ouma, Y.O.; Tetuko, J.; Tateishi, R. Analysis of co-occurrence and discrete wavelet transform textures for differentiation of forest and non-forest vegetation in very-high-resolution optical-sensor imagery. *Int. J. Remote Sens.* **2008**, *29*, 3417–3456. [CrossRef]

55. Chun, H.; Keleş, S. Sparse partial least squares regression for simultaneous dimension reduction and variable selection. *J. R. Stat. Soc. Ser. B (Stat. Methodol.)* **2010**, *72*, 3–25. [CrossRef] [PubMed]

56. Abdel-Rahman, E.M.; Mutanga, O.; Odindi, J.; Adam, E.; Odindo, A.; Ismail, R. A comparison of partial least squares (PLS) and sparse PLS regressions for predicting yield of Swiss chard grown under different irrigation water sources using hyperspectral data. *Comput. Electron. Agric.* **2014**, *106*, 11–19. [CrossRef]

57. Frazer, G.W.; Magnussen, S.; Wulder, M.A.; Niemann, K.O. Simulated impact of sample plot size and co-registration error on the accuracy and uncertainty of LiDAR-derived estimates of forest stand biomass. *Remote Sens. Environ.* **2011**, *115*, 636–649. [CrossRef]

58. Zhao, F.; Yang, X.; Schull, M.A.; Román-Colón, M.O.; Yao, T.; Wang, Z.; Zhang, Q.; Jupp, D.L.; Lovell, J.L.; Culvenor, D.S. Measuring effective leaf area index, foliage profile, and stand height in New England forest stands using a full-waveform ground-based lidar. *Remote Sens. Environ.* **2011**, *115*, 2954–2964. [CrossRef]

59. Cho, M.A.; Skidmore, A.; Corsi, F.; van Wieren, S.E.; Sobhan, I. Estimation of green grass/herb biomass from airborne hyperspectral imagery using spectral indices and partial least squares regression. *Int. J. Appl. Earth Obs. Geoinf.* **2007**, *9*, 414–424. [CrossRef]

60. Skidmore, A.K.; Ferwerda, J.G.; Mutanga, O.; Van Wieren, S.E.; Peel, M.; Grant, R.C.; Prins, H.H.; Balcik, F.B.; Venus, V. Forage quality of savannas—simultaneously mapping foliar protein and polyphenols for trees and grass using hyperspectral imagery. *Remote Sens. Environ.* **2010**, *114*, 64–72. [CrossRef]

61. Haralick, R.M.; Shanmugam, K. Textural features for image classification. *IEEE Trans. Syst. Man Cybern.* **1973**, 610–621. [CrossRef]

62. Zhang, L.; Cheng, Q.; Li, C. Improved model for estimating the biomass of Populus euphratica forest using the integration of spectral and textural features from the Chinese high-resolution remote sensing satellite GaoFen-1. *J. Appl. Remote Sens.* **2015**, *9*, 096010. [CrossRef]

63. Takayama, T.; Iwasaki, A. Optimal Wavelength Selection on Hyperspectral Data with Fused Lasso for Biomass Estimation of Tropical Rain Forest. *ISPRS Ann. Photogramm. Remote Sens. Spat. Inf. Sci.* **2016**, *III-8*, 101–108. [CrossRef]

64. Fujiki, S.; Okada, K.-I.; Nishio, S.; Kitayama, K. Estimation of the stand ages of tropical secondary forests after shifting cultivation based on the combination of WorldView-2 and time-series Landsat images. *ISPRS J. Photogramm. Remote Sens.* **2016**, *119*, 280–293. [CrossRef]

65. Shen, W.; Li, M.; Huang, C.; Wei, A. Quantifying Live Aboveground Biomass and Forest Disturbance of Mountainous Natural and Plantation Forests in Northern Guangdong, China, Based on Multi-Temporal Landsat, PALSAR and Field Plot Data. *Remote Sens.* **2016**, *8*, 595. [CrossRef]

66. Kuplich, T.; Curran, P.J.; Atkinson, P.M. Relating SAR image texture to the biomass of regenerating tropical forests. *Int. J. Remote Sens.* **2005**, *26*, 4829–4854. [CrossRef]

67. Filella, I.; Penuelas, J. The red edge position and shape as indicators of plant chlorophyll content, biomass and hydric status. *Int. J. Remote Sens.* **1994**, *15*, 1459–1470. [CrossRef]

68. Sibanda, M.; Mutanga, O.; Rouget, M. Examining the potential of Sentinel-2 MSI spectral resolution in quantifying above ground biomass across different fertilizer treatments. *ISPRS J. Photogramm. Remote Sens.* **2015**, *110*, 55–65. [CrossRef]

69. Sibanda, M.; Mutanga, O.; Rouget, M. Discriminating Rangeland Management Practices Using Simulated HyspIRI, Landsat 8 OLI, Sentinel 2 MSI, and VENμS Spectral Data. *IEEE J. Sel. Top. Appl. Earth Obs. Remote Sens.* **2016**, *9*, 1–13. [CrossRef]

70. Gara, T.W.; Murwira, A.; Ndaimani, H. Predicting forest carbon stocks from high resolution satellite data in dry forests of Zimbabwe: Exploring the effect of the red-edge band in forest carbon stocks estimation. *Geocarto Int.* **2016**, *31*, 176–192. [CrossRef]

71. Mutanga, O.; Adam, E.; Cho, M.A. High density biomass estimation for wetland vegetation using WorldView-2 imagery and random forest regression algorithm. *Int. J. Appl. Earth Obs. Geoinf.* **2012**, *18*, 399–406. [CrossRef]

72. Tainton, N.; Groves, R.; Nash, R. Time of mowing and burning veld: Short term effects on production and tiller development. *Proc. Annu. Congr. Grassl. Soc. South. Afr.* **1977**, *12*, 59–64. [CrossRef]

73. Kirkman, K.P.; Collins, S.L.; Smith, M.D.; Knapp, A.K.; Burkepile, D.E.; Burns, C.E.; Fynn, R.W.; Hagenah, N.; Koerner, S.E.; Matchett, K.J. Responses to fire differ between South African and North American grassland communities. *J. Veg. Sci.* **2014**, *25*, 793–804. [CrossRef]

74. Mentis, M.; Tainton, N. The effect of fire on forage production and quality. In *Ecological Effects of Fire in South African Ecosystems*; Springer: Berlin/Heidelberg, Germany, 1984; pp. 245–254.

75. Van-Wyk, D. *The Effects of Type, Season and Frequency of Defoliation on Species Diversity, Richness, Evenness and Production of the Mowing-Burning Trials at Ukulinga Research Farm in the Southern Tall Grassveld*; University of Natal: Pietermaritzburg, South Africa, 1998.

remote sensing

MDPI

Article

Fusion of Ultrasonic and Spectral Sensor Data for Improving the Estimation of Biomass in Grasslands with Heterogeneous Sward Structure

Thomas Moeckel *, Hanieh Safari, Björn Reddersen, Thomas Fricke and Michael Wachendorf

Department of Grassland Science and Renewable Plant Resources, University of Kassel, Steinstr. 19, D-37213 Witzenhausen, Germany; safari.hanieh@gmail.com (H.S.); BReddersen@gmx.de (B.R.); thfricke@uni-kassel.de (T.F.); mwach@uni-kassel.de (M.W.)
* Correspondence: thmoeck@uni-kassel.de; Tel.: +49-5542-981337

Academic Editors: Lalit Kumar, Onisimo Mutanga, Lars T. Waser and Prasad S. Thenkabail
Received: 17 November 2016; Accepted: 17 January 2017; Published: 21 January 2017

Abstract: An accurate estimation of biomass is needed to understand the spatio-temporal changes of forage resources in pasture ecosystems and to support grazing management decisions. A timely evaluation of biomass is challenging, as it requires efficient means such as technical sensing methods to assess numerous data and create continuous maps. In order to calibrate ultrasonic and spectral sensors, a field experiment with heterogeneous pastures continuously stocked by cows at three grazing intensities was conducted. Sensor data fusion by combining ultrasonic sward height (USH) with narrow band normalized difference spectral index (NDSI) ($R^2_{CV} = 0.52$) or simulated WorldView2 (WV2) ($R^2_{CV} = 0.48$) satellite broad bands increased the prediction accuracy significantly, compared to the exclusive use of USH or spectral measurements. Some combinations were even better than the use of the full hyperspectral information ($R^2_{CV} = 0.48$). Spectral regions related to plant water content were found to be of particular importance (996–1225 nm). Fusion of ultrasonic and spectral sensors is a promising approach to assess biomass even in heterogeneous pastures. However, the suggested technique may have limited usefulness in the second half of the growing season, due to an increasing abundance of senesced material.

Keywords: pasture biomass; ground-based remote sensing; ultrasonic sensor; field spectrometry; sensor fusion; short grass

1. Introduction

To understand the spatio-temporal changes of forage resources in pasture ecosystems and to support grazing management decisions, an accurate estimation of biomass is needed [1–3]. However, a timely evaluation of biomass is a challenge, as it requires targeted and efficient means to assess numerous data for the creation of continuous maps. Though the traditional "clip-and-weigh" methods of measuring biomass are highly accurate, it is costly, destructive, labor-intensive and time-consuming to obtain biomass properties at a high sampling density. Alternatively, ground-based remote sensing techniques have been used as rapid and non-destructive methods to obtain and map the temporal and spatial variability of vegetation characteristics with high spatial resolution in agricultural and pastoral ecosystems [4–6]. Pastures are highly heterogeneous systems due to variations in sward structure, composition and phenology as well as continuous changes caused by different drivers such as environmental factors and grazing. Therefore, the application of sensors in complex grazing systems is difficult and there are some limitations for each specific sensor used for the prediction of sward characteristics [7,8]. To overcome these constraints, the combination of complementary sensor technologies has been suggested to utilize both the strengths and compensate the weaknesses of individual technologies. Combined sensor systems can support multi-source information acquisition and may provide more accurate property estimates and eventually improved management [9]. Even though some studies have investigated such strategies in different farming fields [10,11], to date, these techniques have not been tested in pastures with complex sward diversity. Thus, an evaluation of sward specific calibration is essential before assessing data on a spatial scale.

Ultrasonic and reflectance sensors are two possible complementary technologies capable of providing comprehensive structural and functional characteristics of vegetation [4,10,12–15]. Sward height measured by ultrasonic distance sensing (referred to as ultrasonic sward height (USH)) has been examined as a possible estimator of biomass in forage vegetation canopies [5,16]. However, the main limitation of this technique is that signals are reflected predominantly from the upper canopy layers, regardless of sward density [4]. Moreover, sonic reflections can be affected by canopy architecture, such as lamina size, orientation, angle and surface roughness of the leaves [5,16,17].

Hyperspectral sensors have also raised considerable interest as a potential tool for prediction of biomass and forage quality in pastures. However, difficulties occur at advanced developmental stages of vegetation, as the ability of the reflectance sensor to detect canopy characteristics could be limited by the presence of a high fraction of senescent material in biomass [18,19] or soil background effects [18], atmospheric conditions [20], grazing impact [21] and heterogeneous canopy structures due to mixed species composition and a wide range of phenological stages [1,22,23]. Remarkably, most studies utilizing remotely sensed data for the estimation of grassland and rangeland biomass were conducted in tropical savannas, since these ecosystems account for 30% of the primary production of all terrestrial vegetation. In contrast, comparable studies on grasslands in temperate climates are rare [24].

The limitations of ultrasonic and hyperspectral reflectance sensors in heterogeneous pastures may be compensated by a combined use of measurement data from both sensors, as shown by [4] for less variable legume/grass-mixtures. Thus, the main objective of the present study was to analyze the potential of ultrasonic and hyperspectral sensor data fusion in pastures with high structural sward diversity to predict biomass, which is a prerequisite for future mapping of spatially heterogeneous grassland.

2. Materials and Methods

2.1. Study Area and Site Characteristics

For data acquisition, a long-term pasture experiment was chosen at the experimental farm Relliehausen of the University of Goettingen (51°46′55″N, 9°42′13″E, 180–230 m above mean sea level; soil type: pelosol-brown earth; soil pH: 6.3; mean annual precipitation: 879 mm; mean annual daily temperature: 8.2 °C). The plant association was a moderately species-rich *Lolio-Cynosuretum* [25]. The pastures exhibited pronounced heterogeneity in sward structure, with short and tall patches and various sward height classes [26,27]. Three levels of grazing intensity were allocated to adjacent pasture paddocks of 1 ha size, which were continuously stocked by cows from the beginning of May to mid-September. Grazing intensities were: (a) moderate stocking, average of 3.4 standard livestock units (SLU, i.e., 500 kg live weight) ha^{-1}; (b) lenient stocking, average 1.8 SLU ha^{-1}; and (c) very lenient stocking, average 1.3 SLU ha^{-1} [25]. To ensure extensive sward variation for data assessment, one representative study plot of 30 × 50 m size was selected within each of the three paddocks using a grazed/ungrazed-classified aerial image to obtain comparable surface proportions.

2.2. Field Measurements

Field measurements were conducted at four sampling dates (designated from now on as Date 1 to Date 4) in 2013: (Date 1) 25 April to 2 May (before grazing), (Date 2) 3 to 5 June, (Date 3) 21 to 23 August and (Date 4) 30 September to 2 October (after final grazing) within each study plot. In each campaign, 18 reference sample plots (each 0.25 m^2) were chosen within each of the 3 study plots, adding up to a total of 54 samples per date which represented the existing range of available biomass levels and sward structures. To verify a representative biomass range, a stratified random sampling was performed. In each study plot, three levels of sward height (low, medium, and high) were sampled randomly to compile all date-specific biomass levels in the data set. A Trimble GeoXH GPS device (Trimble Navigation Ltd., Sunnyvale, California, USA) with DGPS correction from AXIO-net (Hannover, Germany, PED-RTK ±20 mm) was used to avoid repeated sampling at the same location during the growing season.

2.2.1. Ground-Based Remote Sensing Measurements

Sensor measurements took place prior to reference data assessment. Hyperspectral data was measured using a hand-held portable spectro-radiometer (Portable HandySpec Field VIS/NIR, tec5, Germany) in a spectral range of 305–1700 nm. Spectral readings were recorded in 1 nm intervals. Measurements were made from a height of about 1 m above and perpendicular to the soil surface between 10:00 a.m. and 2:00 p.m. (local time) in clear sunshine. The sensor had a field of view of 25°. Spectral calibrations were performed at least after every six measurements using a greystandard (Zenith® Diffuse Reflectance Standard 25%). Ultrasonic sward height (USH) measurements took place subsequent to hyperspectral measurements using an ultrasonic distance sensor of type UC 2000-30GM-IUR2-V15 (Pepperl and Fuchs, Mannheim, Germany). The sensor specific sensing range was from 80 to 2000 mm within a sound cone formed by an opening angle of about 25° [28]. Ultrasonic sward height (mm) was calculated by subtracting the ultrasonic distance measurement value in mm from the sensor mount height using Equation (1).

$$\text{USH (mm)} = \text{Mount height (mm)} - \text{Ultrasonic distance (mm)} \qquad (1)$$

At each sampling plot, five measurements were recorded with the ultrasonic sensors placed at five positions on a frame at a height of about 1 m. Further details of the USH device and methodology can be found in Fricke et al. [5]. In addition to sensor measurements, plant composition of all sampling plots was assessed according to the method of Klapp and Stählin [29] by visually estimating the abundance and dominance of all plant species.

2.2.2. Sampling of Reference Data

The biomass of each sampling plot was cut at ground surface level. Total fresh matter yield was measured and representative sub-samples were either directly dried in the oven for 48 h at 105 °C for the calculation of total dry matter yield or sorted into fractions of grasses, legumes, herbs, mosses and dead material and subsequently also dried at 105 °C for 48 h to determine the proportion of each functional group. These data were used as reference values (dependent variables) in regression analysis procedures.

2.3. Data Analysis

Prior to analysis, an insignificant number of outliers (maximum two were excluded), which appeared as extreme outliers in the box plot analysis [30], were excluded from the dataset due to incorrectly entered or measured data. Moreover, noisy parts of the hyperspectral data (305–360 nm, 1340–1500 nm and 1650–1700 nm) were eliminated, leaving 1126 spectral bands between 360 and 1650 nm. Datasets were combined using a common dataset (n = 214) comprising samples from all study plots (grazing intensities) and all dates, as well as subsets for each date representing a typical phenological status of plants during the vegetation period (n = 52−54). A modified partial least squares regression (MPLSR) was applied as a powerful and full-spectrum based method to analyze the original reflectance values using the WINISI III package (Infrasoft International, LLC. FOSS, State College, PA, version 1.63). To evaluate the potential of a 2-band vegetation index across the available hyperspectral range, the normalized difference spectral index (NDSI) [31] was applied over the range of all single (n = 1126) wavebands using all possible combinations of two-band reflectance ratios based on the NDVI formula [32] according to Equation (2):

$$NDSI\ (b1,\ b2)\ =\ b1-b2/\ b1+b2 \tag{2}$$

where b1 and b2 represent spectral bands of reflection signals with Wavelength b1 > Wavelength b2.

To test the performance of the multispectral approach used in satellites, hyperspectral data were re-combined into 8 broad wavebands according to WorldView-2 satellite images: coastal (400–450 nm), blue (450–510 nm), green (510–580 nm), yellow (585–625 nm), red (630–690 nm), red edge (705–745 nm), near infrared-1 (770–895 nm) and near infrared-2 (869–900 nm) (http://www.landinfo.com/WorldView2.htm).

Ordinary least squares regression analysis was performed using the statistical program R to examine the relationship between the dependent variables (fresh matter yield, dry matter yield and dead material proportion) and USH (Equation (3)), NDSI and satellite bands exclusively (Equations (4) and (5)) and as a combination of USH with variables calculated from hyperspectral data (Equations (6) and (7)) to compare their potential for sensor fusion. After having examined the data and verified that saturation effects could be excluded, it was assumed that squared variables would sufficiently represent possible non-linear effects. Regardless, due to the limited sample size of n ≤ 54, squared satellite band variables were omitted from the regressions to reduce the risk of over-fitting.

Exclusive ultrasonic sward height

$$Y\ =\ USH\ +\ USH^2 \tag{3}$$

Exclusive vegetation index

$$Y\ =\ NDSI\ +\ NDSI^2 \tag{4}$$

Exclusive satellite bands

$$Y\ =\ X1\ +\ X2\ +\ldots+\ Xn \tag{5}$$

Combination of ultrasonic sward height and vegetation index

$$Y\ =\ USH\ +\ NDSI\ +\ USH \times NDSI\ +\ USH^2+\ USH^2 \times NDSI\ +\ NDSI^2\ +\ USH \times NDSI^2\ +\ USH^2 \times NDSI^2 \tag{6}$$

Combination of ultrasonic sward height (USH) and satellite bands

$$Y\ =\ USH\ +\ USH^2\ +\ X1\ +\ X2\ +\ldots+\ Xn\ +\ USH \times X1\ +\ldots+\ USH \times Xn\ +\ USH \times X1\ +\ldots+\ USH^2 \times Xn \tag{7}$$

where Y = fresh matter yield (FMY) (g·m^{-2}), dry matter yield (DMY) (g·m^{-2}) or dead material proportion (DMP) (% of DMY); USH = ultrasonic sward height (mm); NDSI = 2-band combination vegetation index derived from hyperspectral data based on original NDVI formula; and X = WorldView-2 satellite bands.

To determine the best NDSI wavebands in order to maximize R^2, wavelength selection was first conducted according to Equation (4) and (6) for each target parameter. Thus, all possible 2-band NDSI combinations, in all 633,375 indices, were individually used in linear regression models for each sensor combination. The best fit wavelengths for the full models were then used to develop regression models. According to the rules of hierarchy and marginality [33,34], non-significant effects were excluded from the models using a step-wise approach, but were retained if the same variable appeared as part of a significant interaction at α-level of 5%. In order to reduce the risk of over-fitting, all models were validated by a four-fold cross validation method [35]. The prediction accuracy was evaluated using two measures: (a) the cross-validated squared correlation coefficient (R^2_{CV}), which describes the linear relation between the measured dependent variables (i.e., FMY, DMY, and DMP) and the values predicted by the linear model; and (b) the cross-validated root mean square error (RMSE$_{CV}$), which describes the average deviation of the estimated values from the observed ones.

3. Results

3.1. Sward Characteristics

Biomass as FMY and DMY varied from 68.8 to 3207 $g \cdot m^{-2}$ and from 29.2 to 691.9 $g \cdot m^{-2}$ with an overall mean value of 823.9 $g \cdot m^{-2}$ and 276.4 $g \cdot m^{-2}$, respectively, for all sampling dates (Table 1). The sampling date at the beginning of June (Date 2) exhibited the highest biomass (mean value of 1240 $g \cdot m^{-2}$ and 314.5 $g \cdot m^{-2}$ for FMY and DMY, respectively), whereas Date 4 showed the lowest biomass (mean value of 567.5 $g \cdot m^{-2}$ and 237.6 $g \cdot m^{-2}$ for FMY and DMY, respectively). USH ranged from 7 to 646 mm during the growing season and the lowest sward heights were found at Date 1 (mean value = 136 mm). A wide range of DMP (1.4% to 83.6% of DM; sd = 20.5%) was observed throughout the growing season. The highest variability of DMP was observed at more advanced developmental stages of swards (Date 3 and 4; sd = 18.8% and 17.7% of DMY, respectively) which also delivered the highest mean values of DMP (45.7% and 40% of DMY, respectively). The proportion of grass was always considerably higher than proportions of legumes and herbs. The proportion of moss was negligible (overall mean value 1.9%). In total, 48 species were identified in the sampling plots (Table A1). The most important species were *Dactylis glomerata* (Constancy, C = 89.7%) and *Lolium perenne* (C = 70.1%) among the grasses, *Trifolium repens* (C = 39.7%) and *Trifolium pratense* (C = 17.8%) among the legumes, and *Taraxacum officinale* (C = 57.5%) and *Galium mollugo* (C = 40.7%) among the herbs.

Table 1. Descriptive statistics of dry matter yield, fresh matter yield, ultrasonic sward height and proportion of mosses, grasses, legumes, herbs and dead materials for common and date-specific swards.

	N	Min	Max	Mean	Sd	Min	Max	Mean	Sd
		Dry matter yield ($g \cdot m^{-2}$)				Fresh matter yield ($g \cdot m^{-2}$)			
Common	214	29.2	691.9	276.4	145.5	68.8	3207.0	823.9	554.6
Date 1	54	51.9	612.1	248.8	130.0	140.0	1883.0	739.6	416.9
Date 2	54	31.9	691.9	314.5	180.2	107.2	3207.0	1240.0	785.6
Date 3	52	68.2	654.8	305.7	138.1	148.0	1822.0	745.4	337.0
Date 4	54	29.2	468.8	237.6	112.7	68.8	1325.0	567.5	281.7
		Ultrasonic sward height (mm)				Grass proportion (% of DM)			
Common	214	7	646	252	151	8.0	93.7	50.6	23.9
Date 1	54	7	438	136	99	12.9	81.1	44.9	16.8
Date 2	54	31	646	364	174	8.2	93.7	72.2	19.0
Date 3	52	105	615	268	119	8.8	92.9	41.9	24.8
Date 4	54	48	576	240	107	8.0	85.3	43.1	20.6
		Legume proportion (% of DM)				Moss proportion (% of DM)			
Common	214	0.0	39.6	2.9	6.8	0.0	27.5	1.9	4.4
Date 1	54	0.0	36.4	4.7	8.2	0.0	21.3	4.9	6.1
Date 2	54	0.0	39.6	4.1	9.0	0.0	14.7	0.7	2.4
Date 3	52	0.0	31.2	1.9	5.0	0.0	27.5	1.6	4.4
Date 4	54	0.0	7.1	0.6	1.6	0.0	5.8	0.3	0.9
		Herb proportion (% of DM)				Dead material proportion (% of DM)			
Common	214	0.0	63.7	13.1	12.9	1.4	83.6	31.6	20.5
Date 1	54	0.0	44.6	13.6	12.7	2.5	70.3	31.9	14.9
Date 2	54	0.0	63.7	13.9	15.0	1.4	37.6	9.2	6.4
Date 3	52	0.0	47.5	14.6	12.8	3.9	76.3	40.0	18.8
Date 4	54	0.0	42.1	10.3	10.8	10.5	83.6	45.7	17.7

3.2. Exclusive use of Ultrasonic Sward Height

Prediction accuracies for DMY and FMY varied significantly between sampling dates and were predominately low (Figures 1 and 2). Higher accuracies were achieved at Date 1 both for DMY and FMY (R^2_{CV} = 0.73 and 0.80 respectively) when sward heights were much lower than at later dates.

The lowest R^2 values were found at Dates 3 and 4 ($R^2_{CV} < 0.40$). DMP had very weak or no correlation with USH and, thus, data are not shown.

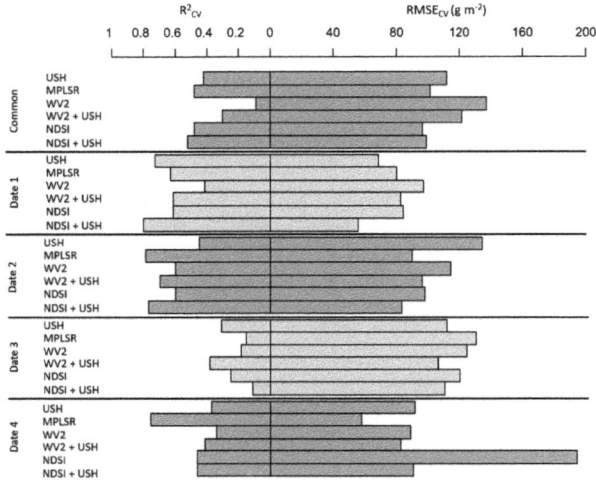

Figure 1. Cross-validation (CV) results for a range of sensor models used for prediction of fresh matter yield (FMY), including exclusive use of ultra-sonic sward height (USH), all hyperspectral wavebands using modified partial least squares regression (MPLSR), normalized difference spectral index (NDSI), and multispectral representation of WorldView-2 wavebands (WV2), as well as models formed from combinations of these sensors.

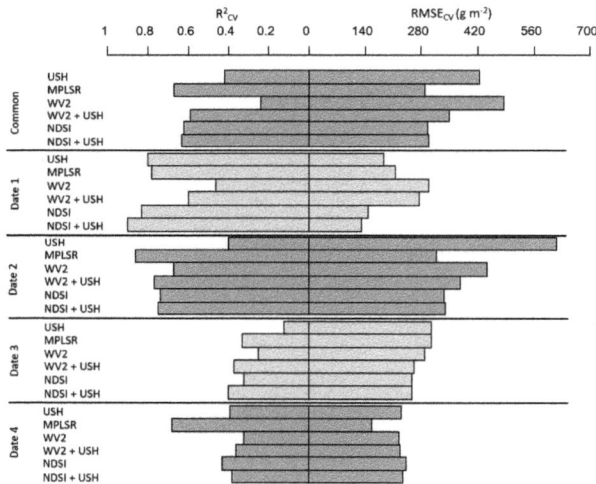

Figure 2. Cross-validation (CV) results for a range of sensor models used for prediction of dry matter yield (DMY), including exclusive use of ultra-sonic sward height (USH), all hyperspectral wavebands using modified partial least squares regression (MPLSR), normalized difference spectral index (NDSI), and multispectral representation of WorldView-2 wavebands (WV2), as well as models formed from combinations of these sensors.

3.3. Exclusive Use of Spectral Data

Maximum prediction accuracy based exclusively on NDSI was found mostly with bands between 1035 and 1139 nm, i.e., the ascending slope of the first water absorption band and the descending slope of the second water absorption band. The ascending slope of the second water absorption band (1188 to 1305 nm) was found to be the most important part of the spectrum for prediction of DMP (Table 2). Among models utilizing sensors exclusively, the MPLSR prediction accuracy was best both for DMY (R^2_{CV} of 0.48 for common and 0.15–0.79 for date-specific models) and FMY (0.67 and 0.33–0.86 respectively) (Figures 1 and 2). For DMP the MPLSR prediction was only best for the common model and date 1 (R^2_{CV} of 0.76 and 0.67), while for the other dates the NDSI showed the best results (R^2_{CV} between 0.43 and 0.68) (Figure 3). This regression approach integrates spectral information from the whole hyperspectral range and its usefulness for measuring grassland properties has been acknowledged by other studies [36–40]. The predictive power of WorldView2 (WV2) bands (R^2 0.13–0.55) was not satisfactory and never outperformed the NDSI or MPLSR approach.

Table 2. Wavelength position of best-fit band combination (b1, b2) for the normalized difference spectral index (NDSI) exclusively and in combination with ultrasonic sward height (USH) predicted target parameter.

	Common (n = 214)		Date 1 (n = 54)		Date 2 (n = 54)		Date 3 (n = 52)		Date 4 (n = 54)	
	b1	b2	b1	b2	b1	b2	b1	b2	b1	b2
	Dry matter yield (g·m⁻²)									
NDSI	1035	1051	389	609	1097	1139	1122	1128	769	778
USH + NDSI	521	578	1215	1225	1024	1031	1116	1118	1622	1633
	Fresh matter yield (g·m⁻²)									
NDSI	1117	1134	1040	1073	1080	1104	1122	1128	751	782
USH + NDSI	1077	1086	996	1005	536	564	1122	1135	1621	1633
	Dead material proportion (% of dry matter yield)									
NDSI	1242	1305	1231	1285	1188	1202	1236	1281	1187	1206

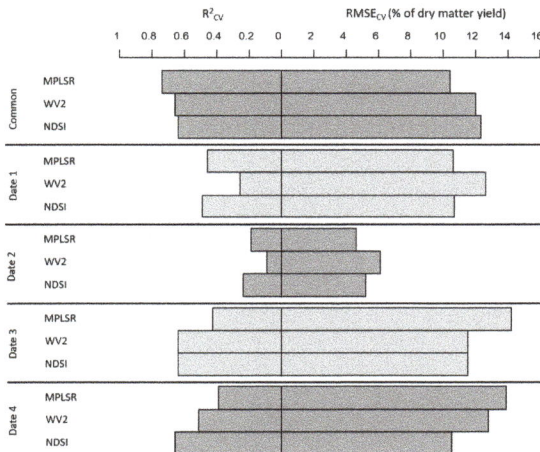

Figure 3. Cross-validation (CV) results for a range of sensor models used for prediction of dead material proportion (DMP), including exclusive use of all hyperspectral wavebands using modified partial least squares regression (MPLSR), normalized difference spectral index (NDSI), and multispectral representation of WorldView-2 wavebands (WV2) as explanatory variables.

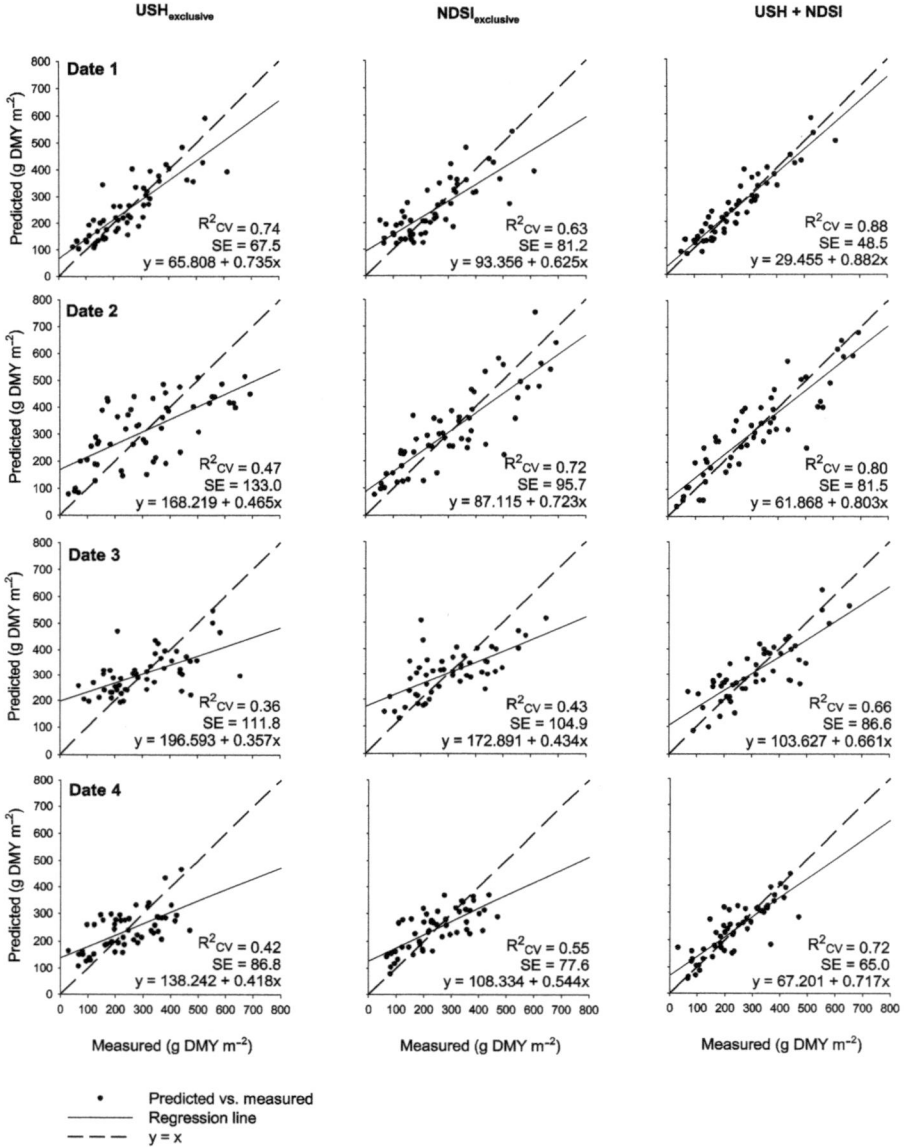

Figure 4. Plots of fit between measured and predicted dry matter yield (DMY) for exclusive use of ultrasonic sward height (USH$_{exclusive}$) and the best fit normalized difference spectral index (NDSI$_{exclusive}$) as well as a combination of USH and NDSI (USH + NDSI) applied in date-specific swards.

3.4. Sensor Data Fusion Using Combinations of USH and Spectral Variables

Combination of USH with the applied spectral variables increased R^2$_{CV}$-values for common swards from 0.42 (USH exclusively) to a maximum of 0.52 (NDSI combined with USH) for DMY and from 0.42 (USH exclusively) to a maximum of 0.63 (NDSI combined with USH) for FMY in common swards (Figures 1 and 2). Irrespective of spectral sensor configuration, date-specific calibrations of yield parameters for Dates 1 and 2 performed better than for Dates 3 and 4. The combination of USH and NDSI consistently produced the best results, both in common and date-specific calibrations. Similar to the model findings with exclusive use of NDSI, the dominant bands of NDSI when in combination with USH were mostly located at water absorption bands, i.e., the ascending slop of the first absorption band (between 996 and 1086 nm) and the ascending slope of the second water absorption band (1215 to 1225 nm) as well as the green region in the visible spectrum (521 to 578 nm) (Table 2). Figure 4 shows example plots of fit for DMY prediction based on USH and NDSI and provides a comprehensive insight into the effects of sensor combination. It becomes clear that with exclusive use of sensors, calibration models led to an overestimation at low levels of DMY, whereas higher values were underestimated. An improvement of fit by combining sensors is obvious for all sampling dates (except Date 3), as demonstrated by higher R^2_{CV} -values and convergence of the regression line to the bisector. Yield predictions in heterogeneous pastures as presented in this study partly show a complex interaction between USH, NDSI and DMP (Figure 5). At higher levels of NDSI (here seen as a measure of, e.g., sward density), DMY and FMY basically follow a linear increase with USH gain (here seen as a measure for sward height), regardless of DMP. In contrast, at low levels of NDSI, DMY and FMY curves show differing trends. While DMY (Figure 5A) just shows a parallel shift to lower yield levels, FMY (Figure 5B) in swards with high DMP shows a saturated curve.

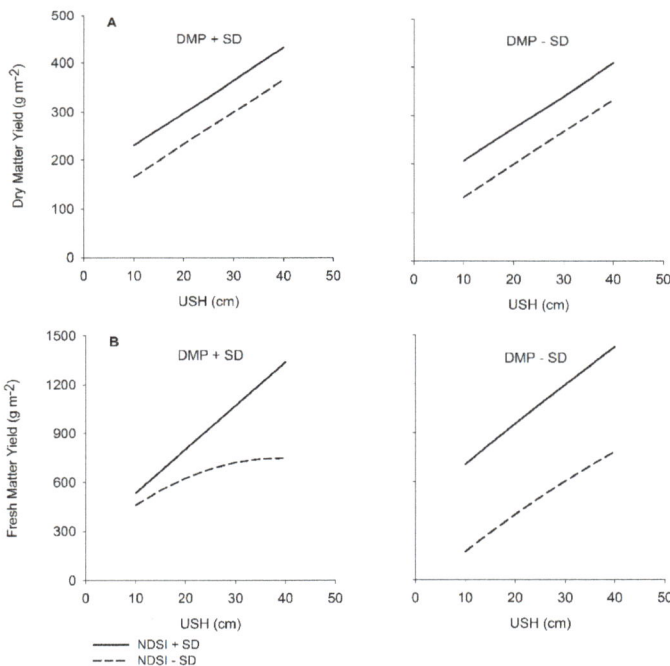

Figure 5. Predictions of dry matter yield (DMY) (**A**); and fresh matter yield (FMY) (**B**) in common swards based on ultrasonic sward height (USH) and the Normalized Difference Spectral Index (NDSI) as influenced by dead material proportion (DMP) in the range of ± standard deviation (SD). NDSI represents narrow-band reflection values selected in combination with USH for each parameter.

4. Discussion

4.1. Exclusive Use of USH

Sward height measured by ultrasonic sensors seems to become a poorer predictor of biomass with progression of the grazing season, as partly utilized patches were short in height but had a dense biomass. In addition, some species such as *Dactylis glomerata* and *Festuca rubra* frequently grow in dense tussocks and produce high biomass at low height, which results in an underestimation of biomass by USH (Figure 1). In some patches rejected by animals, very tall and mature species like *Cirsium arvense*, elongated stems of *Galium mollugo* or very tall and sparse individuals of *Phleum pratensis* at inflorescence stage occurred. Such sward structures may tend to be overestimated (Figure 1) and may have boosted USH measures although the amount of biomass was not particularly high. This effect was also observed by Fricke et al. [5], who further showed that the relationship between forage mass and USH could be influenced by weed proportion, as some weeds grow higher than the sown species. Beside the heterogeneity of canopy structure, variation in leaf angle among plant species and movements of swards during measurement due to wind may have further affected the reflection of ultrasonic signal [16,17]. In summary, exclusive use of USH measurements produced low prediction accuracies for yield parameters in heterogeneous pastures.

4.2. Exclusive Use of Spectral Data

Most spectral variables gave better prediction accuracies than exclusive use of USH measurements. This finding does not match that of Fricke et al. [4] and Adamchuk et al. [41] who reported that exclusive use of USH achieved better results than exclusive use of narrow or broad band spectral vegetation indices for prediction of biomass in more homogeneous grasslands. Contrary to yields, separation of the common dataset into date-specific subsets did not improve prediction accuracy for DMP (Figure 3). Yang and Guo [19] found that the relationship between dead material cover and spectral indices is a function of the amount of dead material, and they concluded that spectral indices could be used for estimating dead material cover which is greater than 50% in mixed grasslands. In this respect, the lower model accuracies for yield at later dates may be partly attributed to the higher amount of dead material at this time. The higher proportion of explained variance in DMP by spectral variables may reflect the impact of dead materials on the canopy reflectance at Date 3 (R^2_{CV} = 0.43–0.64) and, to a lesser degree, at Date 1 (R^2_{CV} = 0.26–0.49) and Date 4 (R^2_{CV} = 0.39–0.66). In contrast, DMP is much lower at Date 2, which corresponds to lower R^2_{CV} values for DMP prediction (0.09–0.24) (Figure 3), but allows higher accuracies for yield prediction, as low levels of DMP are inversely related to higher proportions of green plant material. This is consistent with findings by Chen et al. [42], who pointed out that spectral indicators usually collect data over green vegetation rather than mature and dry vegetation.

Dominant bands of NDSI were mostly located at water absorption bands. This dominance of water absorption bands can be explained by the strong relationship between biomass and canopy water content [43,44]. The importance of water absorption bands for estimating biomass is also confirmed by other investigations [4,45]. Numata et al. [22] found that water absorption features derived from hyperspectral sensors were better measures for estimating pasture biomass compared to spectral vegetation indices, such as Normalized Difference Vegetation Index and Normalized Difference Water Index. In summary, the yield of pastures with complex sward structures could barely be predicted using sensor measurements exclusively.

4.3. Sensor Fusion

Prediction accuracies of the combined measurements were high in the early stages of the grazing season. However, sward structures were so complex at later stages of the grazing season, that even sensor combinations did not produce satisfactory results. Considering the consequences of these limitations for the implementation of sensor data fusion in precision agriculture, it should be noted that the productivity of cool-season pastures is usually highest in the first half of the growing season [46] when the best results with combined sensor data were obtained. Thus, sensor data fusion gains more importance in this particular part of the vegetation period, when efficient and timely estimates of available biomass is most relevant for grazing management decisions. Furthermore, major management measures (e.g., fertilization, evaluation of botanical sward composition) are also typically scheduled before summer, when pasture growth is frequently limited by water scarcity or progressively reduced day lengths.

The fusion of sonar and spectral variables always performed better in predicting yield parameters than the use of each sensor alone. However, the interactions between the two groups of variables with the measured vegetation parameter are complex, particularly for situations with high DMP. Pastures with high cover of dead material might consist of both compacted xeric material leading to higher yield levels at low sward height and sparse high growing mature shoots reaching higher sward layers without much contribution to yield. In contrast, at low DMP, NDSI seems to be more closely linked to pure sward density of green vegetation. The inter-relationship between selective grazing and species phenology creates a broad variation of sward structures posing an enormous challenge for any sensor applications.

Comparable to NDSI, WV2 bands also proved to be an effective spectral tool in combination with USH. This is of particular interest, as this finding points to the potential of the WordView-2 satellite system to provide large-scale images with an acceptable spatial resolution to assess larger pasture areas in farming practice. The relatively high prediction accuracy of WV2 bands, particularly in the major growth period during the first half of the year, opens up a perspective for the development of future management assistant tools. Continuous biomass monitoring based on advanced multispectral satellite images with high spatial resolution like WorldView and GeoEye can be used as support for management decisions such as the planning of grazing time and grazing intervals for cattle on pasture paddocks, site specific re-sowing or targeted cut of less-preferred sub-areas. However, further research is necessary to evaluate the availability of reliable images at a high repetition frequency and their combination with sward height data, as for instance, derived from radar satellites.

5. Conclusions

Mapping the spatio-temporal dynamics of pasture is a necessary prerequisite for making effective grassland management decisions and ensuring timely actions. In order to understand spatio-temporal dynamics, accurate measures of grassland characteristics, such as biomass, are needed, which should preferably be measured in a non-destructive manner. The present study revealed the potential of ultrasonic and hyperspectral sensor data as a non-destructive measurement method for the prediction of biomass in pastures characterized by a high structural diversity.

Our new approach of combining ultrasonic and hyperspectral sensor data improved the precision of biomass estimation when compared to the results gained by each single sensor. In particular, the combination of ultrasonic sensors with a selected subset of hyperspectral bands increased the prediction accuracy significantly. This finding may constitute a promising link to practical use because the identified bands are already implemented on satellite platforms.

However, the inter-relationship between selective grazing and species phenology poses an enormous challenge to sensor applications because it creates highly complex variation in sward structure. More advanced and complex sensor systems are needed to overcome such limitations and future studies should therefore aim at further systematically testing a variety of different sensor applications and their combinations. Purchasing a full range hyperspectral radiometer is still costly and is, therefore, hardly an economically feasible option for grassland managers. This poses another challenge for the practical applicability of the presented methods and should be considered in future studies. However, the increasing use of such sophisticated sensors leads to the assumption that prices will decrease in the future.

Acknowledgments: The authors would like to thank the section of Grassland Science, Georg-August-Universität Göttingen for the provision of experimental pastures and their kind cooperation. This project was supported by a grant of the Research Training Group 1397 "Regulation of soil organic matter and nutrient turnover in organic agriculture" of the German Research Foundation (DFG). We thank the anonymous reviewers for their valuable comments on an earlier version of the manuscript.

Author Contributions: HS, TF, and BR conducted the fieldwork. All authors contributed equally to the data analysis and manuscript preparation.

Conflicts of Interest: The authors declare no conflicts of interest.

Appendix A

Table A1. List of pasture species identified in 214 sampling plots in 2013 with their minimum, maximum and mean values of dry matter contribution estimated according to the Klapp and Stählin method. Constancy (Const.) refers to the relative proportion of plots containing the respective species.

Species	Min	Max	Mean	Const. (%)	Species	Min	Max	Mean	Const. (%)
Grasses					**Herbs**				
Agrostis stolonifera	0.0	79.4	9.22	54.2	*Achillea millefolium*	0.0	85.0	0.92	5.1
Alopecurus pratensis	0.0	95.0	3.83	13.6	*Anthriscus sylvestris*	0.0	28.0	0.13	0.5
Arrhenatherum elatius	0.0	1.0	0.00	0.5	*Bellis perennis*	0.0	59.0	0.31	2.3
Bromus mollis	0.0	7.0	0.10	3.7	*Centaurea jacea*	0.0	1.0	0.00	0.5
Cynosurus cristatus	0.0	59.6	1.77	10.3	*Cerastium holosteoides*	0.0	4.0	0.23	19.6
Dactylis glomerata	0.0	94.0	25.68	89.7	*Cirsium arvense*	0.0	40.0	1.14	9.3
Deschampsia caespitosa	0.0	90.0	0.59	0.9	*Cirsium vulgare*	0.0	15.0	0.30	7.0
Elymus repens	0.0	80.0	5.82	36.9	*Convolvulus arvensis*	0.0	28.6	0.39	6.1
Festuca pratensis	0.0	85.0	0.71	5.6	*Crepis capillaris*	0.0	20.0	0.38	6.1
Festuca rubra	0.0	95.4	4.85	21.0	*Erophila verna*	0.0	4.0	0.04	4.7
Lolium perenne	0.0	88.6	15.64	70.1	*Epilobium spec.*	0.0	16.0	0.20	4.7
Phleum pratense	0.0	4.0	0.06	2.3	*Galium mollugo*	0.0	88.0	9.67	40.7
Poa annua	0.0	1.0	0.01	0.9	*Geranium dissectum*	0.0	13.0	0.20	13.6
Poa pratensis	0.0	45.0	2.32	27.6	*Geum urbanum*	0.0	30.0	0.19	3.3
Poa trivialis	0.0	16.0	1.28	25.2	*Hieracium pilosella*	0.0	0.2	0.00	0.5
					Lamium purpureum	0.0	38.0	0.21	2.3
Legumes					*Leontodon hispidus*	0.0	2.0	0.02	1.9
Medicago lupulina	0.0	5.0	0.03	0.9	*Plantago lanceolata*	0.0	35.0	0.56	10.7
Trifolium campestre	0.0	20.0	0.17	1.9	*Plantago major*	0.0	3.0	0.01	0.5
Trifolium dubium	0.0	25.0	0.18	3.7	*Taraxacum officinale*	0.0	83.0	5.89	57.5
Trifolium pratense	0.0	61.0	1.50	17.8	*Ranunculus acris*	0.0	10.0	0.20	6.5
Trifolium repens	0.0	49.6	2.49	39.7	*Ranunculus repens*	0.0	71.8	1.35	23.8
Vicia cracca	0.0	1.0	0.00	0.5	*Rosa spec.*	0.0	5.0	0.04	0.9
					Rumex acetosa	0.0	4.0	0.03	1.4
					Urtica dioica	0.0	84.0	1.09	2.8
					Veronica chamaedrys	0.0	4.0	0.03	1.9
					Veronica serpyllifolia	0.0	35.0	0.19	1.9

References

1. Cho, M.A.; Skidmore, A.; Corsi, F.; van Wieren, S.E.; Sobhan, I. Estimation of green grass/herb biomass from airborne hyperspectral imagery using spectral indices and partial least squares regression. *Int. J. Appl. Earth Obs. Geoinf.* **2007**, *9*, 414–424. [CrossRef]
2. Fava, F.; Colombo, R.; Bocchi, S.; Meroni, M.; Sitzia, M.; Fois, N.; Zucca, C. Identification of hyperspectral vegetation indices for Mediterranean pasture characterization. *Int. J. Appl. Earth Obs. Geoinf.* **2009**, *11*, 233–243. [CrossRef]
3. Xiaoping, W.; Ni, G.; Kai, Z.; Jing, W. Hyperspectral remote sensing estimation models of aboveground biomass in Gannan rangelands. *Proc. Environ. Sci.* **2011**, *10*, 697–702. [CrossRef]
4. Fricke, T.; Wachendorf, M. Combining ultrasonic sward height and spectral signatures to assess the biomass of legume-grass swards. *Comput. Electron. Agric.* **2013**, *99*, 236–247. [CrossRef]
5. Fricke, T.; Richter, F.; Wachendorf, M. Assessment of forage mass from grassland swards by height measurement using an ultrasonic sensor. *Comput. Electron. Agric.* **2011**, *79*, 142–152. [CrossRef]
6. Lee, H.-J.; Kawamura, K.; Watanabe, N.; Sakanoue, S.; Sakuno, Y.; Itano, S.; Nakagoshi, N. Estimating the spatial distribution of green herbage biomass and quality by geostatistical analysis with field hyperspectral measurements. *Grassl. Sci.* **2011**, *57*, 142–149. [CrossRef]
7. Schellberg, J.; Hill, M.J.; Gerhards, R.; Rothmund, M.; Braun, M. Precision agriculture on grassland: Applications, perspectives and constraints. *Eur. J. Agron.* **2008**, *29*, 59–71. [CrossRef]

8. Pullanagari, R.R.; Yule, I.J.; Hedley, M.J.; Tuohy, M.P.; Dynes, R.A.; King, W.M. Multi-spectral radiometry to estimate pasture quality components. *Precis. Agric.* **2012**, *13*, 442–456. [CrossRef]
9. Adamchuk, V.I.; Sudduth, K.A.; Lammers, P.S.; Rossel, R.A.V. *Sensor Fusion for Precision Agriculture*; In-Tech: Rijeka, Croatia, 2011.
10. Jones, C.L.; Maness, N.O.; Stone, M.L.; Jayasekara, R. Chlorophyll estimation using multispectral reflectance and height sensing. *Trans. ASABE* **2007**, *50*, 1867–1872. [CrossRef]
11. Mazzetto, F.; Calcante, A.; Mena, A.; Vercesi, A. Integration of optical and analogue sensors for monitoring canopy health and vigour in precision viticulture. *Precis. Agric.* **2010**, *11*, 636–649. [CrossRef]
12. Scotford, I.M.; Miller, P. Combination of spectral reflectance and ultrasonic sensing to monitor the growth of winter wheat. *Biosyst. Eng.* **2004**, *87*, 27–38. [CrossRef]
13. Scotford, I.; Miller, P. Estimating tiller density and leaf area index of winter wheat using spectral reflectance and ultrasonic sensing techniques. *Biosyst. Eng.* **2004**, *89*, 395–408. [CrossRef]
14. Farooque, A.A.; Chang, Y.K.; Zaman, Q.U.; Groulx, D.; Schumann, A.W.; Esau, T.J. Performance evaluation of multiple ground based sensors mounted on a commercial wild blueberry harvester to sense plant height, fruit yield and topographic features in real-time. *Comput. Electron. Agric.* **2013**, *91*, 135–144. [CrossRef]
15. Sui, R.; Thomasson, J.A. Ground-Based Sensing System for Cotton Nitrogen Status Determination. *Trans. ASABE* **2006**, *49*, 1983–1991. [CrossRef]
16. Hutchings, N.J.; Phillips, A.H.; Dobson, R.C. An ultrasonic rangefinder for measuring the undisturbed surface height of continuously grazed grass swards. *Grass Forage Sci.* **1990**, *45*, 119–127. [CrossRef]
17. Hutchings, N.J. Factors affecting sonic sward stick measurements: the effect of different leaf characteristics and the area of sward sampled. *Grass Forage Sci.* **1992**, *47*, 153–160. [CrossRef]
18. Boschetti, M.; Bocchi, S.; Brivio, P.A. Assessment of pasture production in the Italian Alps using spectrometric and remote sensing information. *Agric. Ecosyst. Environ.* **2007**, *118*, 267–272. [CrossRef]
19. Yang, X.; Guo, X. Quantifying Responses of spectral vegetation indices to dead materials in mixed grasslands. *Remote Sens.* **2014**, *6*, 4289–4304. [CrossRef]
20. Jackson, R.D.; Huete, A.R. Interpreting vegetation indices. Interpreting vegetation indices. *Prev. Vet. Med.* **1991**, *11*, 185–200. [CrossRef]
21. Duan, M.; Gao, Q.; Wan, Y.; Li, Y.; Guo, Y.; Ganzhu, Z.; Liu, Y.; Qin, X. Biomass estimation of alpine grasslands under different grazing intensities using spectral vegetation indices. *Can. J. Remote Sens.* **2014**, *37*, 413–421. [CrossRef]
22. Numata, I.; Roberts, D.; Chadwick, O.; Schimel, J.; Galvao, L.; Soares, J. Evaluation of hyperspectral data for pasture estimate in the Brazilian Amazon using field and imaging spectrometers. *Remote Sens. Environ.* **2008**, *112*, 1569–1583. [CrossRef]
23. Biewer, S.; Fricke, T.; Wachendorf, M. Determination of dry matter yield from legume–grass swards by field spectroscopy. *Crop Sci.* **2009**, *49*, 1927–1936. [CrossRef]
24. Kumar, L.; Sinha, P.; Taylor, S.; Alqurashi, A.F. Review of the use of remote sensing for biomass estimation to support renewable energy generation. *J. Appl. Remote Sens* **2015**, *9*, 097696. [CrossRef]
25. Wrage, N.; Şahin Demirbağ, N.; Hofmann, M.; Isselstein, J. Vegetation height of patch more important for phytodiversity than that of paddock. *Agric. Ecosyst. Environ.* **2012**, *155*, 111–116. [CrossRef]
26. Scimone, M.; Rook, A.J.; Garel, J.P.; Sahin, N. Effects of livestock breed and grazing intensity on grazing systems: 3. Effects on diversity of vegetation. *Grass Forage Sci.* **2007**, *62*, 172–184. [CrossRef]
27. Jerrentrup, J.S.; Wrage-Mönnig, N.; Röver, K.-U.; Isselstein, J.; McKenzie, A. Grazing intensity affects insect diversity via sward structure and heterogeneity in a long-term experiment. *J. Appl. Ecol.* **2014**, *51*, 968–977. [CrossRef]
28. Pepperl; Fuchs. Sensing your needs: ENU Part No. 200237. 2010. Available online: http://www.pepperl-fuchs.us/usa/downloads_USA/Sensing-your-needs-2010-01-EN.pdf (accessed on 19 January 2017).
29. Klapp, E.; Stählin, A. Standorte, Pflanzengesellschaften und Leistung des Grünlandes. 122 Seiten mit 3 Karten und 20 Abbildungen. Verlag Eugen Ulmer, Stuttgart-S. 1936. *Z. Pflanzenernaehr. Dueng. Bodenk.* **1936**, *43*, 221–222.
30. R Development Core Team. *R: A Language and Environment for Statistical Computing*; R Foundation for Statistical Computing: Vienna, Austria, 2016.

31. Inoue, Y.; Penuelas, J.; Miyata, A.; Mano, M. Normalized difference spectral indices for estimating photosynthetic efficiency and capacity at a canopy scale derived from hyperspectral and CO_2 flux measurements in rice. *Remote Sens. Environ.* **2008**, *112*, 156–172. [CrossRef]
32. Rouse, J.W.; Haas, R.H.; Schell, J.A.; Deering, D.W. Monitoring vegetation systems in the Great Plains with ERTS. In *Third ERTS-1 Symposium*; Fraden, S.C., Marcanti, E.P., Becker, M.A., Eds.; Scientific and Technical Information Office, National Aeronautics and Space Administration: Washington, DC, USA, 1974; pp. 309–317.
33. Nelder, J.A. The statistics of linear models: back to basics. *Stat. Comput.* **1994**, *4*, 221–234. [CrossRef]
34. Nelder, J.A.; Lane, P.W. The computer analysis of factorial experiments: In memoriam—Frank Yates. *Am. Stat.* **1995**, *49*, 382–385. [CrossRef]
35. Diaconis, P.; Efron, B. Computer-intensive methods in statistics. *Sci. Am.* **1983**, *248*, 116–130. [CrossRef]
36. Biewer, S.; Fricke, T.; Wachendorf, M. Development of canopy reflectance models to predict forage quality of legume–grass mixtures. *Crop Sci.* **2009**, *49*, 1917. [CrossRef]
37. Marabel, M.; Alvarez-Taboada, F. Spectroscopic determination of aboveground biomass in grasslands using spectral transformations, support vector machine and partial least squares regression. *Sensors* **2013**, *13*, 10027–10051. [CrossRef] [PubMed]
38. Möckel, T.; Dalmayne, J.; Prentice, H.C.; Eklundh, L.; Purschke, O.; Schmidtlein, S.; Hall, K. Classification of grassland successional stages using airborne hyperspectral imagery. *Remote Sens.* **2014**, *6*, 7732–7761. [CrossRef]
39. Möckel, T.; Dalmayne, J.; Schmid, B.C.; Prentice, H.C.; Hall, K. Airborne hyperspectral data predict fine-scale plant species diversity in grazed dry grasslands. *Remote Sens.* **2016**, *8*, 133. [CrossRef]
40. Möckel, T.; Löfgren, O.; Prentice, H.C.; Eklundh, L.; Hall, K. Airborne hyperspectral data predict Ellenberg indicator values for nutrient and moisture availability in dry grazed grasslands within a local agricultural landscape. *Ecol. Indic.* **2016**, *66*, 503–516. [CrossRef]
41. Reddersen, B.; Fricke, T.; Wachendorf, M. A multi-sensor approach for predicting biomass of extensively managed grassland. *Comput. Electron. Agric.* **2014**, *109*, 247–260. [CrossRef]
42. Chen, P.; Haboudane, D.; Tremblay, N.; Wang, J.; Vigneault, P.; Li, B. New spectral indicator assessing the efficiency of crop nitrogen treatment in corn and wheat. *Remote Sens. Environ.* **2010**, *114*, 1987–1997. [CrossRef]
43. Anderson, M.; Neale, C.; LI, F.; Normann, J.; Kustas, W.; Jayanthi, H.; Chavez, J. Upscaling ground observations of vegetation water content, canopy height, and leaf area index during SMEX02 using aircraft and Landsat imagery. *Remote Sens. Environ.* **2004**, *92*, 447–464. [CrossRef]
44. Mutanga, O.; Skidmore, A.K. Narrow band vegetation indices overcome the saturation problem in biomass estimation. *Int. J. Remote Sens.* **2004**, *25*, 3999–4014. [CrossRef]
45. Psomas, A.; Kneubühler, M.; Huber, S.; Itten, K.; Zimmermann, N.E. Hyperspectral remote sensing for estimating aboveground biomass and for exploring species richness patterns of grassland habitats. *Int. J. Remote Sens.* **2011**, *32*, 9007–9031. [CrossRef]
46. Gherbin, P.; de Franchi, A.S.; Monteleone, M.; Rivelli, A.R. Adaptability and productivity of some warm-season pasture species in a Mediterranean environment. *Grass Forage Sci.* **2007**, *62*, 78–86. [CrossRef]

remote sensing

MDPI

Article

Identifying the Relative Contributions of Climate and Grazing to Both Direction and Magnitude of Alpine Grassland Productivity Dynamics from 1993 to 2011 on the Northern Tibetan Plateau

Yunfei Feng [1], Jianshuang Wu [1,2,*], Jing Zhang [3], Xianzhou Zhang [1] and Chunqiao Song [4]

1   Lhasa Plateau Ecosystem Research Station, Key Laboratory of Ecosystem Network Observation and Modelling, Institute of Geographic Sciences and Natural Resources Research, Chinese Academy of Sciences, Beijing 100101, China; fengyf.13b@igsnrr.ac.cn (Y.F.); zhangxz@igsnrr.ac.cn (X.Z.)
2   Biodiversity—Ecological Modelling, Dahlem Center of Plant Science, Free University of Berlin, 14195 Berlin, Germany
3   College of Global Change and Earth System Sciences, Beijing Normal University, Beijing 100875, China; jingzhang@bnu.edu.cn
4   Department of Geography, University of California, Los Angeles, CA 90095, USA; chunqiao@ucla.edu
*   Correspondence: wujs07s@zedat.fu-berlin.de or wujs.07s@igsnrr.ac.cn; Tel.: +86-10-6488-8176

Academic Editors: Lalit Kumar, Onisimo Mutanga, Parth Sarathi Roy and Prasad S. Thenkabail
Received: 20 November 2016; Accepted: 27 January 2017; Published: 7 February 2017

Abstract: Alpine grasslands on the Tibetan Plateau are claimed to be sensitive and vulnerable to climate change and human disturbance. The mechanism, direction and magnitude of climatic and anthropogenic influences on net primary productivity (NPP) of various alpine pastures remain under debate. Here, we simulated the potential productivity (with only climate variables being considered as drivers; NPP_P) and actual productivity (based on remote sensing dataset including both climate and anthropogenic drivers; NPP_A) from 1993 to 2011. We denoted the difference between NPP_P and NPP_A as NPP_{pc} to quantify how much forage can be potentially consumed by livestock. The actually consumed productivity (NPP_{ac}) by livestock were estimated based on meat production and daily forage consumption per standardized sheep unit. We hypothesized that the gap between NPP_{pc} and NPP_{ac} (NPP_{gap}) indicates the direction of vegetation dynamics, restoration or degradation. Our results show that growing season precipitation rather than temperature significantly relates with NPP_{gap}, although warming was significant for the entire study region while precipitation only significantly increased in the northeastern places. On the Northern Tibetan Plateau, 69.05% of available alpine pastures showed a restoration trend with positive NPP_{gap}, and for 58.74% of alpine pastures, stocking rate is suggested to increase in the future because of the positive mean NPP_{gap} and its increasing trend. This study provides a potential framework for regionally regulating grazing management with aims to restore the degraded pastures and sustainable management of the healthy pastures on the Tibetan Plateau.

Keywords: alpine grassland conservation; anthropogenic disturbance; ecological policies; climate change; grazing exclusion; grazing management; regional sustainability

1. Introduction

Grassland degradation is one of the most important issues closely related to biodiversity conservation, ecological functionality and sustainable development in a rapidly changing world [1–3]. Alpine pasture degradation on the Tibetan Plateau is mainly attributed to overgrazing under the ongoing climate change in the last decades [4]. Grassland degradation on this plateau has been increasingly claimed to not only threaten the livelihood and culture of local residents, but also to widely affect water security in East China and South Asia [5–7]. However, the mechanism, the direction and the magnitude of the relative influences of climatic and anthropogenic drivers are still unknown and under debate, especially in regard to the policy-making at a broader geospatial scale [8–10].

Climate change is believed to primarily impact functions and services of various ecosystems on the Tibetan Plateau. Total precipitation during the plant growing season (GSP) controls both temporal and spatial variabilities in vegetation phenology [9] and biomass production [11] in the context of global warming. Compared with fenced pastures, aboveground biomass and vegetation coverage are reduced by livestock grazing [12,13], so overgrazing by unfenced livestock is very likely to be the most important anthropogenic driver for pasture degradation on this plateau [4]. Furthermore, grazing exclusion with fencing is assumed to be a necessary trajectory for pasture restoration in the heavily degraded grasslands. However, the reduction in aboveground biomass reported in field surveys does not mean a decline in net primary productivity (NPP) because the proportion consumed by large herbivores is unknown. According to the intermediate disturbance hypothesis [14,15], a reasonable stocking rate might be better for maintaining stability in community structure and ecosystem functionality. Therefore, identifying the direction of vegetation dynamics, restoration or degradation, is the keystone of policy-making for alpine pasture conservation [16].

The direct and indirect impacts of either climate change or grazing management, especially on the potential and actual capacity of livestock, are still not clear. It is not necessary for vegetation to linearly respond to either climate change or grazing disturbance [17–19]. Therefore, disentangling and assessing their relative contributions to grassland degradation or restoration is still a challenge but increasingly required [8,20]. In addition, the difference between potential net primary productivity (NPP_P) and actual net primary productivity (NPP_A) [8,20] should be regarded as the proportion of grassland productivity that can be potentially consumed by livestock (NPP_{pc}) rather than the proportion of grassland productivity that has been actually consumed by livestock (NPP_{ac}) (Table 1). This is because NPP_A includes the plant's regrowth after grazing in the same season. In a recent study, Pan et al. [21] proposed a modified framework for assessing the relative impacts of climate and grazing on vegetation productivity at local small villages, and reported a method for estimating the actually consumed productivity by livestock (NPP_{ac}). At a coarser spatial scale, the field investigations generally cost more time and money, while remote sensing provides a more economical and effective data source for assessing historical vegetation dynamics.

To assist regionally specific policy-making concerning livestock regulation and pasture conservation in the future, we hypothesized that the NPP_{gap} (defined as $NPP_{pc} - NPP_{ac}$, Table 1) can effectively indicate the direction of grassland change over a defined period, a positive trend for restoration while a negative one for degradation. Using time-series tendency and correlations between vegetation productivity, climate change, and grazing disturbance (e.g., livestock number and meat production), we also aim to identify their relative contributions to alpine grassland dynamics, and to discuss the potential regulations in livestock management and grassland conservation.

Table 1. Main acronyms of productivity terms and their meanings used in this study.

Acronym	Definition
NPP	net primary productivity
NPP_P	potential net primary productivity, only driven by climatic factors in each grassland type
NPP_A	actual net primary productivity, driven by climatic factors and vegetation index livestock grazing in each grassland type
NPP_{pc}	defined as NPP_P-NPP_A, the proportion of grassland productivity that can be potentially consumed by livestock
NPP_{ac}	the proportion of grassland productivity that has been consumed by livestock. It can be estimated from forage consumed by livestock for body growth and meat output
NPP_{gap}	defined as NPPpc-NPPac. Nine scenarios of the mean and the trend of NPP_{gap} were summarized in Table 2

2. Materials and Methods

2.1. Study Area

The Northern Tibetan Plateau (NTP) is the most traditional and important semi-nomadic region in the Tibetan Autonomous Region, China. In this region, 176,337 herdsmen and 8,941,500 domestic animals live on the 5.2×10^5 square kilometers of available alpine pastures as of 2011 [22]. Livestock husbandry is the dominant economic activity and the major source of income for the herdsman families in this region, generally accounting for 74%–93% of their annual gross income [22]. Livestock grazing is an extensive anthropogenic disturbance in NTP, so this region is increasingly accepted as the most ideal region for studying feedback between vegetation, climate, and grazing on the Qinghai-Tibetan Plateau [23–25]. Across NTP, plants generally start to grow in early May and to senesce in late September, with up to 85% of annual precipitation falling and mean daily temperature being over 5.0 °C during this period [24]. Three zonal alpine grassland types are encountered moving from east to west, from humid alpine meadow (AM) dominated by *Kobresia pygmaea* (a sedge species), to semi-arid alpine steppe (AS) dominated by *Stipa purpurea*, and to arid alpine desert-steppe (ADS) co-dominated by *S. purpurea* and *S. glareosa* [24,26].

2.2. Simulated Potential and Actual Grassland Productivity

In this study, we used the Terrestrial Ecosystem Model (TEM) and the Carnegie–Ames–Stanford Approach (CASA) model, respectively, to simulate the potential and actual net primary productivity (NPP_P and NPP_A) [8]. The former is a process-based ecosystem model and driven by spatially referenced information on vegetation type, climate, elevation, soil water and nutrient availability. The latter is based on remote sensing and climate datasets, with actual influence of human activities being included in remote sensing data. Formulae in TEM and CASA models for NPP calculation are as follows, and detailed parameters can be found in our previous work [8]. In this study, the difference between NPP_P and NPP_A was defined as the productivity that can be potentially consumed (NPP_{pc}) by domestic animals and wild herbivores.

$$GPP = (C_{max}) \frac{PAR}{PAR + k_i} \frac{C_i}{k_c + C_i} (TEMP)(KLEAF) \tag{1}$$

$$NPP_P = GPP - R_a = GPP - (R_m - R_g) \tag{2}$$

$$NPP_A = APAR \times \varepsilon = fPAR \times PAR \times \varepsilon^* \times T_\varepsilon \times W_\varepsilon \tag{3}$$

$$NPP_{pc} = NPP_P - NPP_A \tag{4}$$

2.3. Productivity Actually Consumed by Domestic Herbivores

The number of livestock inventoried at year-end and the quantity of meat output including beef and mutton at the county level over NTP were taken from the Tibet Statistic Yearbooks [22]. The absolute numbers of different domestic animals (yaks, sheep, goats and horses) were firstly converted to standardized sheep units [6]. The actually consumed productivity (NPP_{ac}) included the productivity consumed by the inventoried livestock ($NPP_{livestocck}$) and the productivity consumed for meat output (NPP_{meat}).

$$NPP_{ac} = NPP_{livestock} + NPP_{meat} \tag{5}$$

$$NPP_{livestock} = 0.45 \times daiy\ intake\ per\ sheep\ unit \times Livestock \tag{6}$$

$$NPP_{meat} = 0.45 \times (71.38 \times Yak\ meat + 65.07 \times Mutton) \tag{7}$$

$NPP_{livestock}$ and NPP_{meat} were estimated following the approach of Pan et al. [21]. $NPP_{livestock}$ was estimated from the daily forage intake per standardized sheep unit, about 1.8 kg dry matter per day [27]. NPP_{meat} was estimated by coefficients of gross dry matter consumption per meat weight as reported Pan et al. [28]. In Equations (6) and (7), 0.45 is the coefficient to transform dry matter to carbon. NPP$_{ac}$ was also standardized to g C/m^2, referring to the area of available pastures at the county level (Supplementary Table S1), and finally converted to annual grid surfaces as the ratio of total NPP$_{ac}$ to the summed area of available pasture for each county.

2.4. Precipitation and Temperature Data

Daily mean temperature and total precipitation records between 1993 and 2011 were provided by the National Meteorological Information Center (NMIC) of the China Meteorological Administration (CMA). We aggregated daily climatic records to monthly averages, interpolated and re-aggregated into 8 km × 8 km grids using the ANUSPLIN 4.2 [29] to match the spatio-temporal resolution of the productivity datasets used in this study. The quality of grid climatic surfaces has been demonstrated by the very high correlations to field observation records [8,30]. The average temperature (GST) and sum precipitation (GSP) during the annual plant growth season (generally from May to September) were calculated for time series analysis. Annual and non-growing season average temperatures (MAT and NGST) and sum precipitations (MAP and NGSP) were also provided. In this study, to calculate NPP$_A$, the NDVI data from 1993 to 2000 was obtained from an advanced very high resolution radiometer (AVHRR) dataset, which was developed by the Global Inventory Modeling and Mapping Studies (GIMMS) group (http://glcf.umd.edu/data/gimms/) while the data from 2001 to 2011 was downloaded from the moderate-resolution imaging spectroradimeter (MODIS) product (MYD13A2.5) (https://lpdaac.usgs.gov/get_data/data_pool). Detailed information on data processing methods including resampling and smoothing can be found in Chen et al. [8].

2.5. Time Series Analyses

The method of comparing trends between NPP$_P$ and NPP$_A$ has been widely adopted in identifying the direction of natural and human influences, and in assessing the magnitude of various divers on long-term vegetation trends [8,20,21,31,32]. In each dataset, the temporal trend across the entire study period of 19 years was calculated with Equation (8).

$$Slope_{data} = \frac{n \times \sum_{i=1}^{n}(i \times data_i) - \sum_{i=1}^{n} i \times \sum_{1}^{n} data_i}{n \times \sum_{i=1}^{n} i^2 - \left(\sum_{i=1}^{n} i\right)^2} \tag{8}$$

The significance of the variation tendency was determined by F-test [20]. The calculation for F-statistics is expressed as follows:

$$F = U \times \frac{n-2}{Q} \tag{9}$$

$$U = \sum_{i=1}^{n}(\hat{y}_i - \bar{y})^2 \tag{10}$$

$$Q = \sum_{i=1}^{n}(y_i - \hat{y}_i)^2 \tag{11}$$

$$\hat{y}_i = Slope \times i + b \tag{12}$$

$$b = \bar{y} - Slope \times i \tag{13}$$

where U is the residual sum of the squares; Q is the regression sum; $\hat{y}_i$ is the regression value, which can be calculated by Equations (17)–(19); y_i is the observed value of year I; $\bar{y}_i$ is the mean value over n years; and b is the intercept of the regression formula.

For bivariate analysis at each pixel, the correlations of productivity indicators with GST and GSP, respectively were also explored by the Pearson correlation techniques as shown in Equation (14):

$$r = \frac{n \times \sum_{i=1}^{n}(X_i \times Y_i) - (\sum_{i=1}^{n} X_i)(\sum_{i=1}^{n} Y_i)}{\sqrt{n \times \left(\sum_{i=1}^{n} X_i^2\right) - (\sum_{i=1}^{n} X_i)^2} \sqrt{n \times \left(\sum_{i=1}^{n} y_i^2\right) - (\sum_{i=1}^{n} Y_i)^2}} \tag{14}$$

where n is the sequential year and X_i and Y_i represent productivity and climate variable, respectively.

We did not directly disentangle their relative contributions to vegetation dynamics in a generalized linear model with analysis of variance (co-variance) because of the coarser spatial resolution of the grazing activities and available pasture area datasets compared to plant productivity values. Instead, we used the mean value of NPP_{gap} (termed as $NPP_{pc} - NPP_{ac}$) and its tendency to describe the direction and magnitude of productivity change. Here, we mainly focused on the nine NPP_{gap} variation scenarios as shown in Table 2, to find some potential implications on stocking rate regulation for pasture conservation in the future.

Table 2. The nine scenarios of the mean and the trend of NPP_{gap} (defined as $NPP_{pc} - NPP_{ac}$) at the pixel scale from 1993 to 2011.

Mean	Trend	Vegetation Status	Current Stocking Rate	Future Stocking Rate
=0	>0	Healthy	Reasonable	Can be increased
	=0	Healthy & stable	Reasonable	No regulation
	<0	Healthy	Reasonable	Need to be reduced
>0	>0	Restored	Low	Should be increased
	=0	Restored & stable	Low	No regulation
	<0	Restored	Low	Must not be increased
<0	>0	Degraded	Overgrazed	Should be reduced
	=0	Degraded &stable	Overgrazed	Must be reduced

3. Results

3.1. Trends of Precipitation and Temperature from 1993 to 2011

From 1993 to 2011, temperatures significantly increased across nearly the entire Northern Tibetan Plateau (NTP) (Figures 1 and 2). In the north-central and eastern regions, GST and NGST have significantly risen by 0.08 °C/year–0.09 °C/year and 0.12 °C/year–0.14 °C/year, respectively. In its western, south-central and south-eastern regions, the warming rate during the plant growing season was relatively lower, at approximately 0.04 °C/year–0.06 °C/year. In general, MAT has a similar spatial pattern of long-term trend to GST. Evident spatial variability in precipitation was also observed across the entire NTP (Figure 1). However, significant increasing trends of GSP and MAP were observed only in the north-eastern NTP (Figure 2). In its central and western parts, no significant increase was found for either GSP or MAP. NGSP was observed to significantly decrease in the only three south-eastern counties, Lhari, Biru and Sog (Appendix Figure A1, map (b) for current administrative county boundary).

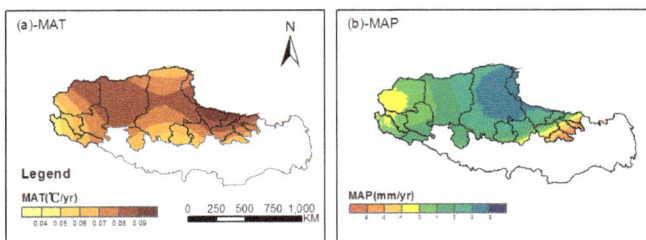

Figure 1. *Cont.*

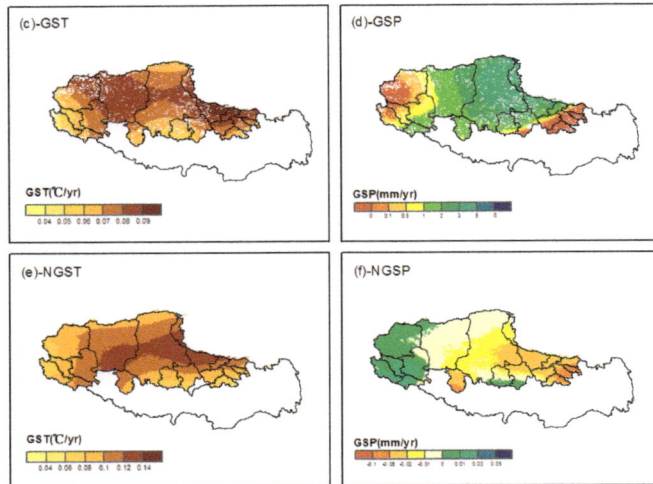

Figure 1. Trends of climatic variables from 1993 to 2011 across the Northern Tibetan Plateau (NTP). (**a**) mean annual temperature (MAT); (**b**) mean annual precipitation (MAP); (**c**) growing season temperature (GST); (**d**) growing season precipitation (GSP); (**e**) non-growing season temperature (NGST); and (**f**) non-growing season precipitation (NGSP).

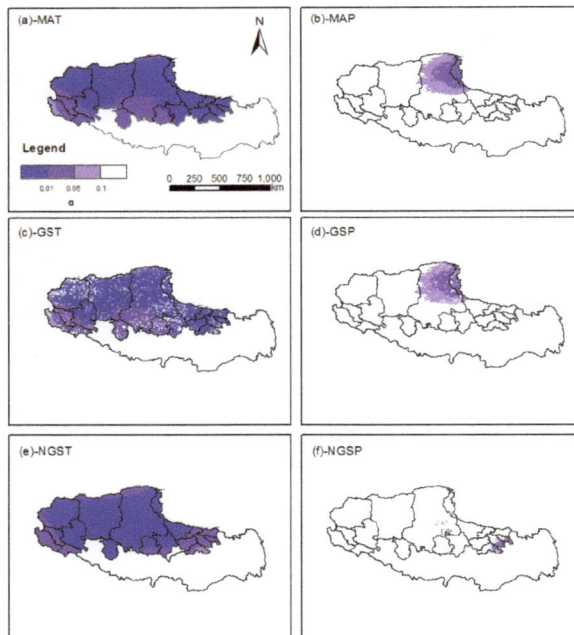

Figure 2. The significance of the corresponding climatic trends from 1993 to 2011 across the Northern Tibetan Plateau (NTP). (**a**) Mean annual temperature (MAT); (**b**) mean annual precipitation (MAP); (**c**) growing season temperature (GST); (**d**) growing season precipitation (GSP); (**e**) non-growing season temperature (NGST); and (**f**) non-growing season precipitation (NGSP).

3.2. Trends of Simulated and Consumed Productivity from 1993 to 2011

Significant increasing trends of NPP_P (Trend > 0, Figure 3a; $P < 0.05$, Figure 4a) were found for 29.63% of alpine grasslands that were mainly distributed in the northern NTP, with an increasing rate up to 5 g C/m^2/year, while in the southern and western regions no evident trend was observed for NPP_P (Figures 3a and 4a). Only 7.84% of pasture patches showed a significant increasing trend in NPP_A, scattering across the entire NTP (Figures 3b and 4b). The increasing trend of NPP_{pc} was observed to be nearly coincident with NPP_P, accounting for 28.8% of pixels across the entire NTP (Figures 3c and 4c). From 1993 to 2011, NPP_{ac} showed significant increasing trends in nearly all the counties except Burang, Biru, Sog and part of Gerze (Figures 3d and 4d. See Appendix Figure A1, map (b) for the current administrative county boundary).

3.3. Correlations of Actual and Potential Productivity with Climate from 1993 to 2011

No significant relation was found between NPP_{pc} and GST, with their correlation coefficient for 87.20% of the pixels being between -0.5 and 0.5 across the entire NTP (Figure 5a). NPP_{pc} was found to be positively correlated with GSP, with their correlation coefficient being over 0.8 for 59.70% of pixels (Figure 5b). There was no evident correlation between NPP_{ac} and climatic variables at the county level across the entire NTP (Figure 5c,d). The interior difference of the correlation of NPP_{ac} with climatic variables within Gerze and Nyima might be due to the historical adjustments of county boundaries (See for the maps of boundary changes between administrative counties before and after 2002 in this region).

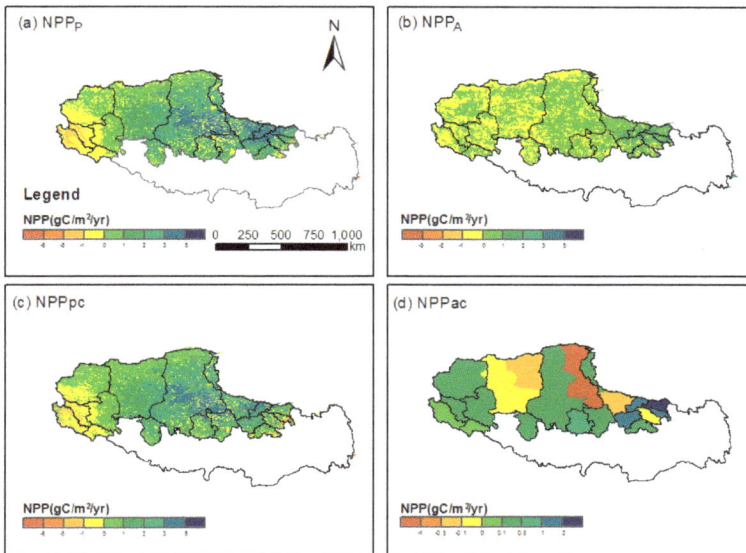

Figure 3. Trends of simulated or calculated grassland productivity from 1993 to 2011. Map (**a**) shows the trend of potential net primary productivity (NPP_P), which was from the Terrestrial Ecosystem Model and driven only by climate variables. Map (**b**) shows the trend of actual net primary productivity (NPP_A), which was simulated by the Carnegie–Ames–Stanford Approach (CASA) model with remote sensing data as the driving variables and can reflect the actual productivity after biomass partly consumed by livestock or wild herbivores. Map (**c**) shows the trend of the difference between NPP_P and NPP_A, which was defined as the productivity proportion that can be potentially consumed by livestock (NPP_{pc}). Map (**d**) shows the trend of NPP_{ac}, which reflected the productivity actually consumed by livestock or converted to meat for human society.

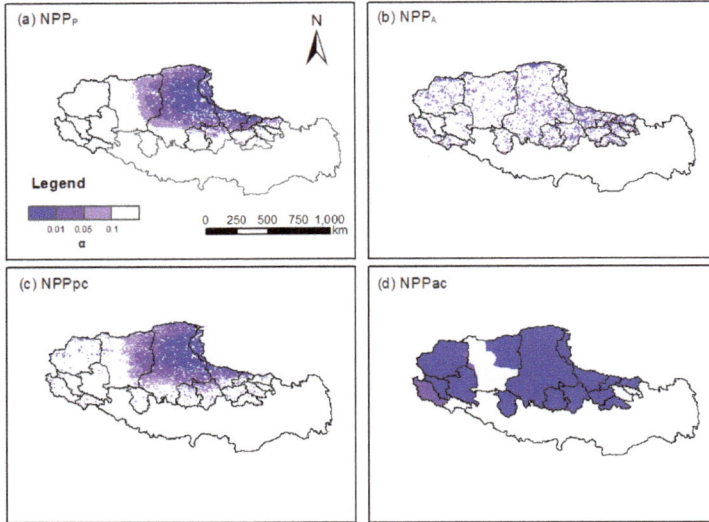

Figure 4. The significance levels of potential net primary productivity (NPP$_P$) from 1993 to 2011 (**a**); Map (**b**) shows the significance levels of actual net primary productivity (NPP$_A$); Map (**c**) shows the significance levels of the difference between NPP$_P$ and NPP$_A$, which was defined as the productivity proportion that can be potentially consumed by livestock (NPP$_{pc}$); Map (**d**) shows the significance levels of NPP$_{ac}$, which reflected the productivity actually consumed by livestock or converted to meat for human society.

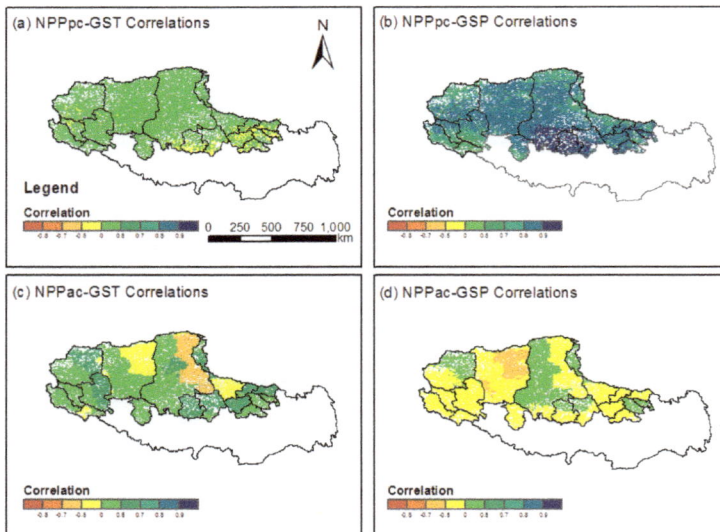

Figure 5. Correlations of net primary productivity that can be potentially consumed (NPP$_{pc}$, maps in (**a**,**b**) or that had been actually consumed by livestock (NPP$_{ac}$, maps in (**c**,**d**)) with average temperatures (GST, growing season temperature, maps in (**a**,**c**) and sum precipitation (GSP, growing season precipitation, maps in (**b**,**d**)).

3.4. Trend, Significance and Climatic Dependency of NPP$_{gap}$ from 1993 to 2011

Implications for stocking rate regulation were summarized in Table 3. Mean annual NPP$_{gap}$ for 29.65% of available alpine grasslands was negative, indicating overgrazing and calling for a reduce in stocking rate. About 69.05% of available alpine pastures are healthy or have been restored, indicating that the current stocking rates are at low or moderate levels. For 58.74% of alpine pastures, therefore, the stocking rates are suggested to increase in future because grasslands there are likely getting better due to the positive mean NPP$_{gap}$ and its increasing trend. On the other hand, 16.37% of alpine grasslands need to be excluded from animals grazing, because both the negative mean NPP$_{gap}$ and the decreasing trend imply that the grassland likely degrades even further (Table 3). NPP$_{gap}$ for 15% of pixels was found to significantly increase in the most north-eastern areas (Figure 6a,b). For 4.4% of pixels NPP$_{gap}$ significantly decreased, but the decrease in the western and south-eastern NTP parts was not significant (Figure 6a,b). In general, NPP$_{gap}$ was not correlated with GST because the correlation coefficient for most pixels was between -0.5 and 0.5 (Figure 6c). However, for 65.9% of pixels NPP$_{gap}$ was highly dependent on GSP, where the correlation coefficient being higher than 0.8 (Figure 6d).

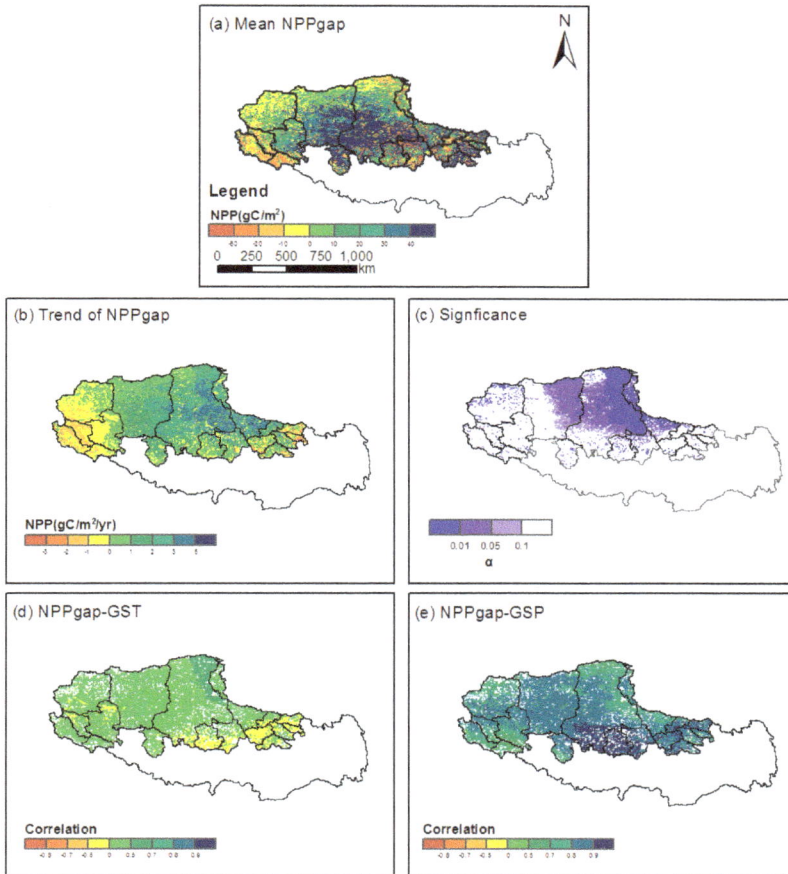

Figure 6. (**a**) Mean values, (**b**) trends and (**c**) the corresponding significance levels of NPP$_{gap}$ from 1993 to 2011. Maps in (**d**) and (**e**) showed the correlation coefficients of NPP$_{gap}$ with air temperature average (GST) and total precipitation (GSP) during the plant growing season, respectively.

Table 3. Summary of the area and the corresponding proportion of NPP_{gap} in different trends (>0, =0, and <0) and different means (>0, =0, and <0) for the nine scenarios proposed as Table 2 in this study.

Mean	Trend	The Target Area (km^2)	The Percentage (%)
=0	>0	0	0
	=0	0	0
	<0	0	0
>0	>0	405,349	58.74
	=0	0	0
	<0	71,162	10.31
<0	>0	91,648	13.28
	=0	0	0
	<0	112,974	16.37

4. Discussion

Alpine grasslands on the Tibetan Plateau are extremely sensitive to climate variation and human activities because of their vulnerability and severe physical conditions in this region [5]. The ongoing climate warming and historical overgrazing are considered to be main drivers for alpine pasture degradation there [4,8]. Adaptive strategies and sustainable policies are increasingly called for with respect to alpine pasture conservation faced in a developing Tibet [6]. In addition, it is a big challenge to identify, disentangle, and assess the relative contribution of climate and anthropogenic variables to variability in ecosystem functionality, but it is also increasingly required [33].

Wessels et al. [34] proposed a method of defining land capability units coupled with the normalized difference vegetation index (NDVI) to distinguish natural variables from anthropogenic influences. Similarly, Li et al. [35] identified and assessed vegetation changes that were mainly induced by human acuities in temperate grasslands in Inner Mongolia using a temporal NDVI residual trend method. In two recent studies, Chen et al. [8] and Wang et al. [20] accepted the difference between potential and actual productivity of alpine grasslands that were simulated by theoretical and remote sensing-based models, respectively, to represent the intensity of human influences. For example, Wang et al. [20] reported that 61.2% of the total grassland area experienced restoration from 2001 to 2013 on the Qinghai-Tibetan Plateau and that human activities, climate variation, and their combined effect accounted for 28.6%, 12.8% and 19.9% of the restored alpine grasslands with increasing productivity.

However, a major shortcoming still remains in the hypotheses of both Chen et al. [8] and Wang et al. [20] in that grassland productivity dynamics are only affected by climate and human activities, with the potential plant regrowth within the same season after livestock grazing being ignored. A multi-site survey with fenced versus grazed paired plots across the northern Tibetan Plateau also found that the differences in aboveground productivity were not only controlled by climate variables but also influenced by grazing management and closely related to plant community properties [13,36,37]. Therefore, the proportion of grassland productivity consumed by livestock should be included in remote-sensing based models for simulating grassland productivity, to represent the actual grassland productivity that has really been influenced by both climate change and human activities. Pan et al. [21] proposed a modified framework for assessing the climate and human impacts on grassland productivity on the Tibetan Plateau, in which the proportion of productivity appropriated by human society was estimated from current-year livestock inventories and meat production with specific transform coefficients for yak and sheep, respectively. Thus, the difference between potential and actual productivity, stimulated by theoretical (mechanism) models and by remote-sensing models, respectively, can be used not only to indicate the relative contribution of natural and anthropogenic factors, but also to reasonably direct the livestock regulation under differential climate change scenarios.

Previous studies suggested that temperature and precipitation showed increasing trends on the Tibetan Plateau, giving an increasing rate of air temperature nearly double that of the last fifty years [38] and three times the global warming rate [5,39]. In contrast, the precipitation exhibited different patterns within the entire Qinghai-Tibetan Plateau, differing among zonal alpine grassland types [8,9], which was accepted to mainly drive the differential response of vegetation phenology to climate changes. We found that only 29.63% of alpine grasslands exhibited significant increasing trends of NPP_P with an average rate up to 5 g·C/m^2/year (Figures 3a and 4a) mainly distributed in the northern NTP, where precipitation during the plant-growing season also significantly increased from 1993 to 2011 (Figures 1d and 2d). Although significant climate warming was observed over the entire northern Tibetan Plateau (Figures 1 and 2), no significant correlation was found between NPP_{pc} and GST for 87.2% of alpine grassland pixels across the entire NTP (Figure 5a). NPP_{pc} was found to be positively correlated with GSP for 59.7% of grassland pixels (Figure 5b). Therefore, our study further confirmed that precipitation is the primary driving force in the edges of the northern Tibetan Plateau, consistent with both remote sensing research [9] and field surveys [11,13,37] in this region. Although NPP_{ac} was found to have no correlation with either GSP or GST, the mosaic pattern of correlation coefficient differs among different counties (Figure 5c,d) and implies that stocking rate and pasture management likely affect the actual productivity of alpine grasslands in this region.

In 2003, the government started to construct metal fences on severely degraded pastures [4]. A new compensatory payment policy was launched in 2011, according to which local herding families can be compensated for alpine pasture conservation if these policies are maintained and effectively administered. Therefore, the degraded grasslands are expected to rapidly recover in the future. Current studies consistently indicate that climate changes, especially in precipitation during the plant growing months, likely limit the spatio-temporal dynamics of forage productivity of alpine grassland on the Tibetan Plateau [9,11,13]. However, these studies cannot offer more mechanistic knowledge or clearer indications on how to make livestock management more reasonable. For example, Chen et al. [8] reported that the area percentage of grassland productivity changes mainly resulting from human activities doubled from 20.16% between 1982 and 2001 to 42.98% between 2002 and 2011. We found the gaps between potential and actual productivity differed regionally, which may reflect differences in alpine grassland types that are dominated by different plant species [37,40]. Similarly, Liang et al. [41] found that alpine grassland biomass in the pastoral area of southern Qinghai Province, a region in the central-eastern Qinghai-Tibetan Plateau, shows considerable spatial heterogeneity because of the geographical, topographical, climatic and biophysical limitations. In this study, NPP_{gap} was included in the assessment framework for evaluating the relative contributions of climate change and grazing activities. The long-term width variation of NPP_{gap} was additionally introduced to identify the directions of vegetation change, restoration or degradation. We even found that the mean NPP_{gap} was positive, negative, zero or a mixture of all over the entire study region, with either increasing, decreasing or invariant trends over the defined period. Here, we point out that uncertainties still remain about the direction of livestock management trends because the spatial scale of data concerning grazing pressure at the county level was coarse.

5. Conclusions

In summary, our study clearly documented recent changes in grassland productivity and analyzed the potential impacts of temperature, precipitation and stocking density on vegetation dynamics. Although precipitation only significantly increased in the northern areas, accounting for a smaller proportion of grasslands over the entire study region, precipitation is still the primary driving force for productivity dynamics during the study period. Due to a decrease in stocking density, the gap between potential and actual consumed productivity increased in the central parts of the northern Tibetan Plateau, suggesting that the stocking rate can be increased in alpine grasslands under restoration processes. Due to the coarse scale of livestock-related variables in this study, some uncertainties remain about the direction of alpine grassland dynamics. However, we found that about 69.05% of available alpine pastures are healthy or have been restored, and are listed as having low or moderate stocking rates in the central parts of the northern Tibetan Plateau.

Supplementary Materials: The following are available online at www.mdpi.com/2072-4292/9/2/136/s1, Table S1: The records of livestock inventory, production of yak meat and mutton, available pastures at the county level on the Northern Tibetan Plateau.

Acknowledgments: This research is supported by the Ministry of Science and Technology of China (2016YFC0502000), the National Natural Sciences Foundation of China (41401070, 41571042), the Chinese Academy of Sciences (XDB03030401). Jianshuang Wu has been awarded an international post-doctoral fellowship by the Alexander von Humboldt Foundation, Germany.

Author Contributions: Jianshuang Wu conceived and designed this research; Xianzhou Zhang provided the datasets; Yunfei Feng analyzed the data and prepared the figures. Jianshuang Wu and Yunfei Feng wrote the manuscript; Jing Zhang, Xianzhou Zhang and Chunqiao Song contributed to the interpretation of the results. All the authors have contributed to this work.

Conflicts of Interest: The authors declare no conflict of interest. The founding sponsors had no role in the design of the study; in the collection, analyses, or interpretation of data; in the writing of the manuscript; and in the decision to publish the results.

Appendix A

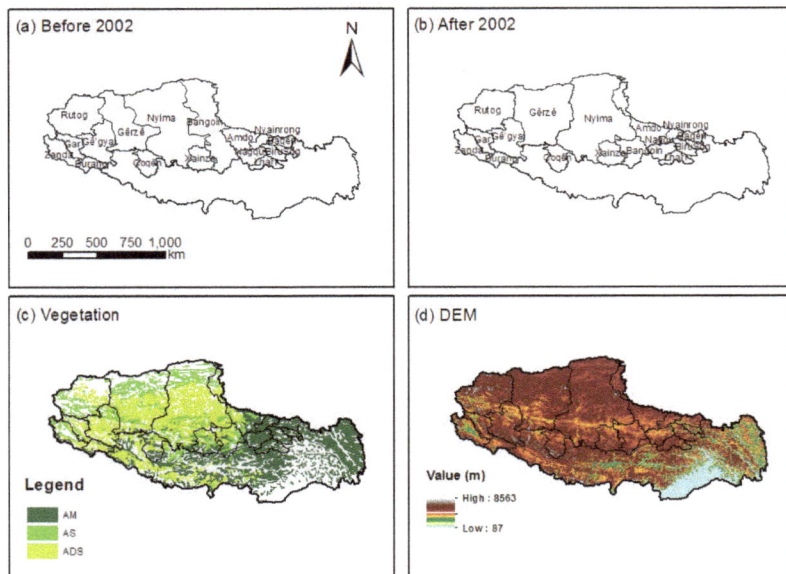

Figure A1. Map (**a**) is for the administrative county boundary before year 2002; Map (**b**) is for the current administrative county boundary after year 2002; Map (**c**) is the vegetation type of Tibet, there are alpine meadow (AM), alpine steppe (AS), and alpine desert steppe (ADS), respectively; Map (**d**) is the elevation of Tibet.

References

1. Wang, X.X.; Dong, S.K.; Yang, B.; Li, Y.Y.; Su, X.K. The effects of grassland degradation on plant diversity, primary productivity, and soil fertility in the alpine region of Asia's headwaters. *Environ. Monit. Assess.* **2014**, *186*, 6903–6917. [CrossRef] [PubMed]
2. Cai, X.B.; Peng, Y.L.; Yang, M.N.; Zhang, T.; Zhang, Q. Grassland degradation decrease the diversity of *Arbuscular mycorrhizal* fungi species in Tibet plateau. *Not. Bot. Horti Agrobot.* **2014**, *42*, 333–339. [CrossRef]

3. Wen, L.; Dong, S.K.; Li, Y.Y.; Li, X.Y.; Shi, J.J.; Wang, Y.L.; Liu, D.M.; Ma, Y.S. Effect of degradation intensity on grassland ecosystem services in the alpine region of Qinghai-Tibetan plateau, China. *PLoS ONE* **2013**, *8*, e58432. [CrossRef] [PubMed]

4. Harris, R.B. Rangeland degradation on the Qinghai-Tibetan plateau: A review of the evidence of its magnitude and causes. *J. Arid Environ.* **2010**, *74*, 1–12. [CrossRef]

5. Yao, T.; Thompson, L.G.; Mosbrugger, V.; Zhang, F.; Ma, Y.; Luo, T.; Xu, B.; Yang, X.; Joswiak, D.R.; Wang, W.; et al. Third pole environment (TPE). *Environ. Dev.* **2012**, *3*, 52–64. [CrossRef]

6. Yu, C.Q.; Zhang, Y.J.; Claus, H.; Zeng, R.; Zhang, X.Z.; Wang, J.S. Ecological and environmental issues faced by a developing Tibet. *Environ. Sci. Technol.* **2012**, *46*, 1979–1980. [CrossRef] [PubMed]

7. Xu, X.D.; Lu, C.G.; Shi, X.H.; Gao, S.T. World water tower: An atmospheric perspective. *Geophys. Res. Lett.* **2008**, *35*, L20815. [CrossRef]

8. Chen, B.X.; Zhang, X.Z.; Tao, J.; Wu, J.S.; Wang, J.S.; Shi, P.L.; Zhang, Y.J.; Yu, C.Q. The impact of climate change and anthropogenic activities on alpine grassland over the Qinghai-Tibet plateau. *Agric. For. Meteorol.* **2014**, *189*, 11–18. [CrossRef]

9. Shen, M.; Piao, S.; Cong, N.; Zhang, G.; Jassens, I.A. Precipitation impacts on vegetation spring phenology on the Tibetan plateau. *Glob. Chang. Biol.* **2015**, *21*, 3647–3656. [CrossRef] [PubMed]

10. Lehnert, L.W.; Wesche, K.; Trachte, K.; Reudenbach, C.; Bendix, J. Climate variability rather than overstocking causes recent large scale cover changes of Tibetan pastures. *Sci. Rep.* **2016**, *6*, 24367. [CrossRef] [PubMed]

11. Shi, Y.; Wang, Y.; Ma, Y.; Ma, W.; Liang, C.; Flynn, D.F.B.; Schmid, B.; Fang, J.; He, J.S. Field-based observations of regional-scale, temporal variation in net primary production in Tibetan alpine grasslands. *Biogeosciences* **2014**, *11*, 2003–2016. [CrossRef]

12. Harris, R.B.; Wenying, W.; Badinqiuying; Smith, A.T.; Bedunah, D.J. Herbivory and competition of Tibetan steppe vegetation in winter pasture: Effects of livestock exclosure and plateau Pika reduction. *PLoS ONE* **2015**, *10*, e0132897.

13. Wu, J.S.; Zhang, X.Z.; Shen, Z.X.; Shi, P.L.; Yu, C.Q.; Chen, B.X. Effects of livestock exclusion and climate change on aboveground biomass accumulation in alpine pastures across the northern Tibetan plateau. *Chin. Sci. Bull.* **2014**, *59*, 4332–4340. [CrossRef]

14. Connell, J.H. Diversity in tropical rain forests and coral reefs. High diversity of trees and corals is maintained only in a nonequilibrium state. *Science* **1978**, *199*, 1302–1310. [CrossRef] [PubMed]

15. Wilkinson, D.M. The disturbing history of intermediate disturbance. *OIKOS* **1999**, *84*, 145–147. [CrossRef]

16. Wessels, K.J.; Prince, S.D.; Reshef, I. Mapping land degradation by comparison of vegetation production to spatially derived estimates of potential production. *J. Arid Environ.* **2008**, *72*, 1940–1949. [CrossRef]

17. Banegas, N.; Albanesi, A.; Pedraza, R.; Dos Santos, D. Non-linear dynamics of litter decomposition under different grazing management regimes. *Plant Soil* **2015**, *393*, 47–56. [CrossRef]

18. Tian, D.; Niu, S.; Pan, Q.; Ren, T.; Chen, S.; Bai, Y.; Han, X. Nonlinear responses of ecosystem carbon fluxes and water-use efficiency to nitrogen addition in Inner Mongolia grassland. *Funct. Ecol.* **2016**, *30*, 490–499. [CrossRef]

19. Wu, J.; Yang, P.; Zhang, X.; Shen, Z.; Yu, C. Spatial and climatic patterns of the relative abundance of poisonous vs. Non-poisonous plants across the northern Tibetan plateau. *Environ. Monit. Assess.* **2015**, *187*, 491. [CrossRef] [PubMed]

20. Wang, Z.; Zhang, Y.; Yang, Y.; Zhou, W.; Gang, C.; Zhang, Y.; Li, J.; An, R.; Wang, K.; Odeh, I.; et al. Quantitative assess the driving forces on the grassland degradation in the Qinghai–Tibet plateau, in China. *Ecol. Inf.* **2016**, *33*, 32–44. [CrossRef]

21. Pan, Y.; Yu, C.; Zhang, X.; Chen, B.; Wu, J.; Tu, Y.; Miao, Y.; Luo, L. A modified framework for the regional assessment of climate and human impacts on net primary productivity. *Ecol. Indic.* **2016**, *60*, 184–191. [CrossRef]

22. Tibet Autonomous Region Bureau of Statistics. *Tibet Statistical Yearbook*; China Statistics Press: Beijing, China, 1993–2011.

23. Li, X.J.; Zhang, X.Z.; Wu, J.S.; Shen, Z.X.; Zhang, Y.J.; Xu, X.L.; Fan, Y.Z.; Zhao, Y.P.; Yan, W. Root biomass distribution in alpine ecosystems of the northern Tibetan plateau. *Environ. Earth Sci.* **2011**, *64*, 1911–1919. [CrossRef]

24. Wu, J.S.; Zhang, X.Z.; Shen, Z.X.; Shi, P.L.; Xu, X.L.; Li, X.J. Grazing-exclusion effects on aboveground biomass and water-use efficiency of alpine grasslands on the northern Tibetan plateau. *Rangel. Ecol. Manag.* **2013**, *66*, 454–461. [CrossRef]

25. Zeng, C.; Wu, J.; Zhang, X. Effects of grazing on above- vs. Below-ground biomass allocation of alpine grasslands on the northern Tibetan plateau. *PLoS ONE* **2015**, *10*, e0135173. [CrossRef] [PubMed]

26. Kattge, J.; Diaz, S.; Lavorel, S.; Prentice, C.; Leadley, P.; Bonisch, G.; Garnier, E.; Westoby, M.; Reich, P.B.; Wright, I.J.; et al. Try - a global database of plant traits. *Glob. Chang. Biol.* **2011**, *17*, 2905–2935. [CrossRef]

27. Ministry of Agriculture of the People's Republic of China. *Calculation of Proper Carrying Capacity of Rangelands*; Ny/t 635–2002; China Zhijian Publishing House: Beijing, China, 2002.

28. Pan, Y.; Wu, J.; Xu, Z. Analysis of the tradeoffs between provisioning and regulating services from the perspective of varied share of net primary production in an alpine grassland ecosystem. *Ecol. Complex.* **2014**, *17*, 79–86. [CrossRef]

29. Hutchinson, M. *Anusplin Version 4.3*; Centre for Resource and Environment Studies, The Australian National University: Canberra, Australia, 2004.

30. Tao, J.; Zhang, Y.J.; Dong, J.W.; Fu, Y.; Zhu, J.T.; Zhang, G.L.; Jiang, Y.B.; Tian, L.; Zhang, X.Z.; Zhang, T.; et al. Elevation-dependent relationships between climate change and grassland vegetation variation across the Qinghai-Xizang plateau. *Int. J. Climatol.* **2015**, *35*, 1638–1647. [CrossRef]

31. Zhang, X.; Lu, X.; Wang, X. Spatial-temporal ndvi variation of different alpine grassland classes and groups in northern Tibet from 2000 to 2013. *Mt. Res. Dev.* **2015**, *35*, 254–263. [CrossRef]

32. Zhang, Y.; Hu, Z.; Qi, W.; Wu, X.; Bai, W.; Li, L.; Ding, M.; Liu, L.; Wang, Z.; Zheng, D. Assessment of effectiveness of nature reserves on the Tibetan plateau based on net primary production and the large sample comparison method. *J. Geogr. Sci.* **2016**, *26*, 27–44. [CrossRef]

33. Yu, C.; Zhang, X.; Zhang, J.; Li, S.; Song, C.; Fang, Y.; Wurst, S.; Wu, J. Grazing exclusion to recover degraded alpine pastures needs scientific assessments across the northern Tibetan plateau. *Sustainability* **2016**, *8*, 1162. [CrossRef]

34. Wessels, K.J.; Prince, S.D.; Frost, P.E.; van Zyl, D. Assessing the effects of human-induced land degradation in the former homelands of northern South Africa with a 1 km AVHRR NDVI time-series. *Remote Sens. Environ.* **2004**, *91*, 47–67. [CrossRef]

35. Li, A.; Wu, J.; Huang, J. Distinguishing between human-induced and climate-driven vegetation changes: A critical application of restrend in Inner Mongolia. *Landsc. Ecol.* **2012**, *27*, 969–982. [CrossRef]

36. Wu, J.S.; Shen, Z.X.; Zhang, X.Z. Precipitation and species composition primarily determine the diversity-productivity relationship of alpine grasslands on the northern Tibetan plateau. *Alp. Bot.* **2014**, *124*, 13–25. [CrossRef]

37. Wu, J.; Wurst, S.; Zhang, X. Plant functional trait diversity regulates the nonlinear response of productivity to regional climate change in Tibetan alpine grasslands. *Sci. Rep.* **2016**, *6*, 35649. [CrossRef] [PubMed]

38. Piao, S.; Cui, M.; Chen, A.; Wang, X.; Ciais, P.; Liu, J.; Tang, Y. Altitude and temperature dependence of change in the spring vegetation green-up date from 1982 to 2006 in the Qinghai-Xizang plateau. *Agric. For. Meteorol.* **2011**, *151*, 1599–1608. [CrossRef]

39. Qiu, J. The third pole. *Nature* **2008**, *454*, 393–396. [CrossRef] [PubMed]

40. Li, S.; Wu, J. Community assembly and functional leaf traits mediate precipitation use efficiency of alpine grasslands along environmental gradients on the Tibetan plateau. *PeerJ* **2016**, *4*, e2680. [CrossRef] [PubMed]

41. Liang, T.; Yang, S.; Feng, Q.; Liu, B.; Zhang, R.; Huang, X.; Xie, H. Multi-factor modeling of above-ground biomass in alpine grassland: A case study in the Three-River Headwaters region, China. *Remote Sens. Environ.* **2016**, *186*, 164–172. [CrossRef]

remote sensing

MDPI

Article

Spectroscopic Estimation of Biomass in Canopy Components of Paddy Rice Using Dry Matter and Chlorophyll Indices

Tao Cheng *, Renzhong Song, Dong Li, Kai Zhou, Hengbiao Zheng, Xia Yao, Yongchao Tian, Weixing Cao and Yan Zhu *

National Engineering and Technology Center for Information Agriculture (NETCIA), Jiangsu Key Laboratory for Information Agriculture, Jiangsu Collaborative Innovation Center for Modern Crop Production, Nanjing Agricultural University, One Weigang, Nanjing 210095, China; 2015101010@njau.edu.cn (R.S.); lidongmath@163.com (D.L.); 2013201073@njau.edu.cn (K.Z.); 2015201019@njau.edu.cn (H.Z.); yaoxia@njau.edu.cn (X.Y.); yctian@njau.edu.cn (Y.T.); caow@njau.edu.cn (W.C.)
* Correspondence: tcheng@njau.edu.cn (T.C.); yanzhu@njau.edu.cn (Y.Z.);
   Tel.: +86-25-8439-6565 (T.C.); +86-25-8439-6598 (Y.Z.)

Academic Editors: Lalit Kumar, Onisimo Mutanga, Jose Moreno and Prasad S. Thenkabail
Received: 28 December 2016; Accepted: 27 March 2017; Published: 28 March 2017

Abstract: Crop biomass is a critical variable for characterizing crop growth development, understanding dry matter partitioning, and predicting grain yield. Previous studies on the spectroscopic estimation of crop biomass focused on the use of various spectral indices based on chlorophyll absorption features and found that they often became saturated at high biomass levels. Given that crop biomass is commonly expressed as the dry weight of canopy components per unit ground area, it may be better estimated using the spectral indices that directly characterize dry matter absorption. This study aims to evaluate a group of four dry matter indices (DMIs) by comparison with a group of four chlorophyll indices (CIs) for estimating the biomass of individual components (e.g., leaves, stems) and their combinations with the field data collected from a two-year rice cultivation experiment. The Red-edge Chlorophyll Index ($CI_{Red-edge}$) of the CI group exhibited the best relationship with leaf biomass ($R^2 = 0.82$) for the whole growing season and with total biomass ($R^2 = 0.81$), but only for the growth stages before heading. However, the Normalized Difference Index for Leaf Mass per Area (ND_{LMA}) of the DMI group showed the best relationships with both stem biomass ($R^2 = 0.81$) and total biomass ($R^2 = 0.81$) for the whole season. This research demonstrated the suitability of dry matter indices and provided physical explanations for the superior performance of dry matter indices over chlorophyll indices for the estimation of whole-season total biomass.

Keywords: rice; biomass; dry matter index; chlorophyll index; $CI_{Red-edge}$; ND_{LMA}

1. Introduction

Rice has a critical role in ensuring food security for the largest population in the world [1]. Timely monitoring of rice growth status is crucial for global food security and agricultural sustainability [2]. Specifically, biomass can be used as an indicator of grain yield, growth status, and gross primary production [3,4]. Furthermore, the information on rice biomass is desired for calculating critical nitrogen (N) concentrations and also the nitrogen nutrition index, which is an important variable for in-season nitrogen management [5]. The traditional approach for measuring rice biomass by manually collecting physical samples is time consuming, labor intensive, and prone to errors. As a non-destructive approach, remote sensing has been successfully used to estimate the biomass of rice and other crops since the late 1990s [6,7].

The majority of previous studies on the remote estimation of crop biomass are based on several methods, including spectral vegetation indices (VIs) [6–9], multivariate regression [9,10], integration of remotely sensed data and crop growth models [11–13] or radiation use efficiency models [14], fusion of optical and radar data [15], and three-dimensional analysis of point cloud data [16,17]. The data-model integration methods are built on the physiological process of crop growth and can be used for estimating crop biomass under various growth and climate conditions, but much effort is required for parameterizing crop models and determining the optimal data assimilating strategy [18]. Crop biomass can be estimated with active remotely sensed data acquired from radar or LiDAR (light detection and ranging) instruments, but those data sources are often expensive and need extensive experiences for data processing [19]. Among all those methods, the use of various spectral VIs has been the most common one due to the simplicity of calculation and the widespread accessibility of spectral data. In the past two decades, most of the VIs for crop biomass estimation are calculated from either multispectral data collected with handheld sensors (e.g., CropScan, GreenSeeker, and Crop Circle) and satellite imagery (e.g., Landsat, RapidEye, and WorldView-2) or hyperspectral data collected with field spectroradiometers (e.g., ASD FieldSpec and Ocean Optics SD2000). The commonly used spectral indices from these data include the Normalized Difference Vegetation Index, NDVI [20,21], the Green NDVI [22], the Modified Chlorophyll Absorption in Reflectance Index, MCARI [23], the Red-edge Chlorophyll Index, $CI_{Red-edge}$ [23], red and red-edge reflectance-based indices [24,25], and near-infrared based indices [26]. In particular, the VIs derived from hyperspectral data are often variable between studies as a result of optimization in the form of NDVI with two new wavelengths for a specific data set, such as $(R_{708} - R_{565})/(R_{708} + R_{565})$ [9], $(R_{1301} - R_{1706})/(R_{1301} + R_{1706})$ [27], and $(R_{752} - R_{549})/(R_{752} + R_{549})$ [28]. Those studies paid considerable attention to various types of indices originally designed for the detection of chlorophyll content, which was based on the chlorophyll absorption features in the red region.

As the crop biomass expressed in most studies is the dry weight of crop components per unit ground area, the physical variable that should be detected directly is actually the dry matter content instead of the chlorophyll content. However, it is still a common practice to use various chlorophyll indices for estimating crop biomass [7–16]. The estimation with these indices was indirect and its performance relied on the relationship between the biomass and chlorophyll content or leaf area index of crops [29,30]. If one uses the appropriate dry matter indices, a direct estimation should become possible and the estimation of crop biomass may be improved. Although a large number of VIs have been reported for estimating foliar chlorophyll content [31,32], only a few narrow-band indices have been developed specifically for detecting dry matter content. They were assessed with experimental and simulated data and proved to work well across a wide range of species [33,34]. These dry matter indices use one or two bands in the shortwave infrared (SWIR) region to characterize the dry matter absorption centered at 1.7 μm [35], and do not use any band in the visible and red edge regions as the chlorophyll indices do. To date, few studies have explicitly evaluated their performance for the estimation of crop biomass and the comparison of them to the commonly used chlorophyll indices. It is unclear whether and in what condition dry matter indices are more appropriate than chlorophyll indices for estimating crop biomass.

In addition, the biomass to be estimated with VIs is often from all the aboveground components of crops, including leaves, stems, and panicles or fruits. The aboveground biomass was found to be nonlinearly related to the chlorophyll indices [9,22,26]. These nonlinear relationships could be due to the poor sensitivity of chlorophyll indices to the aboveground biomass at high biomass levels and could lead to large uncertainties in the biomass estimation. Because of the strong interest in the aboveground biomass, the common practice in the community is still to estimate the biomass of all individual components as a whole using various chlorophyll indices [30]. This problem may be alleviated by exploiting dry matter indices for the spectroscopic estimation and decomposing the aboveground biomass into individual components such as leaf biomass and stem biomass for the evaluation.

Given the smaller amount of mass per unit ground area in leaf biomass than in total biomass, the poor sensitivity of VIs to total biomass at high levels may be better understood by an additional examination of the leaf biomass. The recent research by Kross et al. [25] represented one of the few attempts of this kind and investigated this possibility in corn and soybean crops. Although a number of studies have focused on the estimation of total biomass specifically in rice [6,23,24,27,28], none of them have taken the investigation of individual components in total biomass into consideration. Therefore, particular attention should be paid to the difference in performance between the spectroscopic estimation of total biomass and those of biomass for individual components in rice. Moreover, those pertinent investigations included very few or even no data samples collected after the heading stage and were unable to cover the whole growing season of rice. Considering the high biomass at the post-heading stages of the season, it becomes important to determine whether the models for biomass estimations could be fitted across all critical stages or only for specific stages of the whole season.

The objectives of this study were to evaluate the performance of dry matter indices in comparison with chlorophyll indices for the estimation of leaf biomass, stem biomass, and total biomass in rice and to evaluate the feasibility of fitting a single index-based model across all growth stages of the growing season. Eight spectral indices selected from the literature for such a purpose were evaluated with a large number of samples collected from a two-year experiment for the whole growing season of rice.

2. Materials and Methods

2.1. Experimental Design

The experiment was designed for two consecutive years with the same treatments, involving different rice cultivars, planting densities, and nitrogen (N) rates. The crops were grown in 2014 and 2015 in the same fields at the experimental station of the National Engineering and Technology Center for Information Agriculture (NETCIA), Rugao, Jiangsu, China (120°19′E, 32°14′N). There were four N rate treatments (0, 100, 200, and 300 kg·N·ha^{-1}) with the density of 0.30 m × 0.15 m for the minimum and maximum rates and two densities (0.30 m × 0.15 m and 0.50 m × 0.15 m) for the intermediate rates. The N fertilizers were applied in the form of urea: 40% as basal fertilizer before transplanting, 10% at the tillering stage, 30% at the jointing stage, and 20% at the booting stage. The two rice cultivars involved were Y *liangyou 1* (Indica rice, V1) and *Wuyunjing 24* (Japonica rice, V2). Each plot was 5 m × 6 m in size. A total of 36 plots (12 cultivation conditions with three replications) were grown for the whole study in each year.

2.2. Spectral Measurements

Spectral reflectance was measured with an ASD FieldSpec Pro spectrometer (Analytical Spectral Devices, Boulder, CO, USA) with a 25° field of view at a height of 1.0 m above the rice canopy. The spectral range was 350–2500 nm, with a 1.4 nm sampling interval between 350 and 1050 nm and a 2 nm sampling interval between 1000 and 2500 nm. Spectral measurements were taken from 11:00 a.m. to 1:00 p.m. local time. There were three observation points fixed in each plot and each point was measured five times with the ASD spectrometer. The mean of those measurements was calculated to represent the reflectance spectrum of each plot. Calibration measurements were done with a white reference panel every ten minutes. A summary of the sampling dates is listed in Table 1.

Table 1. Summary of data collection dates for the two-year experiment.

Year	Early Tillering	Late Tillering	Jointing	Early Booting	Late Booting	Heading	Early Filling	Late Filling
2014	10 July	20 July	30 July	/	21 August	2 September	/	21 September
2015	10 July	22 July	30 July	14 August	26 August	/	9 September	27 September

Note: / means no data at that stage due to poor weather conditions.

2.3. Biomass Measurements

The samples of all canopy components at each growth stage were collected within one day of the spectral measurements. For each plot, three hills of plants at the center of the spectral sampling area were cut at the ground surface. All green leaves and panicles when present were separated from the stems. All components were oven-dried at 105 °C for 30 min and then at 80 °C for about 24 h until a constant weight was obtained. A total of 359 leaf samples, 359 stem samples, and 96 panicle samples were collected in the two years at the growth stages of early tillering, late tillering, jointing, early booting, late booting, heading, early filling, and late filling (Table 2).

Table 2. Summary of rice biomass measurements (units of t/ha) for individual components and combinations of components in rice canopies.

Canopy Component	No. of Samples	Mean ± SD	Minimum	Maximum	Growth Stage
Leaf	359	1.66 ± 1.33	0.04	6.75	All stages
Stem	359	3.30 ± 2.93	0.07	12.54	All stages
Panicle	96	5.09 ± 3.11	0.83	12.80	Post-heading
Leaf + stem	359	4.95 ± 4.15	0.11	17.84	All stages
Leaf + stem + panicle (Total)	359	6.32 ± 5.96	0.11	25.94	All stages

2.4. Calculation of Spectral Indices and Estimation of Biomass

Two groups of vegetation indices (VIs) (Table 3) were calculated with the spectral data. One was the group of chlorophyll indices (CIs), including the Red-edge Chlorophyll Index, $CI_{red\ edge}$ [36], the ratio of Transformed Chlorophyll Absorption in Reflectance Index to Optimized Soil-Adjusted Vegetation Index, TCARI/OSAVI [37], the Normalized Difference Vegetation Index, NDVI [38], and the Enhanced Vegetation Index, EVI [39]. They were selected to represent the red-edge based indices, soil-resistant indices, and the two most commonly used vegetation indices. The other was the group of dry matter indices (DMIs), including the Normalized Difference index for the Leaf Mass per Area, ND_{LMA} [33], the Normalized Dry Matter Index, NDMI [34], the Normalized Difference Lignin Index, NDLI [40], and the Normalized Difference Index for leaf canopy biomass, ND_{Bleaf} [41]. The DMIs represented all significant developments in dry matter estimation reported in the literature and were less commonly used in the community due to the use of SWIR bands. To keep the balance between the two groups, this study retained only those four chlorophyll indices although more were available in the literature. The selection of four indices for each group ensured that reasonable representations and adequate attention was paid to their specific relationships with the biomass for the individual and multiple components.

The data collected from the two-year experiment were pooled to examine the relationships between the eight vegetation indices and the biomass of different components or component combinations. Linear and nonlinear (exponential) models were developed to fit those relationships. The predictive capability of those models were assessed by the Root Mean Square Error (RMSE) using a k-fold ($k = 10$) cross-validation procedure.

Table 3. List of vegetation indices used in this study.

Index	Formulation	Reference
Red-edge Chlorophyll Index	$CI_{red\ edge} = \frac{R_{800}}{R_{720}} - 1$	[36]
Ratio of Transformed Chlorophyll Absorption in Reflectance Index to Optimized Soil-Adjusted Vegetation Index	$TCARI/OSAVI =$ $\frac{3[(R_{700}-R_{670})-0.2(R_{700}-R_{550})(R_{700}/R_{670})]}{(1+0.16)(R_{800}-R_{670})/(R_{800}+R_{670}+0.16)}$	[37]
Normalized Difference Vegetation Index	$NDVI = \frac{R_{800}-R_{680}}{R_{800}+R_{680}}$	[38]
Enhanced Vegetation Index	$EVI = 2.5 \frac{R_{800}-R_{680}}{1+R_{800}+6R_{680}-7.56R_{440}}$	[39]
Normalized Difference index for LMA *	$ND_{LMA} = \frac{R_{1368}-R_{1722}}{R_{1368}+R_{1722}}$	[33]
Normalized Dry Matter Index	$NDMI = \frac{R_{1649}-R_{1722}}{R_{1649}+R_{1722}}$	[34]
Normalized Difference Lignin Index	$NDLI = \frac{\log\left(\frac{1}{R_{1754}}\right) - \log\left(\frac{1}{R_{1680}}\right)}{\log\left(\frac{1}{R_{1754}}\right) + \log\left(\frac{1}{R_{1680}}\right)}$	[40]
Normalized Difference Index for leaf canopy biomass	$ND_{Bleaf} = \frac{R_{1540}-R_{2160}}{R_{1540}+R_{2160}}$	[41]

* The 1368 nm band was replaced by the 1320 nm band in this study to avoid the atmospheric water vapor contamination in canopy spectra.

3. Results

3.1. Variation in Biomass of Individual and Multiple Components over the Growing Season

The temporal patterns of biomass measurements across the stages are displayed in Figure 1. The leaf biomass increased gradually from the early tillering stage to the late booting stage and decreased to the minimum at the late filling stage. The stem biomass kept increasing until the early filling stage and also decreased at the late filling stage. The difference between mean leaf biomass and mean stem biomass was greater for the post-heading (heading included) stages than that for the pre-heading (heading excluded) stages. Panicle biomass increased from the heading stage to the late filling stage and exceeded stem biomass at the late filling stage. The biomass of leaves and stems increased rapidly from the early tillering stage to the heading stage and remained almost stable until the decrease from the early filling stage to the late filling stage. The total biomass increased with the growth stage for the whole season.

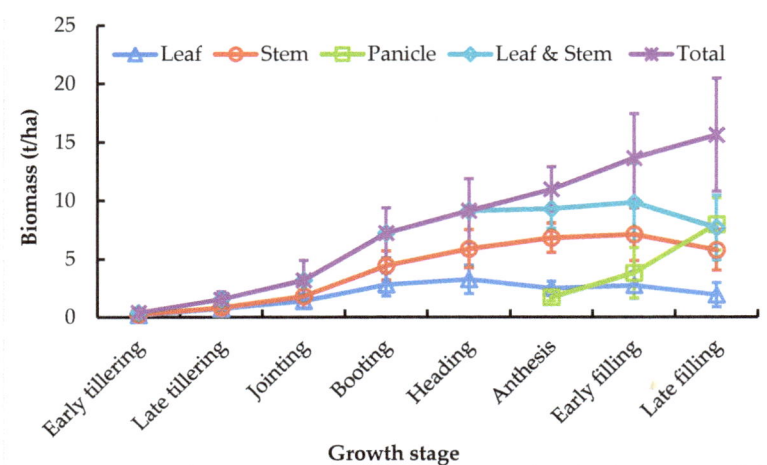

Figure 1. Temporal profiles of biomass for individual components and multiple components (t/ha) over the whole growing season of rice.

3.2. Relationships between Vegetation Indices and Biomass of Individual Components

The best-fit functions for the relationships between the vegetation indices and leaf biomass were mostly nonlinear with R^2 values ranging from 0.68 to 0.82 for chlorophyll indices and from 0.46 to 0.72 for dry matter indices. Only the $CI_{red\ edge}$ from the CI group exhibited linear relationships with the leaf biomass (Figure 2A), even with the best goodness of fit of all indices examined. The TCARI/OSAVI and the NDVI exhibited a strong relationship with the leaf biomass but the sensitivity decreased considerably when the leaf biomass exceeded 1 t/ha (Figure 2B,C). The relationships of the leaf biomass with EVI (Figure 2D) and all indices from the DMI group showed similar asymptotic patterns, but with various degrees of the scattering of data points from the nonlinear fits. Within the DMI group, the ND_{LMA} and the NDMI showed better fits with leaf biomass than the NDLI and the ND_{Bleaf} (Figure 2E–H).

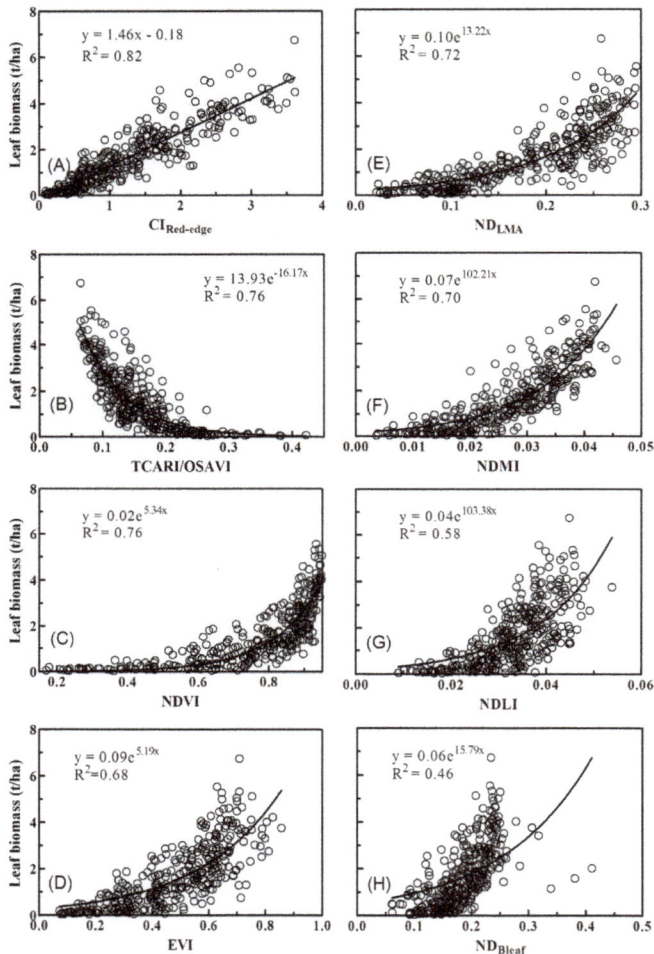

Figure 2. Leaf biomass (t/ha) plotted against vegetation indices: (**A**) $CI_{Red-edge}$; (**B**) TCARI/OSAVI; (**C**) NDVI; (**D**) EVI; (**E**) ND_{LMA}; (**F**) NDMI; (**G**) NDLI and (**H**) ND_{Bleaf}. The solid line is the best-fit function for the data points. All regressions are statistically significant ($p < 0.001$).

Unlike leaf biomass, stem biomass was nonlinearly related to all of the eight spectral indices. The relationships of the stem biomass with $CI_{Red-edge}$ showed an even higher scattering of data points than those with the NDVI and the EVI (Figure 3A–D). The sensitivity of NDVI to the stem biomass became poor when the stem biomass exceeded 2.5 t/ha. The ND_{LMA} from the DMI group showed the strongest relationship ($R^2 = 0.81$, $p < 0.001$) with stem biomass than any other index evaluated (Figure 3E–H). This indicates that the best chlorophyll index examined is more suitable than the best dry matter index for the estimation of leaf biomass, but not for the estimation of stem biomass. In contrast to the leaf biomass and stem biomass, the panicle biomass was not significantly related to any of the spectral indices (data not shown).

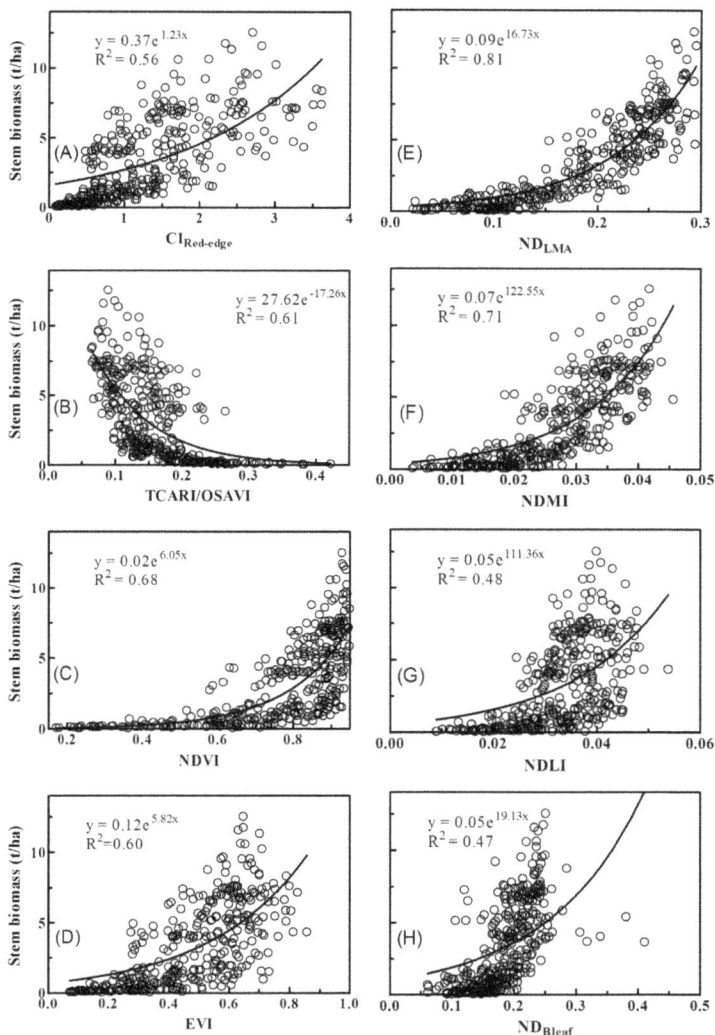

Figure 3. Stem biomass (t/ha) plotted against vegetation indices: (**A**) $CI_{Red-edge}$; (**B**) TCARI/OSAVI; (**C**) NDVI; (**D**) EVI; (**E**) ND_{LMA}; (**F**) NDMI; (**G**) NDLI and (**H**) ND_{Bleaf}. The solid line is the best-fit function for the data points. All regressions are statistically significant ($p < 0.001$).

3.3. Relationships between Vegetation Indices and Biomass of Multiple Components

The relationships of the spectral indices with the biomass of the leaf and stem components showed similar patterns to those with the leaf biomass (Table 4), exhibiting linear best-fits for the $CI_{Red-edge}$ and nonlinear best-fits for other indices. The goodness of fits decreased after the addition of the stem biomass for all chlorophyll indices, particularly for the $CI_{Red-edge}$. In contrast, the goodness of fit for three of the four DMIs increased with the R^2 for the ND_{LMA} being the greatest (0.72 to 0.79).

With regards to the total biomass of the aboveground components, all the best-fit functions were nonlinear and the best fit of all was with the ND_{LMA} (R^2 = 0.81) (Table 4). This suggests that the best dry matter index examined is a better indicator of the biomass across the whole season than the best chlorophyll index for multiple components. Closer examinations of the nonlinear relationships showed that the nonlinearity for the ND_{LMA} across the whole season encompassed two linear fits divided by the growth stage, with one for pre-heading and the other for the post-heading stages. From the transition of the pre-heading phase to the post-heading phase, the change in total biomass appeared faster with the ND_{LMA} (pre-heading: slope = 44.89; post-heading: slope = 99.26) but became slower with the $CI_{Red-edge}$ (pre-heading: slope = 3.62; post-heading: slope = 2.37) as measured by the slopes of the regression lines (Figure 4). The $CI_{Red-edge}$ exhibited the highest correlation (R^2 = 0.81) of all indices with the total biomass for the stages before heading (Figure 4A), followed by the ND_{LMA} (R^2 = 0.75). For the stages after heading, the ND_{LMA} exhibited the highest correlation (R^2 = 0.46) (Figure 4B), which was substantially greater than the correlations with any other chlorophyll index (R^2 ranging from 0.06 to 0.19).

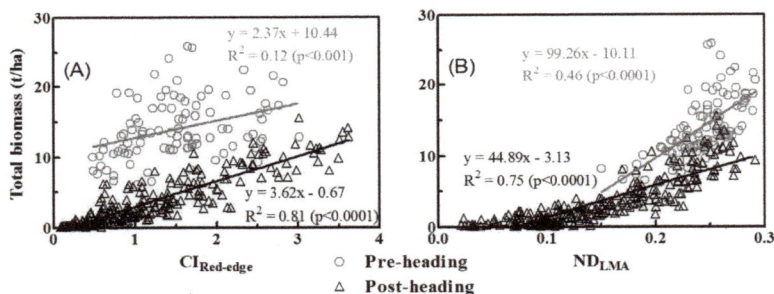

Figure 4. Total biomass (t/ha) plotted against vegetation indices: (A) $CI_{Red-edge}$ and (B) ND_{LMA}. The black triangles and grey circles represent the data from the pre-heading and post-heading stages, respectively.

3.4. Model Validation

For the estimation of leaf biomass with the best-fit models as shown in Figure 2, the $CI_{Red-edge}$ exhibited the lowest RMSE (RMSE = 0.56 t/ha) which was lower than those obtained with the ND_{LMA} (RMSE = 0.75 t/ha) and other indices. For the estimation of stem biomass, the lowest RMSE was produced by the ND_{LMA} among all indices (Table 5). The ND_{LMA} also exhibited the lowest RMSE values for the estimation of the leaf and stem biomass (RMSE = 1.99 t/ha) and of the total biomass across all growth stages (RMSE = 3.07 t/ha). In the CI group, the $CI_{Red-edge}$ performed best with an RMSE of 2.42 t/h for the estimation of the leaf and stem biomass but produced the highest RMSE (8.29 t/ha) for the total biomass. While calibrating linear models by two groups of growth stages, the ND_{LMA} still produced the most accurate estimation of the total biomass for the post-heading group (RMSE = 3.19 t/ha) and the second most accurate estimation for the pre-heading group (RMSE = 1.75 t/ha), which was close to the most accurate estimation with the $CI_{Red-edge}$ (RMSE = 1.51 t/ha).

Table 4. Coefficients of determination (R^2) for the relationships between the vegetation indices and biomass of various components.

Vegetation Index	Leaf Biomass (t/ha)		Stem Biomass (t/ha)		Leaf + Stem Biomass (t/ha)		Total Biomass (t/ha)		
	Linear	Nonlinear	Linear	Nonlinear	Linear	Nonlinear	Nonlinear	Linear (Before Heading)	Linear (After Heading)
NDVI	0.53	**0.76**	0.44	0.68	0.49	0.72	0.68	0.47	0.16
EVI	0.60	0.68	0.47	0.60	0.54	0.63	0.59	0.56	0.19
$CI_{Red\text{-}edge}$	**0.82**	0.67	0.55	0.56	0.66	0.60	0.51	**0.81**	0.12 ***
TCARI/OSAVI	0.58	**0.76**	0.38	0.61	0.46	0.66	0.56	0.55	0.06 *
ND_{LMA}	0.68	0.72	**0.76**	**0.81**	**0.77**	**0.79**	**0.81**	0.75	**0.46**
NDMI	0.68	0.70	0.63	0.71	0.68	0.72	0.69	0.68	0.25
NDLI	0.49	0.58	0.33	0.48	0.39	0.52	0.46	0.45	**0.07 **
ND_{Bleaf}	0.44	0.46	0.41	0.47	0.44	0.48	0.49	0.49	**0.09 **

Note: The number (or numbers) in bold denotes the maximum in each column. Significance level: * $p < 0.05$, ** $p < 0.01$, *** $p < 0.001$, others $p < 0.0001$.

Table 5. Accuracy assessment with the Root Mean Squared Error (RMSE) values for the estimation of rice biomass with vegetation indices. All values were obtained using a 10-fold cross validation procedure.

Vegetation Index	Leaf Biomass (t/ha)		Stem Biomass (t/ha)		Leaf + Stem Biomass (t/ha)		Total Biomass (t/ha)		
	Linear	Nonlinear	Linear	Nonlinear	Linear	Nonlinear	Nonlinear	Linear (Before Heading)	Linear (After Heading)
NDVI	0.92	0.77	2.20	2.13	2.96	2.71	4.79	2.55	3.95
EVI	0.84	0.95	2.13	2.60	2.82	3.45	5.65	2.31	3.92
$CI_{Red\text{-}edge}$	**0.56**	1.68	1.98	4.25	2.42	5.89	8.29	**1.51**	4.04
TCARI/OSAVI	0.87	0.73	2.32	2.35	3.06	2.96	5.48	2.35	4.18
ND_{LMA}	0.75	0.76	**1.45**	**1.49**	**1.99**	**2.03**	**3.07**	1.75	**3.19**
NDMI	0.75	**0.75**	1.78	2.13	2.34	2.71	4.89	1.96	3.72
NDLI	0.95	1.06	2.41	2.85	3.24	3.81	6.11	2.59	4.18
ND_{Bleaf}	1.00	2.88	2.27	9.78	3.12	12.28	19.98	2.49	4.57

Note: The number in bold denotes the minimum in each column.

4. Discussion

4.1. Why Did Dry Matter Indices Work Better Than Chlorophyll Indices?

As the chlorophyll in green leaves is a major absorber of solar radiation within crop canopies, chlorophyll indices are widely used for estimating crop growth parameters such as leaf area index [20] and leaf nitrogen content [9,42] based on their correlations with the leaf chlorophyll content. The $CI_{Red-edge}$ is one such index and relies primarily on the sensitivity of red edge bands to chlorophyll absorption [36]. Our results demonstrated that the chlorophyll indices performed well in the estimation of leaf biomass but not better than the dry matter index ND_{LMA} in the estimation of the total biomass, due to their difficulties in accounting for the variation in stem biomass.

Since the biomass in this study was the mass of dry matter per unit ground area, the relationships of crop biomass with chlorophyll indices were indirect but those with dry matter indices were direct. Although the $CI_{Red-edge}$ exhibited significant relationships with leaf biomass, the strength decreased when the stem biomass was included. The explicit examination of the relationships of stem biomass and total biomass with the $CI_{Red-edge}$ confirmed the breakdown of this indirect connection.

The stable performance of the ND_{LMA} for estimating the biomass of individual and multiple organs in rice canopies suggested that the use of a sensitive dry matter index was a successful choice. On one hand, the ND_{LMA} was originally designed by Féret et al. [33] as an indicator of leaf dry matter content and involved a combination of one NIR band (1320 nm) and one SWIR band (1722 nm). Swain et al. [43] also found that this SWIR band was sensitive to dry matter content as used in their index, NDMI. These dry matter indices expectedly performed better than the chlorophyll indices for the estimation of total biomass. A recent study by Jin et al. [18] showed the better performance of the NDMI for biomass estimation than a few chlorophyll indices, but did not consider the ND_{LMA}. Gnyp et al. [27] determined $(R_{1301} - R_{1706})/(R_{1301} + R_{1706})$, of which the two bands were approximately 20 nm offset from their counterparts in the ND_{LMA}, as their best wavelength combination for the estimation of the total biomass in rice. An analysis of our data demonstrated that their index performed similarly ($R^2 = 0.81$) as did the ND_{LMA}, but exhibited a different model. Although Gnyp et al. [27] did not explicitly link the optimized index to dry matter detection, the successful performance of this index reinforced the suitability of dry matter indices for biomass estimation.

On the other hand, it is common practice to use a spectral index with higher sensitivity to a constituent for detecting low concentrations, but with lower sensitivity to this constituent for detecting high concentrations, as a strategy to avoid optical saturation [36,44]. From the pre-heading phase to the post-heading phase, the total biomass increased to a much higher level (more than doubled) but the $CI_{Red-edge}$ failed to respond to this physiological process (Figure 4A). Compared to the red edge band (700 nm) in the $CI_{Red-edge}$ and the TCARI/OSAVI, the SWIR band (1722 nm) used in the ND_{LMA} and the NDMI exhibited higher reflectance and could be more efficient for detecting dry matter signals from stems that are located deeper in the canopy than the leaves at the top. Although stems could barely be visible from the top of the canopy, dry matter signals could come from the multiple scattering of photons between leaves and stems.

4.2. Partitioning of Aboveground Biomass between Canopy Components

The total biomass of the aboveground components of the canopy is a critical parameter for quantifying nitrogen deficiencies and the harvest index in crops [45–47]. Starting from the booting stage, rice plants transitioned to the reproductive growth phase, which is dominated by grain development with the translocation of dry matter from leaves and stems to panicles. Stem biomass contributed the most to the total biomass for all stages except at the start and the end of the growing season.

To this end, most studies for precision agriculture purposes focus on the remote estimation of total biomass but the estimation of individual components comprising the total biomass is poorly understood. To the best of our knowledge, this study provided the first attempt for the remote estimation for individual components towards a better understanding of the remote estimation of the total biomass. The relationships between the leaf biomass and vegetation indices could be explained by the absorption by dry matter in the leaves or by chlorophyll which is closely related to the leaf biomass of green crops. However, the relationships between stem biomass and vegetation indices could probably be explained by dry matter absorption and the allometric relationships between stem biomass and leaf biomass, given the low exposure of standing stems to the sensor. This leaf vs. stem biomass relationship was strong for the rice plants, but varied with the growth stage (Figure 5).

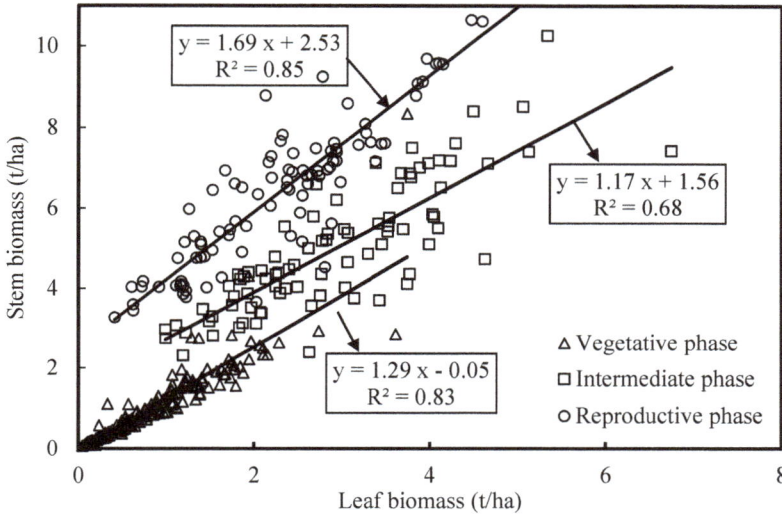

Figure 5. Relationships between leaf biomass and stem biomass of rice for vegetative (i.e., tillering, jointing), intermediate (i.e., booting, heading), and reproductive (i.e., filling) phases.

Our analysis indicated the leaf vs. stem relationships existed separately for at least three periods (i.e., vegetative, intermediate, and reproductive stages) comprising the whole season, with more significant differences in the offset than in the slope between these linear models. The partitioning of aboveground biomass among the leaf, stem, and panicle components of rice is dependent on the growth stage [48], therefore it was unrealistic to apply a single relationship for the whole growing season. This stage-specific relationship could probably explain the worse performance of the $CI_{Red-edge}$ in the estimation of stem biomass than that of leaf biomass. While a single linear function could explain the relationship between the leaf biomass and the $CI_{Red-edge}$ across all stages, even a nonlinear function could not well explain the relationship between the $CI_{Red-edge}$ and stem biomass (Figure 3A). As the ND_{LMA} was used to directly detect the dry matter signals from all aboveground components of the rice plants, the partitioning pattern of dry matter among canopy components did not significantly affect the performance of the ND_{LMA} in the estimation of stem biomass and total biomass. The non-significant relationship between panicle biomass and spectral indices suggested the limited contribution of the rice panicles to canopy spectral reflectance. Since the panicles were located on the upper layer of the crops, it was difficult for them to trap photons and therefore be detected by the sensor from above the canopy.

4.3. Potential for Satellite Observations

Most previous studies on the estimation of crop biomass used vegetation indices constructed with spectral bands in the visible and NIR regions, due to the limitation of the wavelength configuration of satellite instruments [25,28]. With the red edge indices derived from RapidEye image data,

Kross et al. [25] also found linear fitting for the leaf biomass and nonlinear fitting for the total biomass in soybean and corn crops. Their finding from satellite observations was consistent with our results regarding the $CI_{Red-edge}$ derived from ground-based canopy reflectance spectra.

Our results suggested that the red edge indices were better suited for the estimation of leaf biomass than regular NDVI-like indices (without red edge bands) because of the high sensitivity of red edge indices to leaf biomass across all growth stages. For satellite mapping of crop leaf biomass, the red edge indices can be computed from multispectral images acquired from the instruments such as WorldView-2 [28], RapidEye [25], Sentinel-2 [49], and the Medium Resolution Imaging Spectrometer, MERIS [50]. These satellite-based red edge indices also have the potential for accurate estimation of total biomass for pre-heading stages. If the satellite mapping of the total biomass for the whole rice season is required, the bands for the ND_{LMA} are preferred in order to avoid unsatisfactory estimates for the post-heading stages. Once the upcoming hyperspectral satellite missions such as the Environmental Mapping and Analysis Program, EnMap [51] and the Hyperspectral Infrared Imager, HyspIRI [52] are launched into orbit, the ND_{LMA} and other optimal dry matter indices will be available for monitoring the aboveground biomass of rice and even other crops for the whole growing season.

5. Conclusions

This study reports on an investigation of the relationships between vegetation indices and the biomass of individual components and component combinations in rice canopies. Eight indices commonly used for the estimation of chlorophyll and dry matter contents were evaluated with field data collected from a two-year experiment. The $CI_{Red-edge}$ of the chlorophyll index group exhibited the best relationship (linear) of all with the leaf biomass for the whole rice season, and with total biomass, but only for the growth stages before heading due to the poor sensitivity to the large amount of total biomass after heading. The ND_{LMA} of the dry matter index group showed the best relationships (nonlinear) with both stem biomass and total biomass for the whole season. Therefore, the use of canopy sensors that record NDVI or red-edge index data will either be limited to the monitoring of leaf biomass for the whole season or to that of total biomass for the stages before heading. The findings may serve as a guide to choose sensors with appropriate spectral coverages for monitoring the leaf biomass and total biomass for the growing season of rice.

With the detailed analysis of biomass estimation by the components in rice, this research provided physical explanations for the superior performance of the dry matter indices over the chlorophyll indices for the estimation of whole-season total biomass. The dry matter indices, particularly the ND_{LMA}, can serve as useful spectral indicators of biomass for understanding the dry matter or carbon partitioning among aboveground components and the formation of grain yield of rice and other crops. They have great potential for the mapping of crop aboveground biomass for the whole growing season when spectroscopic data from upcoming hyperspectral satellite missions become available.

Acknowledgments: This research was funded by the National Key R&D Program (2016YFD0300601), the Fundamental Research Funds for the Central Universities (KYRC201401), the National Natural Science Foundation of China (31470084), Jiangsu Distinguished Professor Program, Special Program for Agriculture Science and Technology from Ministry of Agriculture in China (201303109), and the Academic Program Development of Jiangsu Higher Education Institutions (PAPD). Field assistance from graduate students Xiang Zhou and Xinqiang Deng is appreciated.

Author Contributions: T.C., X.Y., Y.T., W.C., and Y.Z. conceived and designed the experiments; D.L., K.Z., and H.Z. performed the experiments; T.C., R.S., and D.L. analyzed the data; D.L. provided analysis tools, T.C., R.S., and K.Z. wrote the paper.

Conflicts of Interest: The authors declare no conflict of interest.

References

1.   Cantrell, R.P.; Reeves, T.G. The cereal of the world's poor takes center stage. *Science* **2002**, *296*, 53–53. [CrossRef] [PubMed]

2.   Zhao, G.; Miao, Y.; Wang, H.; Su, M.; Fan, M.; Zhang, F.; Jiang, R.; Zhang, Z.; Liu, C.; Liu, P. A preliminary precision rice management system for increasing both grain yield and nitrogen use efficiency. *Field Crop. Res.* **2013**, *154*, 23–30. [CrossRef]

3.   Harrell, D.; Tubana, B.; Walker, T.; Phillips, S. Estimating rice grain yield potential using normalized difference vegetation index. *Agron. J.* **2011**, *103*, 1717–1723. [CrossRef]

4.   Peng, Y.; Gitelson, A.A. Application of chlorophyll-related vegetation indices for remote estimation of maize productivity. *Agric. For. Meteorol.* **2011**, *151*, 1267–1276. [CrossRef]

5.   Chen, P.; Haboudane, D.; Tremblay, N.; Wang, J.; Vigneault, P.; Li, B. New spectral indicator assessing the efficiency of crop nitrogen treatment in corn and wheat. *Remote Sens. Environ.* **2010**, *114*, 1987–1997. [CrossRef]

6.   Thenkabail, P.S.; Smith, R.B.; De Pauw, E. Hyperspectral vegetation indices and their relationships with agricultural crop characteristics. *Remote Sens. Environ.* **2000**, *71*, 158–182. [CrossRef]

7.   Casanova, D.; Epema, G.F.; Goudriaan, J. Monitoring rice reflectance at field level for estimating biomass and lai. *Field Crop Res.* **1998**, *55*, 83–92. [CrossRef]

8.   Serrano, L.; Filella, I.; Penuelas, J. Remote sensing of biomass and yield of winter wheat under different nitrogen supplies. *Crop Sci.* **2000**, *40*, 723–731. [CrossRef]

9.   Hansen, P.M.; Schjoerring, J.K. Reflectance measurement of canopy biomass and nitrogen status in wheat crops using normalized difference vegetation indices and partial least squares regression. *Remote Sens. Environ.* **2003**, *86*, 542–553. [CrossRef]

10.   Fu, Y.Y.; Yang, G.J.; Wang, J.H.; Song, X.Y.; Feng, H.K. Winter wheat biomass estimation based on spectral indices, band depth analysis and partial least squares regression using hyperspectral measurements. *Comput. Electron. Agric.* **2014**, *100*, 51–59. [CrossRef]

11.   Dong, T.F.; Liu, J.G.; Qian, B.D.; Jing, Q.; Croft, H.; Chen, J.M.; Wang, J.F.; Huffman, T.; Shang, J.L.; Chen, P.F. Deriving maximum light use efficiency from crop growth model and satellite data to improve crop biomass estimation. *IEEE J. Sel. Top. Appl. Earth Obs. Remote Sens.* **2017**, *10*, 104–117. [CrossRef]

12.   He, B.B.; Li, X.; Quan, X.W.; Qiu, S. Estimating the aboveground dry biomass of grass by assimilation of retrieved lai into a crop growth model. *IEEE J. Sel. Top. Appl. Earth Obs. Remote Sens.* **2015**, *8*, 550–561. [CrossRef]

13.   Jego, G.; Pattey, E.; Liu, J.G. Using leaf area index, retrieved from optical imagery, in the stics crop model for predicting yield and biomass of field crops. *Field Crop. Res.* **2012**, *131*, 63–74. [CrossRef]

14.   Liu, J.G.; Pattey, E.; Miller, J.R.; McNairn, H.; Smith, A.; Hu, B.X. Estimating crop stresses, aboveground dry biomass and yield of corn using multi-temporal optical data combined with a radiation use efficiency model. *Remote Sens. Environ.* **2010**, *114*, 1167–1177. [CrossRef]

15.   Jin, X.L.; Yang, G.J.; Xu, X.G.; Yang, H.; Feng, H.K.; Li, Z.H.; Shen, J.X.; Zhao, C.J.; Lan, Y.B. Combined multi-temporal optical and radar parameters for estimating lai and biomass in winter wheat using HJ and RadarSAT-2 data. *Remote Sens.* **2015**, *7*, 13251–13272. [CrossRef]

16.   Bendig, J.; Bolten, A.; Bennertz, S.; Broscheit, J.; Eichfuss, S.; Bareth, G. Estimating biomass of barley using Crop Surface Models (CSMs) derived from UAV-based RGB imaging. *Remote Sens.* **2014**, *6*, 10395–10412. [CrossRef]

17.   Eitel, J.U.H.; Magney, T.S.; Vierling, L.A.; Brown, T.T.; Huggins, D.R. Lidar based biomass and crop nitrogen estimates for rapid, non-destructive assessment of wheat nitrogen status. *Field Crop. Res.* **2014**, *159*, 21–32. [CrossRef]

18.   Jin, X.; Kumar, L.; Li, Z.; Xu, X.; Yang, G.; Wang, J. Estimation of winter wheat biomass and yield by combining the AquaCrop model and field hyperspectral data. *Remote Sens.* **2016**, *8*, 972. [CrossRef]

19.   Wiseman, G.; McNairn, H.; Homayouni, S.; Shang, J.L. Radarsat-2 polarimetric SAR response to crop biomass for agricultural production monitoring. *IEEE J. Sel. Top. Appl. Earth Obs. Remote Sens.* **2014**, *7*, 4461–4471. [CrossRef]

20.   Li, F.; Etc, Y.M.; Li, F. Estimating winter wheat biomass and nitrogen status using an active crop sensor. *Intell. Autom. Soft Comput.* **2010**, *16*, 1221–1230.

21.   Wang, L.A.; Zhou, X.; Zhu, X.; Dong, Z.; Guo, W. Estimation of biomass in wheat using random forest regression algorithm and remote sensing data. *Crop J.* **2016**, *4*, 212–219. [CrossRef]

22. Prabhakara, K.; Hively, W.D.; McCarty, G.W. Evaluating the relationship between biomass, percent groundcover and remote sensing indices across six winter cover crop fields in Maryland, United States. *Int. J. Appl. Earth Obs. Geoinf.* **2015**, *39*, 88–102. [CrossRef]

23. Cao, Q.; Miao, Y.; Wang, H.; Huang, S.; Cheng, S.; Khosla, R.; Jiang, R. Non-destructive estimation of rice plant nitrogen status with crop circle multispectral active canopy sensor. *Field Crop. Res.* **2014**, *154*, 133–144. [CrossRef]

24. Kanke, Y.; Tubaña, B.; Dalen, M.; Harrell, D. Evaluation of red and red-edge reflectance-based vegetation indices for rice biomass and grain yield prediction models in paddy fields. *Precis. Agric.* **2016**, *17*, 507–530. [CrossRef]

25. Kross, A.; Mcnairn, H.; Lapen, D.; Sunohara, M.; Champagne, C. Assessment of rapideye vegetation indices for estimation of leaf area index and biomass in corn and soybean crops. *Int. J. Appl. Earth Obs. Geoinf.* **2015**, *34*, 235–248. [CrossRef]

26. Gnyp, M.L.; Bareth, G.; Li, F.; Lenz-Wiedemann, V.I.S.; Koppe, W.; Miao, Y.; Hennig, S.D.; Jia, L.; Laudien, R.; Chen, X.; et al. Development and implementation of a multiscale biomass model using hyperspectral vegetation indices for winter wheat in the north China plain. *Int. J. Appl. Earth Obs. Geoinf.* **2014**, *33*, 232–242. [CrossRef]

27. Gnyp, M.L.; Miao, Y.X.; Yuan, F.; Ustin, S.L.; Yu, K.; Yao, Y.K.; Huang, S.Y.; Bareth, G. Hyperspectral canopy sensing of paddy rice aboveground biomass at different growth stages. *Field Crop. Res.* **2014**, *155*, 42–55. [CrossRef]

28. Marshall, M.; Thenkabail, P. Advantage of hyperspectral EO-1 Hyperion over multispectral IKONOS, GeoEye-1, WorldView-2, Landsat ETM+, and MODIS vegetation indices in crop biomass estimation. *ISPRS J. Photogramm. Remote Sens.* **2015**, *108*, 205–218. [CrossRef]

29. Babar, M.A.; Reynolds, M.P.; van Ginkel, M.; Klatt, A.R.; Raun, W.R.; Stone, M.L. Spectral reflectance to estimate genetic variation for in-season biomass, leaf chlorophyll, and canopy temperature in wheat. *Crop Sci.* **2006**, *46*, 1046. [CrossRef]

30. Stroppiana, D.; Boschetti, M.; Brivio, P.A.; Bocchi, S. Plant nitrogen concentration in paddy rice from field canopy hyperspectral radiometry. *Field Crop. Res.* **2009**, *111*, 119–129. [CrossRef]

31. Inoue, Y.; Guerif, M.; Baret, F.; Skidmore, A.; Gitelson, A.; Schlerf, M.; Darvishzadeh, R.; Olioso, A. Simple and robust methods for remote sensing of canopy chlorophyll content: A comparative analysis of hyperspectral data for different types of vegetation. *Plant Cell Environ.* **2016**, *39*, 2609–2623. [CrossRef] [PubMed]

32. Le Maire, G.; François, C.; Dufrêne, E. Towards universal broad leaf chlorophyll indices using prospect simulated database and hyperspectral reflectance measurements. *Remote Sens. Environ.* **2004**, *89*, 1–28. [CrossRef]

33. Féret, J.B.; François, C.; Gitelson, A.; Asner, G.P.; Barry, K.M.; Panigada, C.; Richardson, A.D.; Jacquemoud, S. Optimizing spectral indices and chemometric analysis of leaf chemical properties using radiative transfer modeling. *Remote Sens. Environ.* **2011**, *115*, 2742–2750. [CrossRef]

34. Wang, L.; Qu, J.J.; Hao, X.; Hunt, E.R., Jr. Estimating dry matter content from spectral reflectance for green leaves of different species. *Int. J. Remote Sens.* **2011**, *32*, 7097–7109. [CrossRef]

35. Kokaly, R.F.; Asner, G.P.; Ollinger, S.V.; Martin, M.E.; Wessman, C.A. Characterizing canopy biochemistry from imaging spectroscopy and its application to ecosystem studies. *Remote Sens. Environ.* **2009**, *113*, S78–S91. [CrossRef]

36. Gitelson, A.A.; Gritz, Y.; Merzlyak, M.N. Relationships between leaf chlorophyll content and spectral reflectance and algorithms for non-destructive chlorophyll assessment in higher plant leaves. *J. Plant Physiol.* **2003**, *160*, 271–282. [CrossRef] [PubMed]

37. Haboudane, D.; Miller, J.R.; Tremblay, N.; Zarco-Tejada, P.J.; Dextraze, L. Integrated narrow-band vegetation indices for prediction of crop chlorophyll content for application to precision agriculture. *Remote Sens. Environ.* **2002**, *81*, 416–426. [CrossRef]

38. Rouse, J.W., Jr.; Haas, R.H.; Schell, J.A.; Deering, D.W. Monitoring vegetation systems in the great plains with ERTS. In *Third Earth Resources Technology Satellite-1 Symposium-Volume I: Technical Presentations*; NASA SP-351; NASA: Washington, DC, USA, 1974; pp. 309–317.

39. Huete, A.R.; Liu, H.Q.; Batchily, K.; Leeuwen, W.V. A comparison of vegetation indices over a global set of tm images for eos-modis. *Remote Sens. Environ.* **1997**, *59*, 440–451. [CrossRef]

40. Serrano, L.; Peñuelas, J.; Ustin, S.L. Remote sensing of nitrogen and lignin in Mediterranean vegetation from AVIRIS data : Decomposing biochemical from structural signals. *Remote Sens. Environ.* **2002**, *81*, 355–364. [CrossRef]

41. le Maire, G.; Francois, C.; Soudani, K.; Berveiller, D.; Pontailler, J.; Breda, N.; Genet, H.; Davi, H.; Dufrene, E. Calibration and validation of hyperspectral indices for the estimation of broadleaved forest leaf chlorophyll content, leaf mass per area, leaf area index and leaf canopy biomass. *Remote Sens. Environ.* **2008**, *112*, 3846–3864. [CrossRef]

42. Cho, M.A.; Skidmore, A.K. A new technique for extracting the red edge position from hyperspectral data: The linear extrapolation method. *Remote Sens. Environ.* **2006**, *101*, 181–193. [CrossRef]

43. Swain, K.C.; Thomson, S.J.; Jayasuriya, H.P.W. Adoption of an unmanned helicopter for low-altitude remote sensing to estimate yield and total biomass of a rice crop. *Trans. ASAE* **2010**, *53*, 21–27. [CrossRef]

44. Blackburn, G.A. Hyperspectral remote sensing of plant pigments. *J. Exp. Bot.* **2007**, *58*, 855–867. [CrossRef] [PubMed]

45. Ata-Ul-Karim, S.T.; Yao, X.; Liu, X.J.; Cao, W.X.; Zhu, Y. Development of critical nitrogen dilution curve of japonica rice in yangtze river reaches. *Field Crop. Res.* **2013**, *149*, 149–158. [CrossRef]

46. Yao, X.; Ata-Ul-Karim, S.T.; Zhu, Y.; Tian, Y.C.; Liu, X.J.; Cao, W.X. Development of critical nitrogen dilution curve in rice based on leaf dry matter. *Eur. J. Agron.* **2014**, *55*, 20–28. [CrossRef]

47. Lemaire, G.; Jeuffroy, M.-H.; Gastal, F. Diagnosis tool for plant and crop n status in vegetative stage theory and practices for crop N management. *Eur. J. Agron.* **2008**, *28*, 614–624. [CrossRef]

48. Yang, J.C.; Peng, S.B.; Zhang, Z.J.; Wang, Z.Q.; Visperas, R.M.; Zhu, Q.S. Grain and dry matter yields and partitioning of assimilates in japonica/indica hybrid rice. *Crop Sci.* **2002**, *42*, 766–772. [CrossRef]

49. Drusch, M.; Del Bello, U.; Carlier, S.; Colin, O.; Fernandez, V.; Gascon, F.; Hoersch, B.; Isola, C.; Laberinti, P.; Martimort, P. Sentinel-2: ESA's optical high-resolution mission for GMES operational services. *Remote Sens. Environ.* **2012**, *120*, 25–36. [CrossRef]

50. Dash, J.; Curran, P.J. Evaluation of the MERIS terrestrial chlorophyll index (MTCI). *Adv. Space Res.* **2007**, *39*, 100–104. [CrossRef]

51. Guanter, L.; Kaufmann, H.; Segl, K.; Foerster, S.; Rogass, C.; Chabrillat, S.; Kuester, T.; Hollstein, A.; Rossner, G.; Chlebek, C.; et al. The EnMAP spaceborne imaging spectroscopy mission for earth observation. *Remote Sens.* **2015**, *7*, 8830–8857. [CrossRef]

52. Lee, C.M.; Cable, M.L.; Hook, S.J.; Green, R.O.; Ustin, S.L.; Mandl, D.J.; Middleton, E.M. An introduction to the NASA Hyperspectral InfraRed Imager (HyspIRI) mission and preparatory activities. *Remote Sens. Environ.* **2015**, *167*, 6–19. [CrossRef]

remote sensing

MDPI

Article

Comparison and Evaluation of Three Methods for Estimating Forest above Ground Biomass Using TM and GLAS Data

Kaili Liu [1,2], Jindi Wang [1,2,*], Weisheng Zeng [3] and Jinling Song [1,2]

[1]   State Key Laboratory of Remote Sensing Science, Institute of Remote Sensing Science and Engineering, Faculty of Geographical Science, Beijing Normal University, Beijing 100875, China; liukl@mail.bnu.edu.cn (K.L.); songjl@bnu.edu.cn (J.S.)

[2]   Beijing Key Laboratory for Remote Sensing of Environment and Digital Cities, Beijing Normal University, Beijing 100875, China

[3]   Academy of Forest Inventory and Planning, State Forestry Administration, Beijing 100714, China; zengweisheng@forestry.gov.cn

*   Correspondence: wangjd@bnu.edu.cn; Tel./Fax: +86-10-5880-9966

Academic Editors: Lalit Kumar, Onisimo Mutanga, Nicolas Baghdadi and Prasad S. Thenkabail
Received: 21 December 2016; Accepted: 31 March 2017; Published: 2 April 2017

Abstract: Medium spatial resolution biomass is a crucial link from the plot to regional and global scales. Although remote-sensing data-based methods have become a primary approach in estimating forest above ground biomass (AGB), many difficulties remain in data resources and prediction approaches. Each kind of sensor type and prediction method has its own merits and limitations. To select the proper estimation algorithm and remote-sensing data source, several forest AGB models were developed using different remote-sensing data sources (Geoscience Laser Altimeter System (GLAS) data and Thematic Mapper (TM) data) and 108 field measurements. Three modeling methods (stepwise regression (SR), support vector regression (SVR) and random forest (RF)) were used to estimate forest AGB over the Daxing'anling Mountains in northeastern China. The results of models using different datasets and three approaches were compared. The random forest AGB model using Landsat5/TM as input data was shown the acceptable modeling accuracy ($R^2 = 0.95$ RMSE = 17.73 Mg/ha) and it was also shown to estimate AGB reliably by cross validation ($R^2 = 0.71$ RMSE = 39.60 Mg/ha). The results also indicated that adding GLAS data significantly improved AGB predictions for the SVR and SR AGB models. In the case of the RF AGB models, including GLAS data no longer led to significant improvement. Finally, a forest biomass map with spatial resolution of 30 m over the Daxing'anling Mountains was generated using the obtained optimal model.

Keywords: forest above ground biomass (AGB); random forest; mapping

1. Introduction

Forest ecosystems, which are the largest carbon sinks on land, account for about 80% of terrestrial biosphere carbon storage and 40% of underground carbon storage [1] and play a pivotal role in mitigating climate change [2,3]. Biomass, as one of the important parameters of forest environments, is an effective factor for characterizing actual carbon sequestration in the forest ecosystem. Therefore, estimating forest biomass accurately is the basis for terrestrial carbon cycle analysis, and the spatial distribution of forest biomass at regional scale can also reveal spatial variations in carbon sequestration, which can provide a basis for rational carbon reduction targets and forest management programs. Generally, biomass consists of above ground biomass (AGB) and below ground biomass (BGB) [3,4]. Due to the difficulty of collecting and calculating BGB, researchers have focused mainly on AGB, as did this paper.

Remote-sensing technology, which has wide coverage and repeated observation capabilities, has promoted research on the spatial distribution and temporal variation of forest biomass. Biomass models based on remote-sensing data have been shown to be more accurate than other models [5]. The characteristics of the forest can be estimated using the airborne or space-borne multi-spectral remote sensing method [6]. Airborne remote-sensing data, such as aerial photographs, are most useful when fine spatial detail is critical, which are often used for modeling forest canopy structures or tree parameters [4,6]. Compared to airborne remote sensing, satellite imagery can not only capture large areas in a single image but also update information regularly to monitor changes [6].

Three types of remote-sensing data are currently available for biomass estimation: optical sensor data, radar data, and LiDAR data [3,4,7]. Each of these has its own advantages and disadvantages for estimating biomass. Optical remote sensing can be used for continuous estimation of forest biomass due to its long observation time, wide spatial coverage, and multiple bands, which can provide abundant information about the canopy spectrum. Optical remote sensing is limited by its relatively poor penetration. Estimating forest AGB using optical sensor data is based on the close relationship between foliage biomass and forest ecosystem biomass. However, foliage biomass accounts for less than 10% of the total biomass of a mature forest ecosystem [8]. The signal saturation of optical sensor data in dense vegetation is an important factor restricting biomass inversion. The results obtained by Lu et al. [7] confirmed that Thematic Mapper (TM) spectral reflectance changes regularly with increasing AGB in forest sites with low biomass density. As for forest sites with high biomass density, the relationship between AGB and TM spectral reflectance is not obvious. Radar data are also a promising data source for estimating AGB because of their independence of weather and their ability to penetrate the canopy and thereby receive information about trunks and branches [9,10]. Signal saturation is also a problem for radar data [11,12]. LiDAR, an active remote-sensing technology, can acquire forest vertical structure information, which is strongly related to forest biomass. LiDAR data are not affected by signal saturation [13,14]. Incomplete data coverage, short running time, and the effects of clouds and terrain make spatial LiDAR data less than ideal for biomass mapping [3,10,15]. In some studies, LiDAR data were combined with optical images to estimate forest biomass [13].

The techniques for estimating forest biomass can be divided into two categories: parametric and nonparametric algorithms [4,15]. The term "parametric algorithm" refers to common statistical regression. After the model has been developed, the expression relating the dependent variable (AGB) and the independent variables is explicit and easy to calculate [15]. The key is to select suitable variables to represent biomass. In fact, forest biomass is affected by many factors (e.g., forest age, tree species, and tree height), and its relationship with remote-sensing data is difficult to express using a simple linear or nonlinear model. Many researchers have used machine learning and data mining methods (also known as nonparametric algorithms) to estimate forest biomass and have achieved good results [3,16,17].

In the current research, the optimal kind of remote-sensing data and the optimal method for estimating forest AGB remain to be determined. In addition, some issues remain in spatial matching between remote-sensing images and field data. In some studies, the area of a field plot is less than that of a pixel in remote-sensing images [3,4]. In this research, remote-sensing data with a resolution matching the field plot area were chosen as the input data. Three approaches were then developed (stepwise regression, support vector regression, and random forest) to model the relationship between the remote-sensing variables and the measured AGB in the field plots. After comparing the modeling and estimation results using field measurements, the optimal biomass model was used to map regional forest biomass density.

2. Materials

The materials used in this paper included field AGB data measured during 2005–2007, Geospatial Laser Altimeter System (GLAS) data observed during 2003–2008 using laser 2 and laser 3, and Landsat 5 TM data observed in July 2005. The acquisition time of the above data were shown in Table 1.

Table 1. The acquisition time of the materials.

Data		Acquisition Time
Field data		2005, 2006 and 2007
GLAS data	L2A	25 September 2003–19 November 2003
	L2D	25 November 2008–17 December 2008
	L3A	3 October 2004–8 November 2004
	L3B	17 February 2005–24 March 2005
	L3C	20 May 2005–23 June 2005
	L3D	21 October 2005–24 November 2005
	L3F	24 May 2006–26 June 2006
	L3G	25 October 2006–27 November 2006
Landsat5/TM data		July 2005

Note: L2A, L2D, etc. represent the name of the GALS laser campaigns.

2.1. Field Data

Two sources of field measurements were used in this paper. The first was obtained from Sun et al., where GLAS footprints (the red dots in Figure 1) in the Tahe and Changbai Mountain areas were measured in 2006 and 2007 respectively [18,19]. Eighty-six good-quality GLAS data points were obtained in this area (see Section 2.2 for filters). Four sampling plots (the blue solid circles in Figure 2) with a radius of 7.5 m were set within the GLAS footprint after the center of each footprint was located by DGPS (Differential Global Positioning System) [18]. GLAS footprint (the black dots in Figure 1) is elliptical surface, with approximately 65 m in diameter, and the space between footprints is 172 m [3]. The second dataset was obtained from the seventh National Forest Inventory dataset [20], which was obtained in 2005. In this study, 62 forest inventory plots (purple dots in Figure 1) of 0.06 ha each were measured in the Xiaoxing'anling, Daxing'anling, and Changbai Mountains. In addition to the correspondence relationship between the coordinates of remote-sensing data (GLAS data or TM data) and that of field plots, the area near those plots was also forest and was basically homogeneous (see Section 5), which make these plots representative of remote-sensing data. The diameter at breast height (DBH) and tree species were documented for every tree with DBH greater than 5 cm in all these sampling plots.

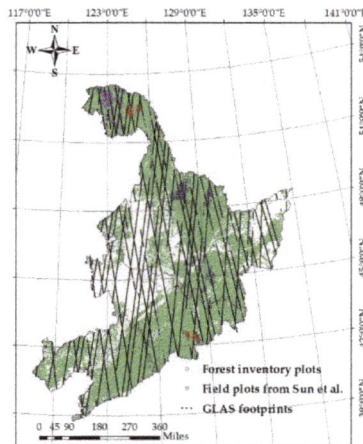

Figure 1. Locations of field plots and GLAS data. The red dots represent field measurements from Sun et al. [18,19]. The purple dots represent data from the Seventh National Forest Inventory. The black dots represent GLAS L3C footprints. The background information is a 30-m forest distribution (in green) map developed by Chen [21].

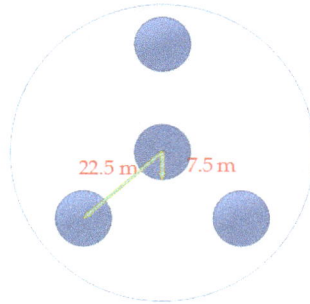

Figure 2. Schematic diagram of field sampling. The solid blue circles represent sampling plots.

In the region studied in the current research, the single-tree biomass for each tree species was estimated using species-specific allometric equations [22–27] (Figure 3) obtained from the literature. The average aboveground biomass of each plot was then obtained by aggregating all single-tree biomass values in this plot and dividing by the area of the sampling plot.

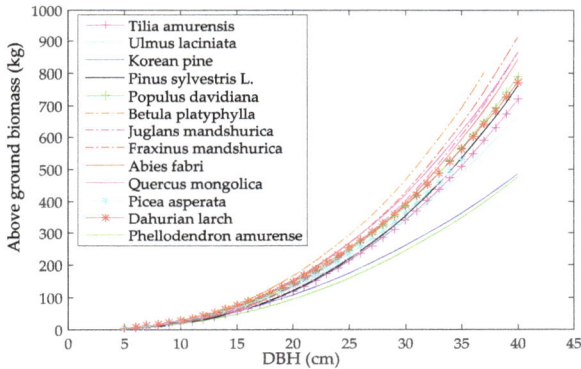

Figure 3. Model predictions from species-specific allometric equations for aboveground biomass. Different lines represent different tree species.

The study region contained 148 field AGB data points. After matching with ICESat/GLAS data, during which the observed time, location, and ICESat/GLAS data quality were considered (see Section 2.2), a total of 108 plot data points were available for modeling (86 from the Sun et al. team and 22 from the Seventh National Forest Inventory dataset). Due to the lack of valid ICESat/GLAS data, 40 data points from the seventh National Forest Inventory dataset were left. The remaining 40 plot data points were used for independent validation of the AGB model using Landsat5/TM as input data.

2.2. ICESat/GLAS Data

The National Aeronautics and Space Administration (NASA) GLAS instrument staged in the Ice, Cloud, and Elevation Satellite (ICESat) is the first space-born full-waveform LiDAR sensor. GLAS emits a pulse waveform in 1064-nm bands, illuminating an elliptical surface footprint approximately 65 m in diameter, and records the returned waveform from the footprint. GLA01 (release 33), recording the transmitted and received waveforms, and GLA14 (release 34), recording the parameters obtained from GLA01 along with the geolocation of the footprint, were used in this study. GLAS shots that were less than 65 m away from field plots collected from the Seventh National Forest Inventory dataset were also used, as well as GLAS data corresponding to field plots from Sun et al. These GLAS data points were downloaded from the National Snow and Ice Data Center (NSIDC) website [28].

To obtain high-quality waveform data, filters are required. With reference to the screening methods proposed by Chi [29], Wu [30] and Baghdadi [31], GLAS data with no cloud and a signal-to-noise ratio (SNR) greater than 60 were retained. Cloudless GLAS data were identified using the cloud detection flag (*i_FRir_qaflag* = 15) in the GLA14 product [29,31]. In addition, the SNR values (Equation (1)) can be calculated using the fields *i_max-RecAmp* and *i_sDevNsOb1* in the GLA14 product [30]. Here, *i_max-RecAmp* represents the peak amplitude of the received echo, and *i_sDevNsOb1* represents the standard deviation of the background noise.

$$\text{SNR} = \frac{i\_max - RecAmp}{i\_sDevNsOb1} \tag{1}$$

After filtering, a total of 108 waveform data points were available. Before calculating the waveform metrics associated with AGB, it was necessary to identify the three crucial locations of the waveform: the signal start location, the signal end location and the ground peak location, which relied on the processing described below.

1.   Filtering the data

The waveform is commonly filtered by a Gaussian filter, which removes high-frequency noise and smooths the data [3,18]. In recent years, some researchers have used wavelet transforms to filter GLAS data [32]. This study compared the denoising effect of the two filters and selected the more effective filtering method. The wavelet transform steps followed in this paper were as follows: first, the signal was decomposed into three layers by a Gaussian wavelet; second, the high-frequency coefficients were denoised using a threshold; and finally, the coefficients underwent an inverse transformation [33]. As for the Gaussian filter, one was created with a width similar to the transmitted pulse and used to filter the original waveform [34].

Three indicators were selected to evaluate the filtering effects: root mean square error (RMSE) [35,36], signal-to-noise ratio (SNR) [35,36], and smoothness (r) [35]. The equations of these indicators can be expressed as follows:

$$\text{RMSE} = \sqrt{\frac{\sum_{i=1}^{N}(s(i) - f(i))^2}{N}} \tag{2}$$

$$\text{SNR} = 10 \times \log \frac{\sum_{i=1}^{N} f(i)^2}{\sum_{i=1}^{N}(s(i) - f(i))^2} \tag{3}$$

$$r = \frac{\sum_{i=1}^{N-1}(f(i+1) - f(i))^2}{\sum_{i=1}^{N-1}(s(i+1) - s(i))^2} \tag{4}$$

where *s* is the original signal, *f* is the filtered signal, and *N* is the length of the signal.

2.   Locating the signal start and end points

Noise was estimated from the signal intensity histogram before the signal start point and after the signal end point. When three consecutive bins were higher than the threshold (the sum of the noise mean and three standard deviations), the signal start and end were located [34].

3.   Gaussian Decomposition

A Gaussian decomposition was applied to the filtered waveform using Levenberg-Marquardt nonlinear least-squares fitting [29,34,37]. To compare this method with the GLA14 product, the

Pearson's correlation coefficient (r) was calculated between the original waveform and the fitted waveforms obtained by the method described above and by the GLA14 product separately.

4.   Identifying the ground peak

By reverse search from the signal endpoint, when the distance between the location of the Gaussian peak and the end of the signal was greater than half the emission pulse width, the location of the Gaussian peak was taken as the ground peak [34].

After processing the GLAS data using this method, waveform metrics sensitive to AGB were extracted according to methods in the available literature. These metrics were divided into two types, height metrics and intensity metrics, and their description and references can be found in Table 2. In addition to these metrics, the results of Gaussian decomposition, including location, amplitude, and width, were also used.

Table 2. Descriptions and references of GLAS metrics derived from GLAS data.

Types	GLAS Metric Abbreviations	Descriptions
The height metrics	Extent	The distance from signal beginning to signal ending [34].
	Treeht	The distance from signal beginning to ground peak [34,38].
	Treeht2 Treeht3	Top tree heights with corrections [34].
	H25 H75	Quartile heights calculated by subtracting the ground elevation from elevation at which 25% or 75% of the returned energy occurs [34,39].
	H10 H20 H100	Decimal heights calculated by subtracting the ground elevation from elevation at which 10% (20%...100%) of the returned energy occurs [34].
	LEE	The distance from the elevation of signal beginning to the first elevation at which the signal strength of the waveform is half of the maximum signal [38].
	TEE	The distance from the last elevation at which the signal strength of the waveform is half of the maximum signal to the elevation of signal ending [38].
	HOME	The height of median energy (HOME) [9].
	Meanh Medh	Mean canopy height, median canopy height [40].
	QMCH	Quadratic mean canopy height (QMCH) calculated from the canopy height profiles [40].
The intensity metrics	Canopy cover	The ratio of the canopy echo area to the total wave area [41].
	AVAW	The area under the waveform from vegetation [41].

After testing the sensitivity to AGB of a number of variables from these GLAS parameters, eight variables (Treeht2, H25, LEE, TEE, HOME, QMCH, AVAW, and gasamp1 (the intensity of the first waveform from Gaussian decomposition)) were retained as predictor variables for GLAS data.

2.3. Landsat5/TM Data

The multispectral data used in this study were TM images with a resolution of 30 m, which matches the area of the field plots. One hundred eight (108) plots were distributed within the range of nine TM scenes. Cloud-free, good-quality images for each scene were downloaded from the United States Geological Survey (USGS) Earth Explorer as close as possible to the peak growing season. The collection duration (day of year (DOY)) of these images was limited to values from 180 to 210. To reduce the influence of spatial mismatch between the plots and the TM images, the mean reflectance was extracted from a 3 × 3 TM pixel window. The validity of the acquired TM images must also be checked by plotting the time-series curves of the spectral reflectances and vegetation indices before extracting variables, and data points that are obviously offset from the curve must be deleted.

The spectral variables extracted in this paper were divided into three categories:

1. Surface reflectance: bands 1, 2, 3, 4, 5 and 7;
2. Spectral indices: normalized difference vegetation index (NDVI) [42], Enhanced Vegetation index (EVI) [43] perpendicular vegetation index (PVI) [44], soil-adjusted vegetation index (SAVI) [45], normalized difference infrared vegetation index (NDIIB6) [46], normalized difference infrared vegetation index (NDIIB7) [46], $\frac{TM4+TM5-TM2}{TM4+TM5+TM2}$ [47], $\frac{TM4}{TM2}$ [47], and $\frac{TM3}{TM7}$ [47];
3. Tasseled Cap indices and their derivatives: Tasseled Cap Brightness (TCB) [48], Tasseled Cap Greenness (TCG) [48], Tasseled Cap Wetness (TCW) [48], Tasseled Cap distance (TCdistance) [49], and Tasseled Cap angle (TCangle) [49].

The formulae used can be found in Table 3.

Table 3. Spectral variables derived from Landsat5/TM data.

Spectral Variables	Formula
NDVI	$(TM4 - TM3)/(TM4 + TM3)$
EVI	$2.5 \times (TM4 - TM3)/(TM4 + 6 \times TM3 - 7.5 \times TM1 + 1)$
PVI	$\sqrt{(0.355 \times TM4 - 0.149 \times TM3)^2 + (0.355 \times TM3 - 0.852 \times TM4)^2}$
SAVI	$(1 + 0.5) \times (TM4 - TM3)/(TM4 + TM3 + 0.5)$
NDIIB6	$(TM4 - TM5)/(TM4 + TM5)$
NDIIB7	$(TM4 - TM7)/(TM4 + TM7)$
TCB	$B \times [TM1, TM2, TM3, TM4, TM5, TM7, 1]^T$
TCG	$G \times [TM1, TM2, TM3, TM4, TM5, TM7, 1]^T$
TCW	$W \times [TM1, TM2, TM3, TM4, TM5, TM7, 1]^T$
TCdistance	$\sqrt{TCB^2 + TCG^2}$
TCangle	$\arctan(TCG/TCB)$

Note: $B = [0.2909, 0.2493, 0.4806, 0.5568, 0.4438, 0.1706, 10.3695]$ $G = [-0.2728, -0.2174, -0.5508, 0.7221, 0.0733, -0.1648, -0.7310]$ $W = [0.1446, 0.1761, 0.3322, 0.3396, -0.6210, -0.4186, -3.3828]$.

For TM data, surface reflectance (band1 and band4), NDVI, $\frac{TM3}{TM7}$, and TCW were retained as predictor variables, after selecting variables sensitive to AGB from the above TM parameters.

3. Methods

The methodology used to estimate forest AGB in this paper is shown in Figure 4.

First, field AGB was calculated based on models of the relationship between the measured data (tree species and DBH) and aboveground tree biomass [22–27] (Figure 3), as described in Section 2.1. The remote-sensing data parameters (GLAS metrics and TM variables) corresponding to field plots were then extracted using the methods described in Sections 2.2 and 2.3

In order to simplify the model and eliminate variables that are not sensitive to AGB and that are collinear with each other. We selected predictor variables for AGB modeling from the GLAS metrics and TM variables using stepwise regression analysis (see Section 3.1).

The dataset included 148 samples; each sample consists of these selected predictor variables and corresponding AGB field data. These samples were divided into two parts, one (108 samples) for modeling, and the other (40 samples) for validation. (1) For 108 samples, the modeling process was as follows: Bootstrapping was used to expand the modeling sample size, creating 300 bootstrap samples from the observations of size 108 (see Section 3.2). AGB models were developed for each bootstrap sample using three methods (stepwise regression (SR), support vector regression (SVR), and random forest (RF)) and three data sources (TM predictor variables, GLAS predictor variables, and TM predictor variables + GLAS predictor variables) (see Section 3.3). After comparing the results from modeling accuracy and cross validation, the optimal AGB model was determined. (2) For the remaining 40 samples, they were used for independent validation of the estimated AGB using the optimal model (see Section 3.4).

Finally, the forest AGB over the Daxing'anling Mountains was mapped using the optimal AGB model.

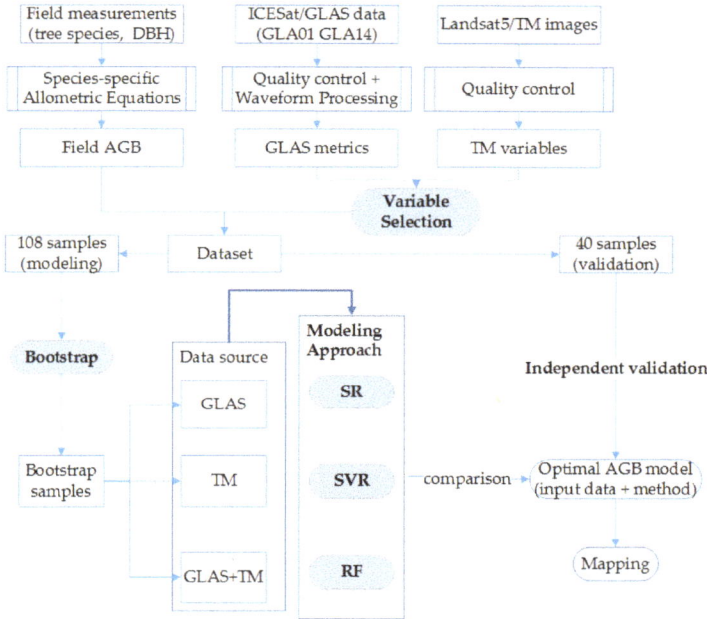

Figure 4. Forest AGB estimation methodology used in this paper.

3.1. Variable Selection

In this study, many potential variables were extracted based on previous studies. Specifically, 42 GLAS variables (the parameters in Table 1 and the results of Gaussian decomposition) and 20 TM variables were available. Therefore, the first step was to determine the predictors to simplify the model and eliminate variables that are not related to AGB and that are collinear with each other. Stepwise regression analysis was used to pare down the potential variables. To test for collinearity between the selected variables, a variance inflation factor (VIF) threshold of 10 was used, with reference to the methods used by Powell [50]. VIF is an indicator of multicollinearity and is calculated as follows:

$$VIF_i = \frac{1}{1 - R_i^2} \tag{5}$$

where VIF_i is the VIF of the i-th variable and R_i^2 is the coefficient of determination of the regression equation between the i-th variable and the remaining variables. To calculate R_i^2, first we run an ordinary least square regression that has Xi (i-th explanatory variable) as a function of all the other explanatory variables. The regression equation would be as follows:

$$X_i = a_1 X_1 + a_2 X_2 + \ldots + a_{i-1} X_{i-1} + a_{i+1} X_{i+1} a_1 X_1 + \ldots a_k X_k + c + e \tag{6}$$

where k is the total number of independent variables, c is a constant and e is the error term. Then the coefficient of determination of the regression Equation (6), R_i^2, is calculated. In this case, we can calculate *k* different VIFs (one for each Xi). Generally, the value of VIF exceeding 10 is regarded as indicating multicollinearity. Particularly, in the process of paring down the variables, the variable with the largest VIF (greater than the selected threshold of 10) was the first one to be removed.

For ICESat/GLAS data, Treeht2, H25, LEE, TEE, HOME, QMCH, AVAW and gasamp1 (the intensity of the first waveform from Gaussian decomposition) were selected as predictor variables. The description of these variables is given in Table 2. For Landsat5/TM data, surface reflectances (band1 and band4), NDVI, $\frac{TM3}{TM7}$ were selected, as well as TCW.

3.2. Bootstrapping

In this study, the number of field data points available for modeling was only 108. To approximate the coefficient distributions and improve modeling accuracy, bootstrapping, which is a resampling technique, was applied to the regression in this paper [51,52].

Bootstrapping, which is a form of a sampling with replacement, initially proposed by Efron in 1797, has been widely used in many fields [52]. Unlike other sampling methods, there is no need to make assumptions about the form of the population [53]. It is a statistical inference method based on a sampling technique that can improve model estimation accuracy by increasing the number of samples [53].

The general bootstrapping process works as follows: a sample of size *X* is drawn from the original sample with replacement, where *X* is the size of the original sample. In this paper, the bootstrapping was combined with stratified sampling, so that bootstrap samples have similar overall properties to those 108 AGB data [52–54]. The steps were as follows:

(1) The original data (AGB field data and corresponding predictor variables) of size 108 was sorted by ascending AGB values.
(2) After that, we divided the dataset into four equal-sized subgroups (size = 27).
(3) For each subgroup, random sampling with replacement was performed and repeated 27 times. Therefore, there were 27 data for each subgroup and a total of 108 data were obtained, which was our first bootstrap sample.
(4) The process (3) was repeated 300 times to obtain 300 bootstrap samples.

In this paper, 300 bootstrap samples were created from the set of 108 observations and modeled separately.

3.3. Modeling Approach

In this paper, three prediction methods were considered: stepwise regression (SR), support vector regression (SVR), and random forest (RF). As shown in Figure 4, specifically, after the remote-sensing predictor variables were retained (see Section 3.1), 300 bootstrap samples were created as described in Section 3.2. Each bootstrap sample included predictor variables (8 variables for GLAS data and 5 variables for TM data) and corresponding AGB field data. For each bootstrap sample, three above modeling approaches (SR, SVR, and RF) and three input datasets (TM predictor variables, GLAS predictor variables, and TM predictor variables + GLAS predictor variables) were used to build AGB models respectively. As a result, nine AGB models were established for each bootstrap sample, namely, three SR AGB models (with TM predictor variables, with GLAS predictor variables, and with TM predictor variables + GLAS predictor variables), three SVR AGB models (with TM predictor variables, with GLAS predictor variables, and with TM predictor variables + GLAS predictor variables), and three RF AGB models (with TM predictor variables, with GLAS predictor variables, and with TM predictor variables + GLAS predictor variables).

SR is a parametric algorithm that is commonly used to estimate AGB [54]. The strength of this approach is that it can select suitable variables for the regression model when many explanatory variables are available. The idea of this algorithm is to introduce all the explanatory variables into the regression equation one by one according to their contributions to the dependent variable and to eliminate the variables whose effects are not significant after the introduction of new variables. In this paper, the underlying regression model used to evaluate the variables in the SR approach is multiple linear. Here, a significance level for deciding when to enter a predictor into the stepwise model is set to 0.15, like many software. Also, a significance level for deciding when to remove a predictor from the stepwise model is set to 0.15.

SVR and RF are two representative non-parametric algorithms that some studies have used to estimate AGB [14,50,54–56]. Unlike parametric algorithms, the strength of non-parametric algorithms is that they do not make assumptions about the form of the model and the distribution of input data, which makes it possible to effectively describe the complex nonlinear relationship between forest AGB and remote-sensing data [55]. SVR transforms a nonlinear regression into a linear regression by mapping the input data into a high-dimensional feature space using a kernel function. The essence of the solution is to find the optimal hyperplane based on the rule of structural risk minimization [56]. In this paper, the radial basis function kernel (RBF), which is the most widely used kernel function, was used because it requires fewer parameters and can reduce the difficulty of numerical calculation [57,58].

RF is an extension of the classification and regression tree (CART) approach. To improve prediction accuracy, random samples and attributes are selected to build multiple independent decision trees [59]. This algorithm is less sensitive to data noise and outliers than others [59]. A flowchart of the RF algorithm is shown in Figure 5. The original data are randomly resampled to yield N samples of size M by bagging repeatedly [59]. In this paper, the value of M (equal to the size of original data) is 108. In addition, the value of N (the number of trees) is 600, determined from the relationship between N and the error, which is also commonly used to determine the number of trees. Then a regression tree is constructed for each dataset. For each regression tree, each node is split using a random subset of size m_{try} (the number of predictors sampled for splitting) from the features, a procedure called "feature bagging". The result is estimated by averaging the predictions of the N regression trees. In this paper, the value of m_{try} was selected based on the RMSE of the data not included in each sample, an approach that is called out-of-bag (OOB) data.

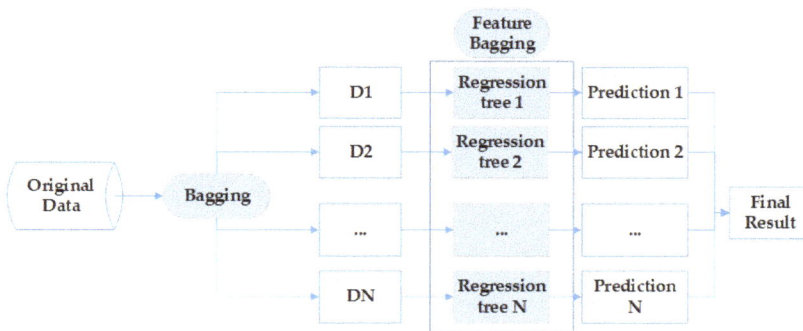

Figure 5. Flowchart of RF algorithm.

In the modeling process, the above algorithms were implemented in R, an open-source software environment [60,61]. For each bootstrap sample, three modeling approaches (SR, SVR, and RF) and three input datasets (TM, GLAS, and TM + GLAS) were used. During which, cross validation was performed with four-fold and five repetitions for each prediction model, which means that 75% of the input data was training data and the rest was test data. After comparison and evaluation, the optimal AGB model was indicated by R-squared (R^2) and root mean square error (RMSE) from both modeling accuracy and cross validation.

3.4. Independent Validation

Due to the lack of corresponding GLAS data, there were 40 field plots that were not used for modeling. To perform further validation of the selected model (the RF AGB model with TM data), these plots were used for independent validation. In addition to R^2 and RMSE, another model evaluation index, the total relative error (TRE) (Equation (7)), which was proposed by Zeng [62], was also used to evaluate the forest AGB model:

$$\text{TRE} = \frac{\sum (y_i - \hat{y}_i)}{\sum \hat{y}_i} \times 100\% \tag{7}$$

where y_i is the *i*-th measured value, and $\hat{y}_i$ is the *i*-th predicted value from the model. TRE is an important indicator reflecting the effect of model fitting and should be controlled within a certain range (such as $\pm 3\%$ or $\pm 5\%$).

4. Results

4.1. ICESat/GLAS Data Processing Results

In order to calculate the waveform metrics associated with AGB, it was necessary to pre-process the GLAS raw data. In the process of Filtering and Gaussian Decomposition, the results of different methods were compared and analyzed.

Comparison of the filtering effects of wavelet transform and Gaussian filter (Figure 6 and Table 4) showed that the RMSE and SNR of the wavelet transform were better than those of the Gaussian filter, but that the smoothness was not significantly different. As a result, the wavelet transform was chosen to filter the waveform of the GLAS footprint.

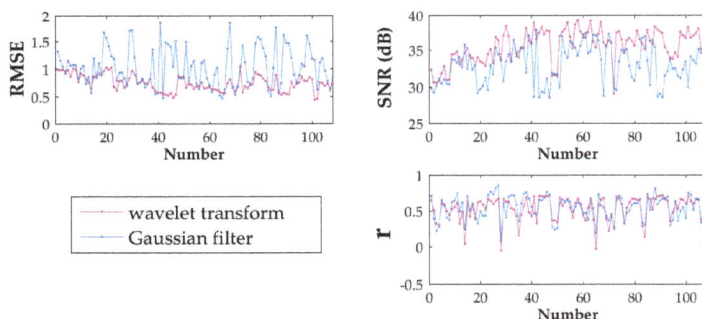

Figure 6. RMSE, SNR, and r from different filters. The pink lines represent results from the wavelet transform and the blue line represents results from the Gaussian filter. The X-axis (Number) represents the serial number of the 108 GLAS data points.

Table 4. Mean values of RMSE, SNR, and r from different filters.

Method	RMSE	SNR(dB)	r
wavelet transform	0.73	35.67	0.53
Gaussian filter	1.02	33.15	0.54

As mentioned in Section 2.2, the Pearson's correlation coefficient (r) was calculated between the original waveform and the fitted waveforms obtained by the proposed method and by the GLA14 product separately. A comparison of these two correlations is shown in Figure 7. Clearly, the overall correlation obtained using the proposed method is superior to that obtained using the GLA14 product.

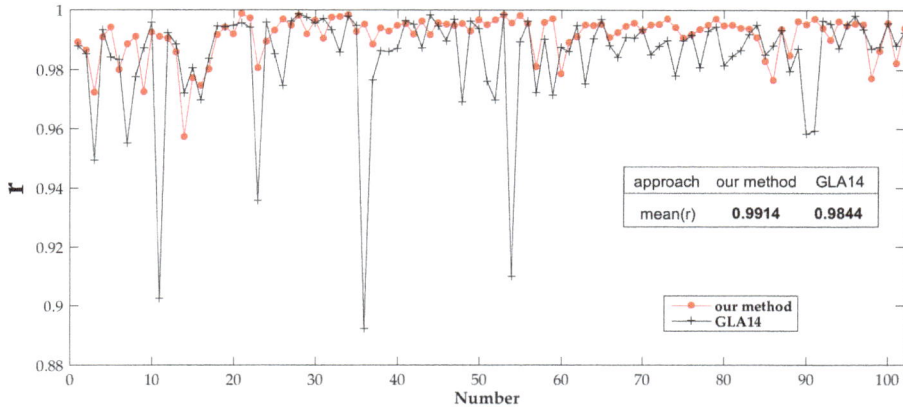

Figure 7. Pearson's correlation coefficient (r) between the fitted waveforms and the raw waveforms. The red line represents the results from the proposed method, and the black line represents the results from GLA14. The *X*-axis (Number) represents the serial number of the 108 GLAS data points.

After processing the GLAS data using the above method, three crucial locations of the waveform (the signal start location, the signal end location and the ground peak location) were successfully identified. In addition, then the waveform metrics in Table 2 were extracted.

4.2. AGB Model Results

The performances of all AGB models, evaluated in terms of R^2 and RMSE, is shown in Figures 8 and 9. Figure 8 summarizes the modeling accuracy results from regression using different approaches and input data combined with bootstrapping, and Figure 9 shows the results from repeated cross validation. RF outperformed the other two approaches in all three cases: with TM data alone, with GLAS data alone, and with GLAS data and TM data together. RF AGB models generally led to higher R^2 and smaller RMSE, both in modeling accuracy (R^2_{max} = 0.96, $RMSE_{min}$ = 17.73 Mg/ha) and cross validation (R^2_{max} = 0.76, $RMSE_{min}$ = 39.60 Mg/ha). The performance of the SR AGB models was the worst in terms of R^2 and RMSE for both modeling accuracy and cross validation. The presence of GLAS data significantly improved AGB predictions for SVR and SR AGB models by decreasing RMSE and increasing R^2. As for the RF AGB models, inclusion of GLAS data no longer led to significant improvement. There was little difference in terms of R^2 and RMSE between the RF AGB model with TM alone and that with GLAS or TM + GLAS. Considering that the GLAS footprints are spatially discontinuous, this model needs to be extrapolated at regional scale by adding more data, which will introduce new errors at the same time. Therefore, in this paper, the Landsat5/TM dataset was used as input data with RF as the prediction method to estimate forest AGB at regional scale.

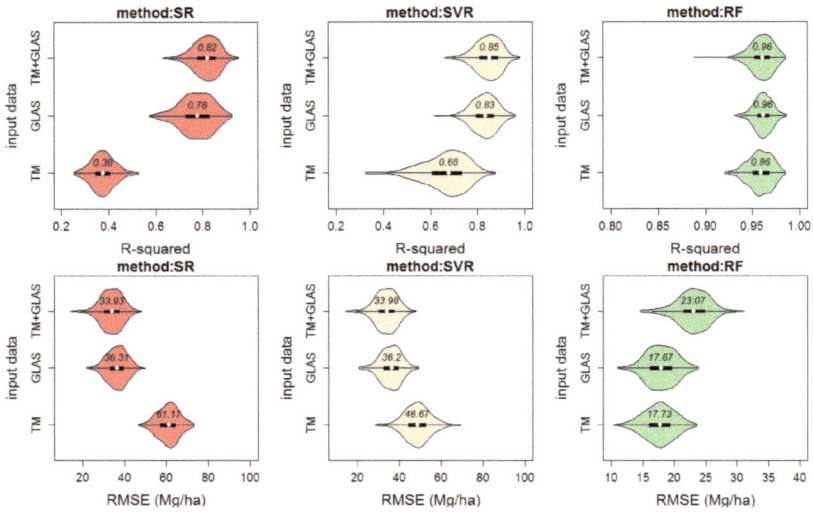

Figure 8. Modeling accuracy results from regression using different input data and prediction approaches. The distribution of RMSE and R^2 is shown as a violin plot [63], which is the combination of a box plot and a density plot. The white point represents the median, and the black box indicates the interquartile range.

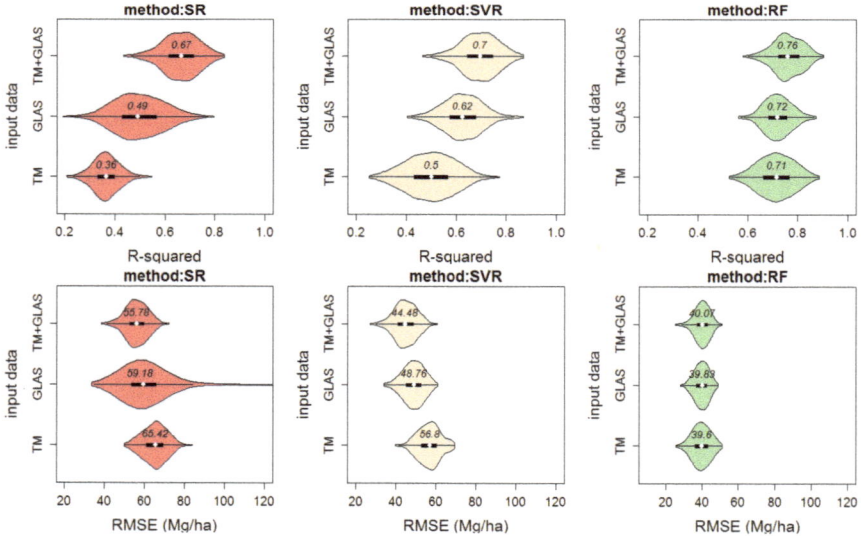

Figure 9. Results from cross validation using different input data and prediction approaches. The distribution of RMSE and R^2 is shown as a violin plot [63].

The performance of the RF AGB model with TM data was further investigated. Scatter plots of field AGB against predicted biomass from RF models with 300 bootstraps is shown in Figure 10. The distribution of scatter points is concentrated near the 1:1 line, but this model underestimated forest AGB at high AGB levels (200–400 Mg/ha) and overestimated it at low AGB levels (0–200 Mg/ha).

The modeling accuracy results of RF AGB models created with different sample sizes (Figure 11) show that increasing the sample size led to an increase in R^2, a decrease in RMSE, and a reduction in the range of variation, implying that the established model is more stable.

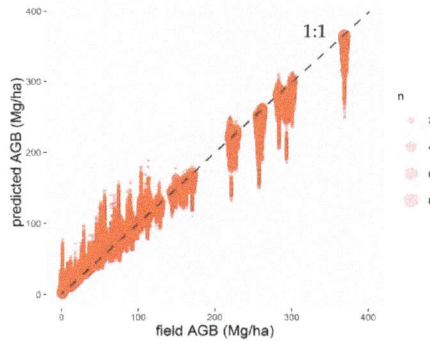

Figure 10. Predicted AGB vs. field AGB. The size of the point, n, represents the number of repetition points. The color of the point is transparent pink. The black dotted line represents the 1:1 line.

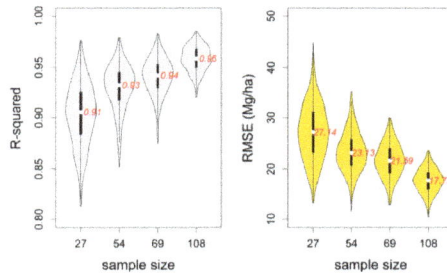

Figure 11. Modeling accuracy results of regression from different sample size using the RF AGB model. The numbers represent median values.

Sixty-nine samples, which were randomly selected from the 108 datasets, were used for RF AGB modeling with TM and TM + GLAS respectively. The modeling accuracy results (Figure 12) further confirmed the previous finding that the presence of GLAS data did not lead to a significant increase in R^2. In this case, inclusion of GLAS data resulted in an increase in RMSE.

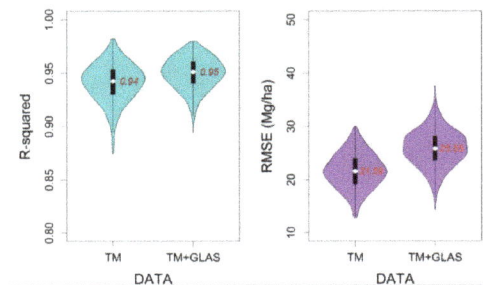

Figure 12. Modeling accuracy results of regression from RF AGB models with different input data from 69 samples. The numbers represent median values.

To perform further validation of the selected model (the RF AGB model with TM data), the remaining 40 plots were used for independent validation (Figure 13). The predicted forest AGB values were the medians of the 300 bootstrap estimates. The results show an R² of 0.54, an RMSE of 20.5 Mg/ha and a TRE of 4.97%, which was within the acceptable range.

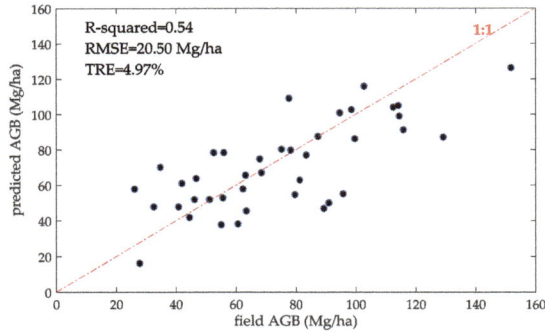

Figure 13. Independent validation results from RF AGB model with TM data from 40 datasets. The red dotted line represents the 1:1 line.

4.3. Wall-to-Wall AGB Prediction over the Daxing'anling Mountains in Heilongjiang Province

The spatial distribution of forest AGB density in 2005 over the Daxing'anling Mountains is shown in Figure 14, using the optimal AGB model established in Section 4.1. The predicted forest AGB density values were the medians of the 300 bootstrap estimates. The forest AGB density over the Daxing'anling Mountains was distributed mainly in the 60–90 Mg/ha range, and the highest value was 304 Mg/ha. The average forest AGB density over the Daxing'anling Mountains was 83.13 Mg/ha. This value is close to the average AGB density, 83.50–102.49 Mg/ha, estimated by Zhang et al. [3] in northeastern China, who also found that the forest AGB of the Daxing'anling Mountains was less than those of the Changbai and Xiaoxing'anling Mountains. The result obtained here is slightly larger than the 80.18 Mg/ha provided by Huang and Xia [64] using the Dong model in northeastern China.

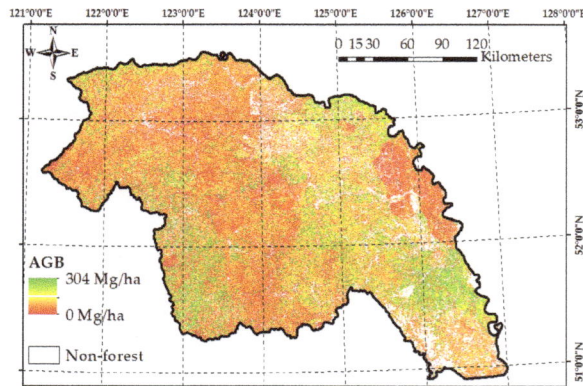

Figure 14. Forest AGB density map from RF AGB model with TM data over the Daxing'anling Mountains for 2005. The background information is a 30-m forest distribution map developed by Chen [21].

5. Discussion

In this paper, the results of models using different datasets and three approaches were compared. The random forest AGB model using Landsat5/TM as input data has the acceptable modeling accuracy (R^2 = 0.95 RMSE = 17.73 Mg/ha) and it was also shown to estimate AGB reliably by cross validation (R^2 = 0.71 RMSE = 39.60 Mg/ha). We also compared our results with other similar research works. Powell [50] modeled aboveground tree biomass using field data and Landsat satellite imagery in Minnesota and Arizona by comparing different statistical techniques. The RMSE of the modeling accuracy results ranged from 32.19 to 44.43 Mg/ha. Zhang et al. [3] developed forest AGB models in northeastern China based on GLAS data, achieving an R^2 of modeling accuracy results for field-measured points of 0.86 and an RMSE of 26.76 Mg/ha. Compared with other published studies, the forest AGB model in this paper achieved better performance in terms of modeling accuracy (R^2 and RMSE).

5.1. Spatio-Temporal Matching between GLAS Data and Measured Data

When matching the measured field data and the GLAS waveforms, it was assumed that no significant change in forest AGB in the field plot had occurred within the previous three years. In terms of geographical location, especially for matching forest inventory data with GLAS data, the authors believe that a central location difference in the 65-m range is acceptable. The above assumptions are due to the difficulty of matching the two datasets, which is caused by the short observation time of GLAS, the small number of repeated observations, and the spatial discontinuities of GLAS. To make these hypotheses reasonable, the spectral difference of the plots between the GLAS observation time and the measurement time were examined, and the spectral variance from a 3 × 3 TM pixel window corresponding to the GLAS data was also examined. Data points with an abnormal deviation were deleted.

The modeling accuracy and cross validation results showed that selection of representative GLAS data for the field plots is an important step towards effective modeling and improved modeling accuracy.

5.2. GLAS Data and Terrain Effects

LiDAR waveforms are susceptible to ground slopes. When the ground slope is greater than 20 degrees, information from the ground and from the canopy are intermixed, making the extracted metrics no longer accurate [65,66]. Before modeling, slope values were calculated for all field plots, and all were found to be less than 20 degrees, with most less than 15 degrees.

In addition, an effort was made to add auxiliary data in the form of a digital elevation model (DEM) to the model, but the results were not improved. Therefore, terrain effects were not taken into account in this paper, but when the GLAS model is applied to complex terrain, terrain effects must be eliminated.

5.3. Influence of Regional Coverage Types on Estimation

The field measurements used in this paper consisted of two parts, one from measurements of the GLAS footprints, where three sampling plots were established within each footprint, and the other from National Forest Inventory data, where each tree with DBH greater than 5 cm in the range of 0.06 ha was measured. The difference in sampling methods between these two datasets may have introduced errors to the results. Only the tree biomass was involved in field measurements, ignoring shrubs and herbaceous plants. Therefore, the resulting estimates of forest AGB were lower than the actual values.

5.4. Effects of TM Data on Regional Biomass Mapping

In this modeling exercise, the acquisition time of the TM images was close to the peak of the growing season, and the time differences among scenes were less than 30 days. The results were highly affected by TM data quality when the AGB model was applied at regional scale. In this research, the TM data for the study area close to the growing season in 2005 were of good quality, and most areas were clear. However, the model will be limited when good-quality TM data are unavailable for the entire growing season, which is highly probable in some areas. In this case, TM image reconstruction methods can be used to compensate for the lack of data.

5.5. Effects of range of AGB values on validation

The RMSE from independent validation using the 40-sample dataset is significantly lower than the RMSE from cross validation, which may due to the range of AGB values measured on the field plots. In particular, the AGB values of 40-sample dataset for independent validation were within the smaller AGB (AGB < 160 Mg/ha) (see Figure 13), while the range of AGB used in the modeling was 0–400 Mg/ha, with the main distribution values ranging from 0–160 Mg/ha. Therefore, the RMSE from independent validation is smaller. In addition, the larger RMSE result from cross validation is mainly affected by the large-values of AGB data.

6. Conclusions

To map the distribution of forest AGB density at regional scale, two types of remote-sensing data matching were selected for a group of field plots: optical remote-sensing data (Landsat5/TM) with a resolution of 30 m, and LiDAR data (ICESat/GLAS) with a footprint approximately 65 m in diameter. AGB models were built using these field measurements and remote-sensing datasets. The results showed that including GLAS data improved AGB predictions for the SR and SVR AGB models. However, for the RF AGB models, there was little difference between the results from the three input datasets. Therefore the combination of data type and prediction method is important, and LiDAR data (e.g., GLAS data) may not be a necessary option for estimating forest AGB. After comparing and analyzing the effects of the various AGB models using the three modeling approaches and three remote-sensing datasets combined with bootstrapping, it was found that the RF AGB model with TM data was optimal for mapping. Finally, forest AGB density with spatial resolution of 30 m over the Daxing'anling Mountains was mapped. Compared with some other researches, the estimated forest AGB at the regional scale is acceptable.

Acknowledgments: This research was supported by the National Basic Research Program of China under grant No. 2013CB733403. The authors would also like to express gratitude to Sun Guoqing, who provided field measurements for this research, and Zhang Yuzhen, who provide much help and instruction in data processing.

Author Contributions: Kaili Liu designed the framework of this research work and wrote the manuscript; Jindi Wang proposed the main idea, provided important guidance on the work, and checked the writing; Weisheng Zeng provided the field data and guidance; and Jinling Song gave guidance on this work.

Conflicts of Interest: The authors declare no conflict of interest.

References

1. Olson, J.S.; Watts, J.A.; Allison, L.J. *Carbon in Live Vegetation of Major World Ecosystems;* Oak Ridge National Laboratory: Oak Ridge, TN, USA, 1983.
2. Brown, S.; Sathaye, J.; Cannell, M.; Cannell, M.; Kauppi, P.E. Mitigation of carbon emissions to the atmosphere by forest management. *Commonw. For. Rev.* **1996**, *75*, 80–91.
3. Zhang, Y.; Liang, S.; Sun, G. Forest biomass mapping of northeastern China using GLAS and MODIS data. *IEEE J. Sel. Top. Appl. Earth Obs. Remote Sens.* **2014**, *7*, 140–152. [CrossRef]
4. Lu, D. The potential and challenge of remote sensing–based biomass estimation. *Int. J. Remote Sens.* **2006**, *27*, 1297–1328. [CrossRef]

5. McRoberts, R.E.; Næsset, E.; Gobakken, T. Inference for lidar-assisted estimation of forest growing stock volume. *Remote Sens. Environ.* **2013**, *128*, 268–275. [CrossRef]
6. Ahamed, T.; Tian, L.; Zhang, Y.; Ting, K.C. A review of remote sensing methods for biomass feedstock production. *Biomass Bioenergy* **2011**, *35*, 2455–2469. [CrossRef]
7. Lu, D.; Batistella, M.; Moran, E. Satellite estimation of aboveground biomass and impacts of forest stand structure. *Photogramm. Eng. Remote Sens.* **2005**, *71*, 967–974. [CrossRef]
8. Dobson, M.C.; Ulaby, F.T.; LeToan, T.; Beaudoin, A. Dependence of radar backscatter on coniferous forest biomass. *IEEE Trans. Geosci. Remote Sens.* **1992**, *30*, 412–415. [CrossRef]
9. Drake, J.B.; Dubayah, R.O.; Clark, D.B.; Knox, R.G.; Blair, J.B.; Hofton, M.A.; Chazdone, R.L.; Weishampelf, J.F.; Prince, S. Estimation of tropical forest structural characteristics using large-footprint lidar. *Remote Sens. Environ.* **2002**, *79*, 305–319. [CrossRef]
10. Yu, Y.; Yang, X.; Fan, W. Estimates of forest structure parameters from GLAS data and multi-angle imaging spectrometer data. *Int. J. Appl. Earth Obs. Geoinform.* **2015**, *38*, 65–71. [CrossRef]
11. Mougin, E.; Proisy, C.; Marty, G.; Fromard, F.; Puig, H.; Betoulle, J.L.; Rudant, J.P. Multifrequency and multipolarization radar backscattering from mangrove forests. *IEEE Trans. Geosci. Remote Sens.* **1999**, *37*, 94–102. [CrossRef]
12. Sandberg, J.; Tsoukas, H. Grasping the logic of practice: Theorizing through practical rationality. *Acad. Manag. Rev.* **2011**, *36*, 338–360. [CrossRef]
13. Hajj, M.E.; Baghdadi, N.; Fayad, I.; Vieilledent, G.; Bailly, J.S.; Minh, D.H.T. Interest of integrating spaceborne LiDAR data to improve the estimation of biomass in high biomass forested areas. *Remote Sens.* **2017**, *9*, 213. [CrossRef]
14. Fayad, I.; Baghdadi, N.; Guitet, S.; Bailly, J.S.; Hérault, B.; Gond, V.; Hajj, M.E.; Minh, D.H.T. Aboveground biomass mapping in French Guiana by combining remote sensing, forest inventories and environmental data. *Int. J. Appl. Earth Obs. Geoinform.* **2016**, *52*, 502–514. [CrossRef]
15. Lu, D.; Chen, Q.; Wang, G.; Liu, L.; Li, G.; Moran, E. A survey of remote sensing-based aboveground biomass estimation methods in forest ecosystems. *Int. J. Digit. Earth* **2016**, *9*, 63–105. [CrossRef]
16. Dixon, R.K.; Brown, S.; Houghton, R.E.A.; Solomon, A.M.; Trexler, M.C.; Wisniewski, J. Carbon pools and flux of global forest ecosystems. *Science* **1994**, *263*, 185–190. [CrossRef] [PubMed]
17. Saatchi, S.; Malhi, Y.; Zutta, B.; Buermann, W.; Anderson, L.O.; Araujo, A.M.; Phillips, O.L.; Peacock, J.; Steege, H.T.; Gonzalez, G.L.; et al. Mapping landscape scale variations of forest structure, biomass, and productivity in Amazonia. *Biogeosci. Discuss.* **2009**, *6*, 5461–5505. [CrossRef]
18. Pang, Y.; Lefsky, M.; Sun, G.; Miller, M.E.; Li, Z. Temperate forest height estimation performance using ICESat GLAS data from different observation periods. *Int. Arch. Photogramm. Remote Sens. Spat. Inf. Sci.* **2008**, *37*, 777–782.
19. Sun, G.; Ranson, K.; Masek, J.; Guo, Z.; Pang, Y.; Fu, A.; Wang, D. Estimation of tree height and forest biomass from GLAS data. *J. For. Plan.* **2008**, *13*, 157–164.
20. Chen, X.; Zeng, W.; Xiong, Z.; Zhang, M. New development of China National Forest Inventory (NFI)—On revision of NFI technical regulations. *For. Resour. Manag.* **2004**, *5*, 40–45.
21. Chen, J.; Chen, J.; Liao, A.; Cao, X.; Chen, L.; Chen, X.; He, C.; Han, G.; Peng, S.; Lu, M.; et al. Global land cover mapping at 30 m resolution: A POK-based operational approach. *ISPRS J. Photogramm. Remote Sens.* **2014**, *103*, 7–27. [CrossRef]
22. Zeng, W. Methodoly on Moeling of Single-Tree Biomass Equations for National Biomass Estimation in China. Ph.D. Thesis, Chinese Academy of Forestry, Beijing, China, 2011.
23. Zeng, W.S.; Zhang, H.R.; Tang, S.Z. Using the dummy variable model approach to construct compatible single-tree biomass equations at different scales—A case study for Masson pine (Pinus massoniana) in southern China. *Can. J. For. Res.* **2011**, *41*, 1547–1554. [CrossRef]
24. Dong, L.; Zhang, L.; Li, F. Developing additive systems of biomass equations for nine hardwood species in Northeast China. *Trees* **2015**, *29*, 1149–1163. [CrossRef]
25. Wang, H. Dynamic Simulating System for Stand Growth of Forest in Northeast China. Ph.D. Thesis, Northeast Forest University, Harbin, China, 2012.
26. Wang, C. Biomass allometric equations for 10 co-occurring tree species in Chinese temperate forests. *For. Ecol. Manag.* **2006**, *222*, 9–16. [CrossRef]

27. Dong, L.; Zhang, L.; Li, F. A compatible system of biomass equations for three conifer species in northeast, China. *For. Ecol. Manag.* **2014**, *329*, 306–317. [CrossRef]

28. National Snow & Ice Data Center. Available online: http://nsidc.org/ (accessed on 2 April 2017).

29. Chi, H.; Sun, G.; Huang, J.; Guo, Z.; Ni, W.; Fu, A. National forest aboveground biomass mapping from ICESat/GLAS data and MODIS imagery in China. *Remote Sens.* **2015**, *7*, 5534–5564. [CrossRef]

30. Wu, D.; Fan, W. Synergistic use of ICESat/GLAS and MISR data for estimating forest aboveground biomass. *Bull. Bot. Res.* **2015**, *3*, 397–405. (In Chinese)

31. Baghdadi, N.; El Hajj, M.; Bailly, J.S.; Fabre, F. Viability statistics of GLAS/ICESat data acquired over tropical forests. *IEEE J. Sel. Top. Appl. Earth Obs. Remote Sens.* **2014**, *7*, 1658–1664. [CrossRef]

32. Pourrahmati, M.R.; Baghdadi, N.N.; Darvishsefat, A.A.; Namiranian, M.; Fayad, I.; Bailly, J. S.; Gond, V. Capability of GLAS/ICESat data to estimate forest canopy height and volume in mountainous forests of Iran. *IEEE J. Sel. Top. Appl. Earth Obs. Remote Sens.* **2015**, *8*, 5246–5261. [CrossRef]

33. Sihag, R. Wavelet thresholding for image de-noising. In Proceedings of the International Conference on VLSI, Communication & Instrumentation (ICVCI), Kottayam, Kerala, India, 7–9 April 2011; Volume 201, pp. 20–23.

34. Sun, G.; Ranson, K.; Kimes, D.; Blair, J.B.; Kovacs, K. Forest vertical structure from GLAS: An evaluation using LVIS and SRTM data. *Remote Sens. Environ.* **2008**, *112*, 107–117. [CrossRef]

35. Yang, H.; Zhang, D.; Huang, W.; Gao, Z.; Yang, X.; Li, C.; Wang, J. Application and evaluation of wavelet-based denoising method in hyperspectral imagery data. In Proceedings of the International Conference on Computer and Computing Technologies in Agriculture, Beijing, China, 29–31 October 2011; Springer: Berlin/Heidelberg, Germany, 2011; pp. 461–469.

36. Yi, T.-H.; Li, H.-N.; Zhao, X.-Y. Noise smoothing for structural vibration test signals using an improved wavelet thresholding technique. *Sensors* **2012**, *12*, 11205–11220. [CrossRef] [PubMed]

37. Ranson, K.J.; Kimes, D.; Sun, G.; Nelson, R.; Kharuk, V.; Montesano, P. Using MODIS and GLAS data to develop timber volume estimates in central Siberia. In Proceedings of the 2007 IEEE International Geoscience and Remote Sensing Symposium, Barcelona, Spain, 23–28 July 2007; pp. 2306–2309.

38. Lefsky, M.A.; Keller, M.; Pang, Y.; De Camargo, P.B.; Hunter, M.O. Revised method for forest canopy height estimation from Geoscience Laser Altimeter System waveforms. *J. Appl. Remote Sens.* **2007**, *1*, 013537.

39. Kimes, D.S.; Ranson, K.J.; Sun, G.; Blair, J.B. Predicting lidar measured forest vertical structure from multi-angle spectral data. *Remote Sens. Environ.* **2006**, *100*, 503–511. [CrossRef]

40. Lefsky, M.A.; Harding, D.; Cohen, W.B.; Parker, G.; Shugart, H.H. Surface lidar remote sensing of basal area and biomass in deciduous forests of eastern Maryland, USA. *Remote Sens. Environ.* **1999**, *67*, 83–98. [CrossRef]

41. Harding, D.J.; Carabajal, C.C. ICESat waveform measurements of within-footprint topographic relief and vegetation vertical structure. *Geophys. Res. Lett.* **2005**, *32*, 322. [CrossRef]

42. Rouse, J.W.; Haas, R.H.; Schell, J.A.; Deering, D.W. *Monitoring Vegetation Systems in the Great Plains with ERTS*; NASA Special Publication; NASA: Washington, DC, USA, 1974; Volume 351, p. 309.

43. Huete, A.; Didan, K.; Miura, T.; Rodriguez, E.P.; Gao, X.; Ferreira, L.G. Overview of the radiometric and biophysical performance of the MODIS vegetation indices. *Remote Sens. Environ.* **2002**, *83*, 195–213. [CrossRef]

44. Richardson, A.J.; Wiegand, C.L. Distinguishing vegetation from soil background information. *Photogramm. Eng. Remote Sens.* **1978**, *43*, 1541–1552.

45. Huete, A.R. A soil-adjusted vegetation index (SAVI). *Remote Sens. Environ.* **1988**, *25*, 295–309. [CrossRef]

46. Hunt, E.R.; Rock, B.N. Detection of changes in leaf water content using near-and middle-infrared reflectances. *Remote Sens. Environ.* **1989**, *30*, 43–54.

47. Cai, T.; Ju, C.; Yao, Y. Quantitative estimation of vegetation coverage in Mu Us sandy land based on RS and GIS. *Chin. J. Appl. Ecol.* **2005**, *16*, 2301–2305. (In Chinese).

48. Crist, E.P. A TM Tasseled Cap equivalent transformation for reflectance factor data. *Remote Sens. Environ.* **1985**, *17*, 301–306. [CrossRef]

49. Duane, M.V.; Cohen, W.B.; Campbell, J.L.; Hudiburg, T.; Turner, D.P.; Weyermann, D.L. Implications of alternative field-sampling designs on landsat-based mapping of stand age and carbon stocks in oregon forests. *For. Sci.* **2010**, *21*, 77–79.

50. Powell, S.L.; Cohen, W.B.; Healey, S.P.; Kennedy, R.E.; Moisen, G.G.; Pierce, K.B.; Ohmann, J.L. Quantification of live aboveground forest biomass dynamics with Landsat time-series and field inventory data: A comparison of empirical modeling approaches. *Remote Sens. Environ.* **2010**, *114*, 1053–1068. [CrossRef]

51. Juan, S.; Lantz, F. Application of bootstrap techniques in econometrics: The example of cost estimation in the automotive industry. *Oil Gas Sci. Technol.* **2001**, *56*, 373–388. [CrossRef]

52. Wehrens, R.; Putter, H.; Buydens, L.M.C. The bootstrap: A tutorial. *Chemom. Intell. Lab. Syst.* **2000**, *54*, 35–52. [CrossRef]

53. Freedman, D.A. Bootstrapping regression models. *Ann. Stat.* **1981**, *9*, 1218–1228. [CrossRef]

54. Fassnacht, F.E.; Hartig, F.; Latifi, H.; Berger, C.; Hernández, J.; Corvalán, P.; Koch, B. Importance of sample size, data type and prediction method for remote sensing-based estimations of aboveground forest biomass. *Remote Sens. Environ.* **2014**, *154*, 102–114. [CrossRef]

55. Avitabile, V.; Baccini, A.; Friedl, M.A.; Schmullius, C. Capabilities and limitations of Landsat and land cover data for aboveground woody biomass estimation of Uganda. *Remote Sens. Environ.* **2012**, *117*, 366–380. [CrossRef]

56. Zhao, K.; Popescu, S.; Meng, X.; Pang, Y.; Agca, M. Characterizing forest canopy structure with lidar composite metrics and machine learning. *Remote Sens. Environ.* **2011**, *115*, 1978–1996. [CrossRef]

57. Müller, K.-R.; Mika, S.; Rätsch, G.; Tsuda, K.; Schölkopf, B. An introduction to kernel-based learning algorithms. *IEEE Trans. Neural Netw.* **2001**, *12*, 181–201. [CrossRef] [PubMed]

58. Crone, S.F.; Lessmann, S.; Pietsch, S. Parameter sensitivity of support vector regression and neural networks for forecasting. In Proceedings of the IEEE International Conference on Data Mining (DMIN), Hongkong, China, 18–22 December 2006; pp. 396–402.

59. Breiman, L. Random forests. *Mach. Learn.* **2001**, *45*, 5–32. [CrossRef]

60. Ihaka, R.; Gentleman, R. R: A language for data analysis and graphics. *J. Comput. Grph. Stat.* **1996**, *5*, 299–314. [CrossRef]

61. Dalgaard, P. Introductory statistics with R. *J. R. Stat. Soc. Ser. A Stat. Soc.* **2009**, *172*, 274–275.

62. Zeng, W.; Tang, S. Goodness evaluation and precision of tree biomass equations. *Sci. Silvae Sin.* **2011**, *47*, 106–113.

63. Potter, K.; Hagen, H.; Kerren, A.; Dannenmann, P. Methods for presenting statistical information: The box plot. *Vis. Larg. Unstruct. Data Sets S* **2006**, *4*, 97–106.

64. Huang, G.; Xia, C. MODIS-based estimation of forest biomass in northeast China. *For. Resour. Manag.* **2005**, *4*, 40–44. (In Chinese)

65. Pang, Y.; Li, Z.; Lefsky, M.; Sun, G.; Yu, X. Model based terrain effect analyses on ICEsat GLAS waveforms. In Proceedings of the 2006 IEEE International Symposium on Geoscience and Remote Sensing, Vancouver, BC, Canada, 28–30 August 2006; pp. 3232–3235.

66. Wang, X.; Huang, H.; Gong, P.; Liu, C.; Li, C.; Li, W. Forest canopy height extraction in rugged areas with ICESAT/GLAS data. *IEEE Trans. Geosci. Remote Sens.* **2014**, *52*, 4650–4657. [CrossRef]

remote sensing

MDPI

Article

Evaluation of Remote Sensing Inversion Error for the Above-Ground Biomass of Alpine Meadow Grassland Based on Multi-Source Satellite Data

Baoping Meng [1], Jing Ge [1], Tiangang Liang [1,*], Shuxia Yang [1], Jinglong Gao [1], Qisheng Feng [1], Xia Cui [2], Xiaodong Huang [1] and Hongjie Xie [3]

[1]   State Key Laboratory of Grassland Agro-ecosystems, College of Pastoral Agriculture Science and Technology, Lanzhou University, Lanzhou 730020, China; mengbp09@lzu.edu.cn (B.M.); gej12@lzu.edu.cn (J.G.); yangshx2014@lzu.edu.cn (S.Y.); rslabjinlong@163.com (J.G.); fengqsh@lzu.edu.cn (Q.F.); huangxd@lzu.edu.cn (X.H.)
[2]   Key Laboratory of Western China's Environmental Systems (Ministry of Education), College of Earth and Environmental Sciences, Lanzhou University, Lanzhou 730000, China; xiacui@lzu.edu.cn
[3]   Laboratory for Remote Sensing and Geoinformatics, University of Texas at San Antonio, San Antonio, TX 78249, USA; hongjie.xie@utsa.edu
*   Correspondence: tgliang@lzu.edu.cn; Tel.: +86-931-981-5306; Fax: +86-931-891-0979

Academic Editors: Lalit Kumar, Onisimo Mutanga, Lars T. Waser and Prasad S. Thenkabail
Received: 21 January 2017; Accepted: 13 April 2017; Published: 16 April 2017

Abstract: It is not yet clear whether there is any difference in using remote sensing data of different spatial resolutions and filtering methods to improve the above-ground biomass (AGB) estimation accuracy of alpine meadow grassland. In this study, field measurements of AGB and spectral data at Sangke Town, Gansu Province, China, in three years (2013–2015) are combined to construct AGB estimation models of alpine meadow grassland based on these different remotely-sensed NDVI data: MODIS, HJ-1B CCD of China and Landsat 8 OLI (denoted as $NDVI_{MOD}$, $NDVI_{CCD}$ and $NDVI_{OLI}$, respectively). This study aims to investigate the estimation errors of AGB from the three satellite sensors, to examine the influence of different filtering methods on MODIS NDVI for the estimation accuracy of AGB and to evaluate the feasibility of large-scale models applied to a small area. The results showed that: (1) filtering the MODIS NDVI using the Savitzky–Golay (SG), logistic and Gaussian approaches can reduce the AGB estimation error; in particular, the SG method performs the best, with the smallest errors at both the sample plot scale (250 m × 250 m) and the entire study area (33.9% and 34.9%, respectively); (2) the optimum estimation model of grassland AGB in the study area is the exponential model based on $NDVI_{OLI}$, with estimation errors of 29.1% and 30.7% at the sample plot and the study area scales, respectively; and (3) the estimation errors of grassland AGB models previously constructed at different spatial scales (the Tibetan Plateau, Gannan Prefecture and Xiahe County) are higher than those directly constructed based on the small area of this study by 11.9%–36.4% and 5.3%–29.6% at the sample plot and study area scales, respectively. This study presents an improved monitoring algorithm of alpine natural grassland AGB estimation and provides a clear direction for future improvement of the grassland AGB estimation and grassland productivity from remote sensing technology.

Keywords: alpine meadow grassland; above-ground biomass; inversion model; error analysis; applicability evaluation

1. Introduction

As the largest terrestrial biome on the Earth's surface [1], the grassland biome occupies approximately 40% of the total land area [2], and its net primary productivity accounts for approximately 20% of the production capability of the entire land biome [3]. In China, the grassland biome accounts for approximately 1/3 of the total national territory. Grasslands play a critical role not only in animal husbandry, but also in energy exchange, carbon sequestration and the biogeochemical cycle. In particular, the effect of the grassland biome is more prominent in the vast central and western regions of China [4,5].

Grassland vegetation is defined as permanent vegetation composed of an herbaceous plant community [2]; the above-ground biomass (AGB) of grassland provides the basis for estimating the net primary productivity of grassland [6]. Overall, AGB can be used to directly estimate grassland productivity [7,8]. In general, AGB is defined as the weight of dry grass in the above-ground portion within one unit area [9]. Monitoring using remote sensing data is the most effective method for collecting continuous spatial and temporal data on the regional or global scale [10,11], because satellite remote sensing can provide large-scale, frequent, low cost and massive information [12]. Therefore, it has gradually replaced traditional methods of ground biomass monitoring, which are inefficient and expensive. Since NDVI was first applied to study natural grasslands in the 1970s, research on the linkage between vegetation indices and AGB has had a history extending over several decades [13–17]. Because grassland AGB estimations can directly guide livestock production, evaluating the accuracy of estimation models is highly important. Numerous studies have indicated that factors such as the representativeness of ground sampling sites, the temporal and spatial resolutions of satellite images, the types of sensors and the methods of remote sensing image processing [18] are responsible for the large differences between the grassland AGB models developed by different scholars even for the same region or the same type of grassland. These differences exist not only in the form of the models, but also in the estimation errors. At present, grassland AGB estimation is often based on the statistical models on a specific remote sensing vegetation index [19–22]. For alpine meadow grassland, which type of remote sensing estimation model is the most suitable is currently unknown. Existing studies indicate that using both high spatial resolution imagery and filtering the vegetation index can increase the accuracy of grassland AGB estimation [18,23–25]. However, it is not yet clear whether there is any difference in using remote sensing data of different spatial resolutions and filtering methods to improve the AGB estimation accuracy of alpine meadow grassland. Therefore, further comparisons and studies must be conducted.

In this study, considering the factors discussed above, a region of alpine meadow grassland in the Tibetan Plateau area is used to perform the following investigations: (1) by comparing and analyzing the AGB estimation models and their accuracies based on the NDVI of multi-source remote sensing data, the influences of different remote sensing data and filtering methods on the error of grassland AGB estimation are revealed; (2) the data from the sample plots observed inside the study area are used to validate the applicability of previous grassland models based on MODIS data and to investigate the reasons for errors from different models; and (3) based on the above research results, we propose a method to improve the accuracy of grassland AGB estimation.

2. Data and Methods

2.1. Study $A = \pi r^2$

The study area ($102°23'$–$102°26'$ E, $35°5'$–$35°7'$ N) is located in the Yangji Community of Sangke Town in Xiahe County, Gansu Province, with a size of approximately 3.86 km (N–S) × 2.77 km (E–W) and a mean elevation of 3050 m (Figure 1). The natural grassland type in the study area is alpine meadow. The dominant pasture plants include alpine *Kobresia pygmaea* C B Clarke, *Elymus nutans* Griseb, *Festuca ovina* L, *Poa annua* L, *Koeleria litwinowii* Domin var and *Astragalus membranaceus* Bunge, while the primary malignant weeds include *Ligularia virgaurea* Mattf, *Leontopodium japonicum*, *Potentilla chinensis* Ser and *Pedicularis resupinata* L. The dominant grazing livestock are yak and Ganjia sheep. The study area is cold and wet throughout the year, and it belongs to the continental monsoon climate of the temperate plateau. The annual average temperature is 2.1 °C; the annual average rainfall is 580 mm.

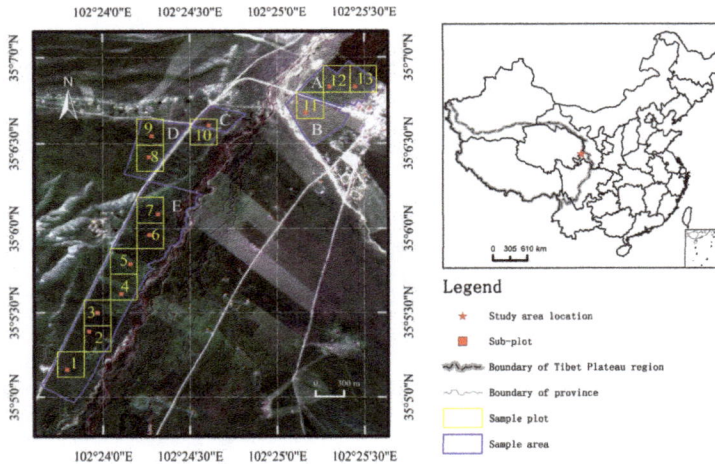

Figure 1. Distributions of experimental sample areas (A–E), with sample plots (yellow squares) and sub-plots (red small squares with their identification numbers as yellow), in Xiahe County, Gansu Province. Each sample area has a similar situation of grass growth; each sample plot is a MODIS pixel of 250 m × 250 m; each sub-plot is a 30 m × 30 m plot with five sample quadrats (see the details in Section 2.2).

The study area consists of five fenced experimental sample areas (A–E) of natural alpine meadow grassland (Figure 1), with a total area of 161.36 ha. Specifically, Area A (19.38 ha) was used to conduct natural grassland reseeding tests; Area B (16.06 ha) was used for grazing utilization experiments; Area C (7.52 ha) acted as a non-treatment testing control area; Area D (19.30 ha) was an experimental area for grassland enclosure; and Area E (99.10 ha, accounting for 61.55% of the total experimental area) was used for artificial fertilization experiments. The grassland growth in these five experimental areas differed because of the above different usages.

2.2. Sampling Strategy and Data Collection

A total of 13 sample plots of 250 m × 250 m were set up inside the five experimental sample areas. In each sample plot, a 30 m × 30 m sub-plot was set up for data collection. In each sub-plot, five quadrats (1.5 m × 1.5 m each) were set up as shown in Figure 2, taking the central point and the four corner points to represent the entire sub-plot. To reduce the artificial sampling error of biomass measurements, a strategy of 9 sub-quadrats was also used for avoiding repeat sampling in each year (Figure 2). The locations of the 13 sample plots were selected based on the following factors: (1) the growth status of the grassland was relatively uniform and was spatially representative; and (2) each sample plot (250 m × 250 m) corresponded to one MODIS pixel. However, in each sample plot, the sub-plot of 30 m× 30 m was randomly selected, corresponding to a Landsat 8 OLI and HJ-1B CCD pixel.

During the grassland growth seasons from 2013 to 2015, observations were performed approximately every 10 d, and a total of 20 field investigations was conducted over the three years. Due to factors such as weather conditions and satellite image time phases, only five times of field measurement data were matched well with satellites and were selected. They were August 2013, July 2014 and July, August and October 2015; with a total of 325 quadrat observations (see the details in Section 2.4) used to construct and analyze the biomass model.

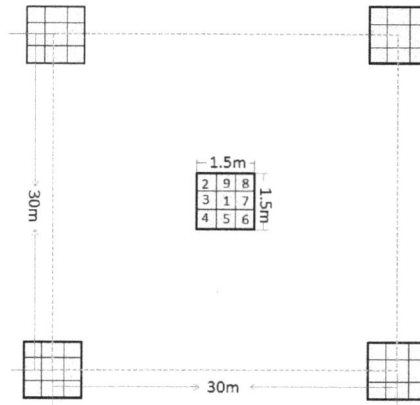

Figure 2. Distributions of the five quadrats (1.5 m × 1.5 m each) in each sub-plot of 30 m × 30 m. Each quadrat consists of 9 sub-quadrats of 0.5 m × 0.5 m. The sub-quadrat identification number (1–9) is the order that we used to sample grass each time, e.g., Sub-quadrat 1 was used the first time of any year, 2 was used in second time of the same year, etc.

Grass samples in each sub-quadrat (0.5 m × 0.5 m) were collected and recorded in the same way through the years. It was cut to the ground level using shears and then packed and weighed, after removing all residual litter and other non-plant materials. The sampling record included latitude, longitude, elevation, grassland types, dominant species, grassland vegetation coverage, grass height, the fresh weight and dry weight (after being dried in the lab at 64 °C until the weight remained constant).

In the field, we also collected spectral data of grassland, in a total of 18 out of the 20 field campaigns at 1170 sub-quadrats. These data were used to analyze the influence of the three filtering methods on MODIS NDVI. The spectral data were acquired using an AvaField-3 portable spectrometer (made by Holland Avantes Company). The spectral range of AvaField-3 is 300–2500 nm, particularly with a resolution of 1.4 nm and a sampling interval of 0.6 nm in the 300–1100 nm and a resolution of 15 nm and a sampling interval of 6 nm in 1100–2500 nm. The fieldwork was conducted on several consecutive sunny days. Measurements were taken on clear sunny days between 11:30 a.m. and 2:00 p.m. The fiber optic sensor, with a field view of 25°, was pointed on the target at nadir position from about a 1-m height above the ground surface. In order to increase the stability and precision of the instrument as advised in the operational manual, we preheated the spectrometer for half an hour before measurements. For each target measurement, the downwelling radiance was first measured by pointing to the white reference panel. The spectral reflectivity was directly recorded by using the upwelling radiance (i.e., reflective radiance) to the target divided by the downwelling radiance. Ten spectral reflectivity curves for each target were collected for later processing in the lab. An example of the mean spectrum of 13 sample plots on 28–29 July 2014 is shown in Figure 2. The artifacts seen (in the Figure 3) around 1150 nm could be mostly due to the mixed effect of different objects in the sample plots, including grass, soil, gravel and dead grass.

In addition, a portable GPS device for 20 ground control points (GCP) was used in the study to record the longitude, latitude and elevation, for later precise geometric corrections of the Landsat and HJ-1B satellite images.

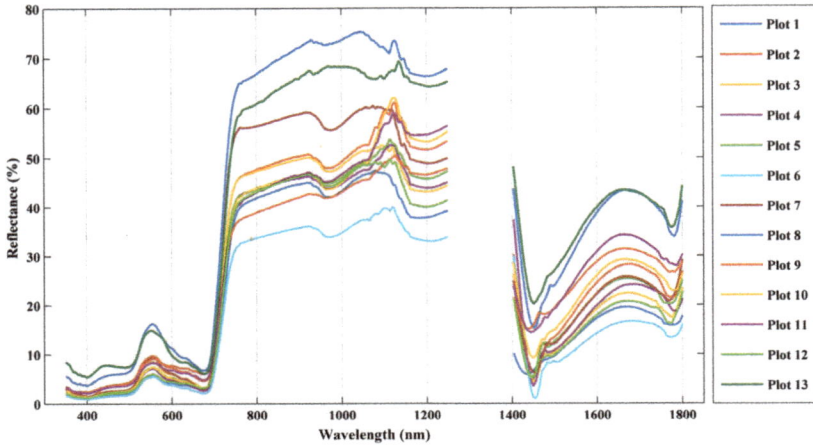

Figure 3. The spectral curves of 13 sample plots on 28–29 July 2014.

2.3. Preprocessing of MODIS Vegetation Index Data

The MODIS vegetation index data were selected from the MODIS 16-d maximum composite NDVI vegetation index product (MOD13Q1) of the United States National Aeronautics and Space Administration. In total, 69 images spanning the years 2013–2015 taken with a spatial resolution of 250 m from orbit number h26v05 were selected. Three filtering methods, the asymmetric Gaussian function (GA), double-logistic curve (LO) and Savitzky–Golay (SG), were used to reconstruct the time series data of NDVI. The four vegetation indices were denoted as $NDVI_{MOD}$ (MOD13Q1 NDVI), $NDVI_{GA}$ (after GA filtering), $NDVI_{LO}$ (after LO filtering) and $NDVI_{SG}$ (after SG filtering). The performance of the three filtering methods was examined using the NDVI values derived from spectrum measurements in the 18 out of 20 field campaigns, as shown in Table 1.

Table 1. The dates between MODIS images (h26v05) and spectral measurements in 13 sample plots in the study area.

Date of MODIS	Measurement Time
2013.08.30–09.14	2013.09.12–09.13
2014.05.26–06.10	2014.05.30–05.31
2014.06.11–06.26	2014.06.14–06.16
2014.06.27–07.12	2014.06.28–06.29
2014.07.13–07.28	2014.07.11–07.13
2014.07.13–07.28	2014.07.26–07.28
2014.08.14–08.29	2014.08.14–08.15
2014.08.30–09.14	2014.09.01–09.02
2014.09.15–09.30	2014.09.26–09.28
2014.10.17–11.01	2014.10.20–10.22
2015.05.10–05.25	2015.05.20–05.22
2015.07.13–07.28	2015.07.14–07.15
2015.07.13–07.28	2015.07.24–07.25
2015.07.29–08.13	2015.08.10–08.11
2015.08.14–08.29	2015.08.20–08.23
2015.08.30–09.14	2015.09.11–09.13
2015.10.01–10.16	2015.10.10–10.11
2015.10.17–11.01	2015.10.20–10.22

For the period of field sampling in August 2013, July 2014 and July, August and September 2015, the NDVI$_{MOD}$, NDVI$_{GA}$, NDVI$_{LO}$ and NDVI$_{SG}$ were used to construct the grassland AGB models, and the performance of the models was assessed using the field measurements.

2.4. Data Processing of Landsat 8 OLI and HJ-1B CCD and Calculation of the Vegetation Index

The Landsat 8 OLI data were obtained from the website of the United States Geological Survey (USGS). The OLI imager includes eight multi-spectral bands with a spatial resolution of 30 m and one panchromatic band with a spatial resolution of 15 m. The imaging range is 185 km × 185 km, and the revisit period is 16 d. In this study, five scenes of Landsat 8 OLI satellite images with no cloud cover in the study area in Xiahe County, Gansu Province, during 2013–2015 were downloaded. The HJ-1B CCD data were acquired from the China Center for Resource Satellite Data and Applications (http://www.cresda.com/n16/index.html). The Environment and Disaster Monitoring Satellite B (HJ-1B) carries two CCD cameras. Each CCD creates four multi-band images with a spatial resolution of 30 m; the frame width of a single CCD image is 360 km. Together, the frame width of the two CCD images spans 700 km, and the revisit period is 4 d. Based on the time of the ground surveys, the revisiting periods of the Landsat 8 and the HJ-1B satellites and cloud cover in the images, a total of five scenes of the HJ-1B CCD images and five scenes of Landsat 8 images close to the field investigation times that cover the entire study area without clouds (Table 2) were selected.

Table 2. The date between satellite images and field measurements in the study area.

Date of Satellite Images	Satellite	Sensor Type	Path	Row	Sampling Time
2013.08.08	Landsat8	OLI	131	36	2013.08.06–09
2013.08.09	HJ-1B	CCD2	12	73	2013.08.06–09
2013.07.29–08.13	MODIS	Terra	26	5	2013.08.06–08.09
2014.07.26	Landsat8	OLI	131	36	2014.07.27–31
2014.07.28	HJ-1B	CCD2	13	72	2014.07.27–31
2014.07.29–08.13	MODIS	Terra	26	5	2014.07.27–07.31
2015.07.13	Landsat8	OLI	131	36	2015.07.11–17
2015.07.13	HJ-1B	CCD1	20	72	2015.07.11–17
2015.07.13–07.28	MODIS	Terra	26	5	2015.07.11–07.17
2015.08.14	Landsat8	OLI	131	36	2015.08.10–11
2015.08.12	HJ-1B	CCD2	16	72	2015.08.10–11
2015.07.29–08.13	MODIS	Terra	26	5	2015.08.10–08.11
2015.09.15	Landsat8	OLI	131	36	2015.09.14–18
2015.09.14	HJ-1B	CCD1	19	72	2015.09.14–18
2015.09.15–09.30	MODIS	Terra	26	5	2015.09.14–09.18

The OLI and CCD data were both processed using ENVI 5.0 software, and the Radiometric Calibration module, the Image to Image module in the registration and the FLAASH Atmospheric Correction module were used for converting the original DN value to atmospheric surface reflectance, precise geometric correction and atmospheric correction of OLI and CCD images, respectively. The image projection was defined as WGS_1984_UTM_ZONE_47N [26]. Then, the Band Math module was used to calculate the NDVI values for the Landsat 8 OLI and HJ-1B CCD images. The NDVI values were extracted from the Landsat 8 OLI and HJ-1B CCD images in the 13 sample sub-plots (250 m × 250 m). These results were used as the NDVI values (namely, NDVI$_{CCD}$ and NDVI$_{OLI}$) corresponding to the ground sampling sites.

2.5. Spectral Data Processing and Accuracy Evaluation of MODIS NDVI

The spectral data were processed by using the Viewer 7 software of AvaField-3. First, we previewed the spectral curves in the Viewer and removed the abnormal curves. Then, we calculated the mean value of the normal curves for each plot. Third, according to the spectral response function of the MODIS sensor, the reflectance of the red and near-infrared bands corresponding to the MODIS sensor (near-infrared band: 841–876 nm; red band: 620–670 nm) was calculated (Equation (1)). Finally, the NDVI was calculated as Equation (2).

$$\rho_i = \frac{\sum_{\lambda_{ai}}^{\lambda_{bi}} \varphi(\lambda_i)\psi(\lambda_i)}{\sum_{\lambda_{ai}}^{\lambda_{bi}} \varphi(\lambda_i)} \tag{1}$$

where ρ_i represents the reflectivity of band i, λ_{ai} represents the starting wavelength of band i, λ_{bi} represents the termination wavelength of band i, $\psi(\lambda_i)$ represents the reflectivity value at wavelength λ and $\varphi(\lambda_i)$ represents the spectral response factor at wavelength λ of band i.

$$\text{NDVI} = (\rho_{Nir} - \rho_{Red})/(\rho_{Nir} + \rho_{Red}) \tag{2}$$

where ρ_{Nir} represents the reflectance of the near-infrared band and ρ_{Red} represents the reflectance of the red band.

According to the measured spectrum of grassland in the sub-quadrat, the NDVI values of the grassland were calculated by taking the average NDVI of the five quadrats as the ground-measured NDVI value for that sub-plot and then calculating the root mean square error (RMSE) with NDVI_{MOD}, NDVI_{GA}, NDVI_{LO} and NDVI_{SG} (Equation (3)) and their correlation coefficients (r). The smaller the RMSE and the larger the r value are, the better the vegetation index.

$$\text{RMSE} = \sqrt{\frac{\sum_{i=1}^{n}(y_i - y_i')^2}{n}} \tag{3}$$

where y_i represents the ground-measured value, y_i' is the estimated value, i represents a sampling plot and n stands for the number of sample plots.

2.6. Construction of Grassland Biomass Monitoring Model and Accuracy Evaluation

The average dry weight of AGB in all of the sub-quadrats of each sub-plot was estimated as the biomass for that sample plot. Taking the grassland biomass of different sample plots as the dependent variable and taking NDVI_{MOD}, NDVI_{GA}, NDVI_{LO}, NDVI_{SG}, NDVI_{CCD} and NDVI_{OLI} as the independent variables, the leave-one-out cross validation (LOOCV) method, RMSE and relative estimate error (REE) (Equation (4)) were used to evaluate the accuracy and estimation errors of the linear, exponential, logarithmic and power regression models for the 6 vegetation indices at the sample plot level.

$$\text{REE} = \sqrt{\frac{\sum_{i=1}^{n}\left[(y_i - y_i')/y_i'\right]^2}{n}} \tag{4}$$

where y_i represents the ground-measured value, y_i' is the estimated value, i represents a sampling plot and n stands for the number of sample plots.

In addition, to study the relative estimation error of the alpine meadow grassland AGB monitoring model on the regional scale (i.e., the experimental area of this study), the average AGB dry weight for all of the sample plots in each sample area was adopted as the ground-measured biomass in this sample area. Then, the RMSE and REE were used to assess the performance of each model for estimating total yield and yield per unit area.

3. Results and Analysis

3.1. Statistical Analysis of Ground Observation AGB and the Corresponding Multi-Source Satellite NDVI

Table 3 shows the statistical results of AGB in the surveyed sample plots and the NDVI calculated using the corresponding remote sensing data from 2013–2015. There are considerable differences in the biomass during the grass growing season in the 13 sample plots of the experimental area. The average values range from 1280–1887 kg DW/ha over the past three years, while the coefficient of variation ranges from 0.16–0.73. The largest biomass from the 13 sample plots is 3963 kg DW/ha, and the minimum is 745.5 kg DW/ha. The magnitude of the variation in the $NDVI_{CCD}$, $NDVI_{OLI}$ and $NDVI_{MOD}$ values is relatively small, and their dispersion degree is similar; the standard deviation and coefficient of variation range from 0.03–0.13 and 0.05–0.18, respectively. In terms of mean NDVI value ranges, they are very similar, from 0.59–0.75 and 0.57–0.74, respectively, for $NDVI_{CCD}$ and $NDVI_{MOD}$ and 0.62–0.83 for $NDVI_{OLI}$, slightly larger than the former two.

For the entire experimental area, the average total biomass in August 2013–2015 is 285×10^3 kg, and the average grass yield is 1813.5 kg DW/ha. In the experimental area, the average biomass of the five times in July–September of 2013–2015 is 1808.4 kg DW/ha, with the standard deviation of 847.3 kg DW/ha and the variation coefficient of 0.47. Similar to the sample plots level, the $NDVI_{CCD}$ and $NDVI_{MOD}$ means are very close (0.68–0.69), slightly smaller than the 0.78 of $NDVI_{OLI}$.

3.2. Influence of Different Filtering Methods on MODIS NDVI

The NDVI calculated using the measured spectrum data on the ground is treated as the ground-measured value to examine the accuracy of $NDVI_{MOD}$ and the three NDVI time series after the filters are applied, i.e., $NDVI_{GA}$, $NDVI_{LO}$ and $NDVI_{SG}$. The results are shown in Table 4, indicating that the three filtering methods perform better overall than the original MOD13Q1 NDVI, and the SG filtering is the best. At the sample plot level, the SG filtering method achieves the best results for Sample Plots 1, 4–8, 10, 12 and 13, with RMSE ranging from 0.028–0.098 and r values ranging from 0.78–0.91. The logistic filtering method achieved the best results for Sample Plots 3 and 11; with RMSE of 0.050–0.076 and r values of 0.88–0.89. The Gaussian filtering method achieved the best results for Sample Plots 2 and 9; with RMSE of 0.078–0.094 and r values of 0.81–0.89. Among all of the sample plots, all three filtering methods achieved relatively good results at Plot 1, with RMSE of 0.028–0.029 and r of 0.90–0.91. In contrast, all three filtering methods achieve relatively poor results at Plot 13, with RMSE values ranging from 0.098–0.113 and r values from 0.85–0.91.

Table 3. Descriptive statistics of grassland above-ground biomass (AGB) and corresponding multi-source satellite NDVI during July–September of 2013–2015 in the study area, Xiahe County.

Index	Statistics	Plot													
		E1	E2	E3	E4	E5	E6	E7	D8	D9	C10	B11	A12	A13	All
Biomass (10³ kg/ha)	Maximum	2.20	2.28	2.70	2.08	2.52	2.94	3.20	2.47	1.77	2.03	2.41	3.96	2.67	3.96
	Minimum	1.12	1.15	0.83	0.86	0.79	1.13	0.97	0.94	1.18	0.95	0.93	0.83	0.75	0.75
	Average	1.77	1.89	1.86	1.29	1.48	1.71	1.75	1.82	1.51	1.37	1.44	1.68	1.28	1.81
	Standard deviation	0.47	0.52	0.81	0.534	0.74	0.84	1.00	0.68	0.25	0.46	0.67	1.52	0.93	0.85
	Cv	0.27	0.27	0.44	0.42	0.50	0.50	0.57	0.37	0.16	0.34	0.47	0.12	0.73	0.47
	n	25	25	25	25	25	25	25	25	25	25	25	25	25	325
HJ-CCD NDVI	Maximum	0.68	0.74	0.80	0.76	0.80	0.78	0.73	0.74	0.68	0.71	0.71	0.67	0.67	0.80
	Minimum	0.57	0.64	0.67	0.62	0.68	0.64	0.65	0.65	0.62	0.58	0.57	0.53	0.51	0.51
	Average	0.65	0.71	0.75	0.67	0.73	0.72	0.70	0.70	0.65	0.64	0.64	0.61	0.59	0.68
	Standard deviation	0.04	0.05	0.05	0.07	0.05	0.06	0.04	0.04	0.03	0.05	0.06	0.06	0.07	0.07
	Cv	0.07	0.07	0.06	0.10	0.07	0.08	0.06	0.05	0.05	0.08	0.09	0.09	0.12	0.10
	n	5	5	5	5	5	5	5	5	5	5	5	5	5	65
Landsat-8 OLI DVI	Maximum	0.77	0.86	0.90	0.86	0.87	0.86	0.85	0.87	0.82	0.78	0.82	0.79	0.71	0.97
	Minimum	0.60	0.70	0.74	0.65	0.73	0.73	0.75	0.76	0.63	0.66	0.69	0.64	0.55	0.55
	Average	0.70	0.79	0.84	0.755	0.81	0.80	0.81	0.82	0.72	0.72	0.75	0.70	0.62	0.78
	Standard deviation	0.07	0.07	0.07	0.10	0.06	0.06	0.05	0.05	0.08	0.06	0.06	0.07	0.07	0.10
	Cv	0.10	0.09	0.08	0.13	0.08	0.08	0.06	0.06	0.11	0.08	0.08	0.10	0.11	0.13
	n	5	5	5	5	5	5	5	5	5	5	5	5	5	65
MOD13Q1 NDVI	Maximum	0.82	0.84	0.82	0.82	0.81	0.81	0.82	0.79	0.73	0.75	0.79	0.69	0.67	0.84
	Minimum	0.66	0.69	0.68	0.65	0.61	0.64	0.64	0.58	0.62	0.46	0.49	0.49	0.44	0.44
	Average	0.73	0.74	0.73	0.72	0.70	0.71	0.72	0.69	0.68	0.60	0.60	0.57	0.58	0.69
	Standard deviation	0.06	0.07	0.07	0.07	0.08	0.07	0.08	0.08	0.05	0.12	0.13	0.09	0.10	0.10
	Cv	0.09	0.09	0.09	0.10	0.12	0.10	0.11	0.12	0.07	0.20	0.21	0.15	0.18	0.15
	n	5	5	5	5	5	5	5	5	5	5	5	5	5	65

Note: Cv depicts the coefficient of variation, equal to the standard deviation divided by the mean; n depicts the number of observations.

Table 4. The result of NDVI$_{MOD}$, NDVI$_{GA}$, NDVI$_{LO}$ and NDVI$_{SG}$ as compared with the field measurements in 13 sample plots of the study area.

Vegetation Index	Index	Plot													Average
		E1	E2	E3	E4	E5	E6	E7	D8	DB9	C10	B11	A12	A13	
NDVI$_{MOD}$	RMSE	0.035	0.083	0.081	0.079	0.087	0.089	0.105	0.115	0.110	0.087	0.051	0.120	0.111	0.102
	R	0.83**	0.85**	0.86**	0.83**	0.81**	0.81**	0.78**	0.70	0.74**	0.82**	0.87**	0.82	0.77**	0.77**
NDVI$_{GA}$	RMSE	0.029	0.078	0.077	0.062	0.077	0.070	0.087	0.094	0.094	0.079	0.051	0.113	0.107	0.091
	R	0.90**	0.89**	0.88**	0.90**	0.86**	0.89**	0.86**	0.82*	0.81**	0.86**	0.89**	0.85**	0.79**	0.83**
NDVI$_{LO}$	RMSE	0.028	0.078	0.076	0.062	0.077	0.072	0.088	0.095	0.095	0.080	0.050	0.112	0.107	0.091
	R	0.90**	0.89**	0.88**	0.90**	0.86**	0.88**	0.86**	0.81*	0.80**	0.86**	0.89**	0.86**	0.78**	0.83**
NDVI$_{SG}$	RMSE	0.028	0.079	0.080	0.060	0.075	0.070	0.085	0.093	0.095	0.078	0.051	0.098	0.107	0.090
	R	0.91**	0.88**	0.87**	0.91**	0.87*	0.89**	0.86**	0.82*	0.81**	0.88**	0.88**	0.91**	0.78**	0.84**

** $p < 0.01$, * $p < 0.05$.

3.3. Grassland Biomass Monitoring Model in the Study Area and Evaluation of Its Accuracy at the Sample Plot Level

The results of the accuracy evaluation validated by LOOCV for the grassland biomass models constructed on the three satellites, $NDVI_{MOD}$, $NDVI_{CCD}$ and $NDVI_{OLI}$, as well as the three filtering methods applied to MODIS NDVI only are listed in Table 5. It is found that in the four types of grassland AGB estimation models (linear, logarithmic, power and exponential), the exponential model performs the best, with the smallest RMSE and REE, while the linear or power model performs the second best. Among the six vegetation indices, the exponential model based on $NDVI_{OLI}$ is the best, with the smallest RMSE of 511.6 kg DW/ha and REE of 29.1% and the highest estimation accuracy of the AGB (Figure 4d). The exponential model based on $NDVI_{CCD}$ is the second best (Figure 4c). The exponential model based on $NDVI_{MOD}$ is worse (Figure 4a). All of the filtering methods perform better than the $NDVI_{MOD}$, with $NDVI_{SG}$ the best (Figure 4b).

Table 6 shows the parameters of the best fit model for each of the six vegetation indices. As shown, all six models pass the *T* test and *F* test at a significance level of 0.01, with $NDVI_{OLI}$ the best, followed by $NDVI_{CCD}$, $NDVI_{SG}$ and $NDVI_{MOD}$.

Table 5. The validation results by leave-one-out cross validation for the grassland biomass models based on multi-source satellite data. REE, relative estimate error.

Vegetation Index	Model	Accuracy Evaluation	
		RMSE (kg/ha)	REE (%)
$NDVI_{MOD}$	Linear	594.5	47.8
	Exponential	574.6	35.3
	Logarithm	619.2	61.4
	Power	598.8	36.7
$NDVI_{SG}$	Linear	573.3	45.0
	Exponential	549.7	33.9
	Logarithm	594.6	57.7
	Power	571.4	35.0
$NDVI_{LO}$	Linear	581.1	44.1
	Exponential	560.4	34.1
	Logarithm	602.4	53.8
	Power	582.0	35.2
$NDVI_{GA}$	Linear	583.8	44.2
	Exponential	562.0	34.1
	Logarithm	605.8	53.9
	Power	585.2	35.3
$NDVI_{CCD}$	Linear	557.1	45.2
	Exponential	548.4	31.6
	Logarithm	567.6	74.0
	Power	552.6	32.1
$NDVI_{OLI}$	Linear	516.8	33.7
	Exponential	511.6	29.1
	Logarithm	528.9	41.2
	Power	512.1	29.4

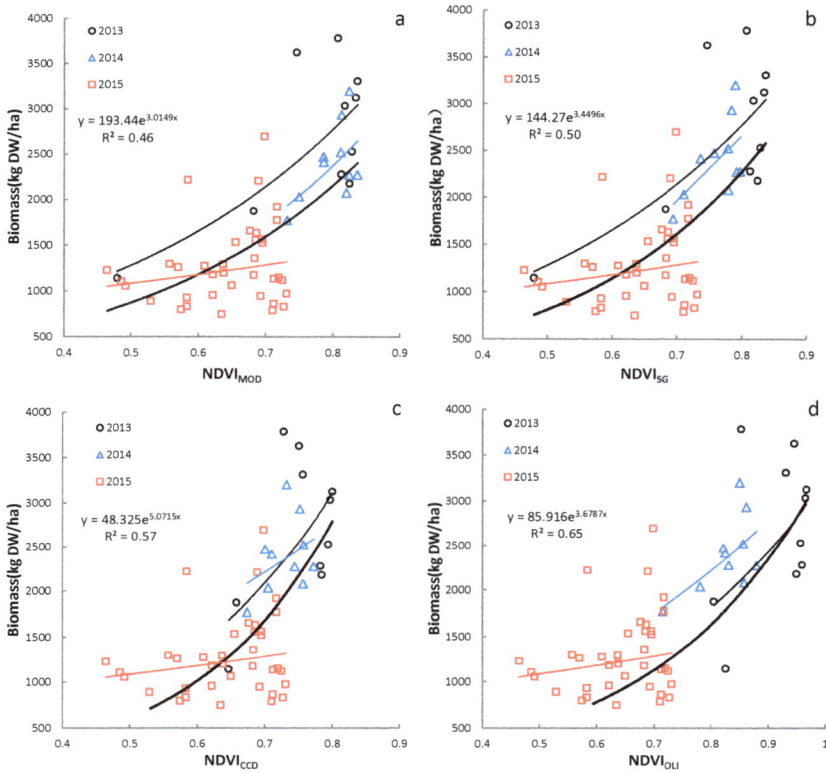

Figure 4. The best fit models constructed based on NDVI$_{MOD}$ (**a**), NDVI$_{SG}$ (**b**), NDVI$_{CCD}$ (**c**) and NDVI$_{OLI}$ (**d**).

Table 6. The results of model fits with the optimum inversion models based on multi-source satellite data.

Vegetation Index	Parameter Estimation and T Test			Regression Significance Test	
	Parameter	Estimated Value	T	R^2	F
NDVI$_{MOD}$	b	3.0149	6.817 **	0.46	46.478 **
	a	193.442	3.229 **		
NDVI$_{SG}$	b	3.4496	7.400 **	0.50	54.759 **
	a	144.265	3.078 **		
NDVI$_{LO}$	b	3.496	7.030 **	0.47	49.413 **
	a	140.404	2.888 **		
NDVI$_{GA}$	b	3.487	6.889 **	0.46	47.464 **
	a	141.383	2.839 **		
NDVI$_{CCD}$	b	5.0715	8.581 **	0.57	73.634 **
	a	48.325	2.457 **		
NDVI$_{OLI}$	b	3.6787	10.017 **	0.65	100.341 **
	a	85.916	3.427 **		

Note: ** represents $p < 0.01$; a and b represent the constant and exponential term of the model, respectively; T and F are the significant values according to the T and F tests.

4. Discussion

4.1. Influence of Different Remote Sensing Data on the Estimation Error of Grassland Biomass

In this study, $NDVI_{MOD}$, $NDVI_{CCD}$ and $NDVI_{OLI}$ data were used in combination with ground observation data during 2013–2015 to construct a grassland biomass inversion model. The errors based on $NDVI_{MOD}$, $NDVI_{CCD}$ and $NDVI_{OLI}$ reached 35.3%, 31.6% and 29.1%, respectively, at the sample plot level (Table 5). Our results indicate that the yield per unit area estimated using the exponential model based on $NDVI_{OLI}$ (1518 kg DW/ha) is closest to the ground-measured value, and its estimation error is the lowest (30.7%), followed by $NDVI_{CCD}$ (32.4%) and $NDVI_{MOD}$ (39.6%) (Table 7). This result suggests that the inverse models based on 30-m resolution (namely, $NDVI_{CCD}$ and $NDVI_{OLI}$) are better than that based on MODIS data with 250-m resolution, and the model based on $NDVI_{OLI}$ is better than that on $NDVI_{CCD}$.

Table 7. The estimation error of the grassland AGB evaluated by different remote sensing data and filtering methods in study area.

Data Type	Vegetation Index	Formula	Yield (AGB) (kg DW/ha)	REE (%)
Different remote sensing data	$NDVI_{OLI}$	$y = 85.916e^{3.6787x}$	1518.0	30.7
	$NDVI_{CCD}$	$y = 48.325e^{5.0715x}$	1472.7	32.4
	$NDVI_{MOD}$	$y = 193.442e^{3.0149x}$	1431.0	39.6
Different filtering methods	$NDVI_{SG}$	$y = 132.146e^{3.584x}$	1564.1	34.9
	$NDVI_{LO}$	$y = 140.404e^{3.496x}$	1422.3	38.6
	$NDVI_{GA}$	$y = 141.383e^{3.478x}$	1408.1	39.3

There are multiple reasons for the above results. Existing studies indicate that the accuracy of the grassland biomass inversion model is not only affected by the temporal and spatial resolutions of the satellite images, sensor type and method of image processing [18,27], but is also subject to the influence of the size, number and representativeness of ground sampling sites. In this study, the five-point method is adopted to determine the sub-plot (30 m × 30 m), which matches the spatial resolution of the Landsat 8 OLI and HJ-1B CCD images. Although the Landsat or HJ-1B pixel could not always exactly match a sub-plot (30 m × 30 m) on the ground due to the orbit drift, we consider this a minor problem, since the relatively uniform distribution of grass in each 250 m × 250 m sample plot was one of the selection criteria. A slight mismatch between sub-plot and satellite pixel does not really make much difference in our case. The average grassland biomass based on five quadrats can reasonably represent the variation of grassland biomass within a full 30 m × 30 m sub-plot; therefore, the accuracy is relatively high for the grassland biomass monitoring model constructed using the remote sensing image data with 30-m resolution (namely, $NDVI_{OLI}$ and $NDVI_{CCD}$). However, the spatial matching with the MODIS vegetation index, which uses a 250-m spatial resolution, has some limitations; consequently, the accuracy of the simulation model is relatively low. In addition, there is an obvious difference between the estimation errors of the biomass models based on $NDVI_{OLI}$ and $NDVI_{CCD}$, which have the same spatial resolution (30 m). The error of the AGB estimation model based on $NDVI_{OLI}$ is smaller than $NDVI_{CCD}$. This difference occurred mainly because the near-infrared (NIR) and red bands (RED) used to calculate their NDVI values come from different sensors that have different band ranges. Specifically, the NIR and RED band ranges used to calculate $NDVI_{CCD}$ are 0.76–0.90 μm and 0.63–0.69 μm, respectively, while for $NDVI_{OLI}$, they are 0.845–0.885 μm and 0.630–0.680 μm, respectively. The narrow band receives less signal disturbances. In addition, the OLI NIR also excludes the influence of water vapor absorption at 0.825 μm, and therefore, the OLI NIR's contribution to retrieve grassland information is more prominent [28].

4.2. Influence of Three Filtering Methods on the Error of Grassland AGB Estimation Based on MODIS NDVI

The exponential models (based on the accuracy of the best inversion mode) for grassland AGB of the three filtering methods including $NDVI_{SG}$, $NDVI_{LO}$ and $NDVI_{GA}$, can better simulate the grassland AGB than the $NDVI_{MOD}$, in which the error decreased by approximately 1.40, 1.14 and 1.13%, respectively (Table 5). The yield per unit area for grassland AGB estimated by $NDVI_{SG}$ (1564.1 kg DW/ha) is the closest to the ground-measured value, and its estimation error was the lowest (34.9%), followed by $NDVI_{LO}$ and $NDVI_{GA}$ (Table 7). This result indicates that the SG filtering method is better at eliminating abnormal values from the MOD13Q1 NDVI; consequently, it yields biomass estimates closer to the ground-measured value. This finding is similar to the research results of Chen et al. (2004) and Geng et al. (2014) Their studies in Southeast Asia indicated that SG filtering can partially eliminate atmospheric interference and the influence of mixed pixels on the grassland vegetation index and can improve the representativeness of MODIS NDVI for grassland vegetation [23,29].

4.3. Assessment of Previously-Established Biomass Inversion Models Based on the MODIS Vegetation Index over the Tibetan Plateau

Many scholars have used the MODIS vegetation index to conduct numerous successful studies on grassland biomass in the Tibetan Plateau area. However, the existing studies show considerable differences in model errors for alpine grassland in the Tibetan Plateau area at the sample plot scale (Table 8). For example, the grassland biomass inversion model constructed by Xu et al. (2007) used field measurement data in the Tibetan Plateau area and the 10-d maximum composite MODIS NDVI (with a 1-km resolution) from the end of July to the end of September in 2007, to construct an exponential function (Model I), with R^2 of 0.75 and overall estimation accuracy up to 80% [19]. In the grassland biomass regression models constructed by Feng et al. (2011), which combined measurement data from the Tibetan Plateau during July–August of 2005 and 2006 with the monthly maximum composition 1-km resolution MODIS NDVI, the exponential function model is the optimum inversion model (Model II). However, its accuracy is lower than that of Model I; with R^2 of 0.49 and RMSE up to 671.8 kg DW/ha [30]. Cui et al. (2011), Wang et al. (2010) and Bao et al. (2010) used the MODIS vegetation index at 500-m, 1-km and 250-m resolution, respectively, to construct biomass regression models for the Gannan prefecture. Their results are similar. The optimum inversion model (in the order of Models III, IV and V) is a power function based on MODIS EVI, with the R^2 values of 0.63, 0.62 and 0.63, respectively [31–33]. Clearly, there are large differences among models. Here, we intend to examine the feasibility of these models all developed at large scales to be applied to our small study area with detailed field data.

Table 8. The existing AGB estimation models for alpine grassland in the Tibetan Plateau region.

Model	Study Area	Area (10⁴ ha)	MODIS	Formula	R^2	Literature
I	Tibetan Plateau	25,724	NDVI	$y = 225.42 \times e^{4.4368x}$	0.75	[19]
II	Tibetan Plateau	25,724	NDVI	$y = 268.810 \times e^{2.398x}$	0.49	[30]
III	The northeast of Tibetan Plateau (Gannan Prefecture)	380	EVI	$y = 3738.073x^{1.553}$	0.63	[31]
IV	The northeast of Tibetan Plateau (Gannan Prefecture)	380	EVI	$y = 5320.7x^{1.9776}$	0.62	[32]
V	The northwest of Gannan Prefecture (Xiahe County)	62.74	EVI	$y = 1719.1x^{2.2588}$	0.63	[33]

Table 9 shows the results of statistical analysis using the biomass data measured on the ground surface during 2013–2015 in our study area to examine the error of the aforementioned five models. As shown, the Model IV error is the smallest (47.2%), followed by V, III, II and I. Among the five models, Models I and II based on the entire Tibetan Plateau perform the worse. Although Model I achieved the lowest error in the Tibetan Plateau (Table 8), its estimation error of grassland AGB in the study area is highest (71.7%). Models III, V and IV are built on the regional area (Xiahe and Gannan) of the Tibetan Plateau; their estimation errors are much smaller, around 47%.

Table 9. The estimation error for the grassland AGB based on existing inversion models at the sample plot and the entire study area scales.

Model	Sample Plot Scale	Study Area Scale	
	REE (%)	Yield (kg DW/ha)	REE (%)
I	71.7	5748.8	68.9
II	58.6	1352.9	48.3
III	47.8	1397.8	46.2
IV	47.2	1470.8	48.0
V	47.3	1551.3	44.6

By analyzing the differences in biomass estimation among Models I–V in the study area (Table 8), we can see that the yield per unit area of grassland biomass (1551.3 kg DW/ha) estimated by Model V in Xiahe County is closest to the ground-measured value of the grassland AGB in the experimental area, and its estimation error for yield per unit area is the lowest among the five models (44.6%). The yield per unit area of grassland AGB estimated by Model I for the experimental area (5748.8 kg DW/ha) is much larger than the ground-measured value, and its estimation error is also the largest (68.9%); the error of the yield per unit area estimated by the other three models ranked from small to large as III, IV and II.

Compared with this study, either at the sample plot scale or the regional scale (namely, the study area of Xiahe County), the estimation accuracy of grassland AGB based on NDVI$_{MOD}$ is the highest, and its estimation error for the yield of grassland AGB per unit area is the smallest in the experimental area (39.3%) (Table 7). Although the previous inversion Models I–V can reflect the overall variation trend of grassland biomass, their errors are higher when applied to our study area. The reason is that in large areas (the Tibetan Plateau area or an entire prefecture or county), the types of natural grassland pasture are complicated, the geographical distribution is wide, the spatial heterogeneity is strong and the vegetation index value is subject to influences from many features and factors (e.g., the ecological environment of the grassland). Therefore, the models built based on a large scale, when applied to small local areas (such as this experimental area), would result in large errors and low accuracy, poor stability and large spatial variations. This conclusion is similar to the results of other studies in other areas [16,34–37].

4.4. Limitations and Prospects of Remote Sensing Monitoring Biomass

In this study, although the optimum models for the remote sensing monitoring of grassland biomass in the study area are determined, due to the limitations of factors, such as the duration of sampling on the ground (only the growing seasons of the grassland from 2013–2015 were sampled), there is still some error and uncertainty for these inversion models, especially the grassland biomass inversion model based on the MODIS NDVI data. There are clearly two areas that we may improve in the future. One is the NDVI saturation problem, as is clearly seen in Figure 4 when the biomass is larger than 3000 kg DW/ha. One way to solve this problem is to use the wide dynamic range vegetation index (WDRVI) [38–40], which was developed mostly to simulate the biomass for crops, usually with very high biomass. It is worthy to introduce WDRVI for the grassland biomass study in the future, especially to explore its sensitivity to the high end of grassland biomass. The other one is the bidirectional reflectance distribution function (BRDF) effects on all remotely-sensed NDVI. As the BRDF effects are already considered in the MODIS NDVI products, our derived NDVI from the Landsat and HJ-1B are not. Although it is not the intention of this study to evaluate the methods for deriving BRDF from these two satellites, we notice that quite a few studies have already reported their methods [41–44]. For example, one study [45] found that there are 0.02–0.06 differences for Landsat images between regular reflectivity and BRDF, and it is worthy to explore and use BRDF-derived NDVI for modeling purposes in our future efforts.

The area of grassland resources is distributed broadly worldwide (accounting for approximately 40% of the global land biome) [2], and the variation on the temporal and spatial scales is dramatic [46]. Satellite data using the relatively high resolutions of Landsat MSS, TM and SPOT show relatively high accuracy in the monitoring of grassland AGB over extended periods; however, because of influences from the transit period, frame width and cloud cover, it is difficult to obtain long time series of high-quality images. Although MODIS has relatively low spatial resolution and relatively large estimation error of grassland AGB, it is still particularly suitable for monitoring widely distributed grassland AGB, mostly due to its high temporal resolution (daily) and large spatial coverage (2330 km). Therefore, it is important to explore new research approaches to grassland AGB monitoring based on MODIS data to improve the accuracy of grassland remote sensing inversion over large regions in the future. These approaches mainly involve the following aspects: (1) enhance the spatial representativeness of ground sampling sites, for example by increasing their number and area and improving the range observed by the ground sampling site and the corresponding spatial matching problem given the size of satellite image pixels; (2) improve the temporal matching between ground sampling sites and remote sensing data, for example by better scheduling the times of field investigations to reduce the time differences between ground surveys and satellite image acquisitions; (3) incorporate new remote sensing observation techniques (e.g., hyperspectral imagery and the UAV remote sensing technique) and strengthen research on the spectral characteristics of the grassland vegetation community and the applications of the narrow band remote sensing vegetation index in monitoring grassland AGB [47,48]; and (4) construct multi-factor grassland AGB estimation models based on statistical analysis and machine learning techniques. These multiple factors include climatic factors (e.g., sunlight, temperature and rainfall), soil factors (e.g., soil nutrients, soil structure and fertility), biological factors (e.g., grassland type, species richness and distribution of malignant weeds) and management factors (e.g., pasture, fencing enclosures and rotational grazing) [49–51]. For example, Li et al. (2013) used neural networks to build an AGB model based on multiple MODIS vegetation indices and showed higher accuracy than a model based on a single index, by decreased RMSE of 433 kg/ha and increased R^2 of 0.35 [49]. A multi-factor model by Liang et al. (2016) showed decreased RMSE by 14.5% as compared with the optimum single-factor model [50]. Diouf et al. (2016) studied the semi-arid grassland in the Sahel region and indicated that a combined photosynthetic radiation and meteorological data model had better performance ($R^2 = 0.69$ and RMSE = 483 kg DW/ha) than the single-factor model of photosynthetic radiation or meteorological data ($R^2 = 0.63$ and 0.55 and RMSE = 550 kg DW/ha and 585 kg DW/ha, respectively) [51].

5. Conclusions

In this study, based on MOD13Q1, HJ-1B CCD and Landsat 8 OLI remote sensing data, grassland observation data in the Sangke grassland of Xiahe County during 2013–2015 are combined to construct a grassland AGB estimation model based on different remote sensing data, and the influence of different filtering approaches for MODIS NDVI on the biomass estimation error of alpine meadow grassland is investigated. The simulation errors of several grassland AGB models in the study area are compared and analyzed, and the applicability of the models is evaluated. The following primary conclusions have been reached.

(1) There is a significant difference in the estimation errors of alpine meadow grassland AGB using remote sensing data from the Chinese HJ-1B CCD, Terra MODIS and Landsat 8 OLI. In this study, the grassland AGB optimum inversion model of the experimental area is the exponential model based on NDVI$_{MOD}$, NDVI$_{OLI}$ and NDVI$_{CCD}$, but different models show considerable differences in the error of grassland AGB inversion. The errors for the estimation of grassland AGB for the optimum models based on NDVI$_{MOD}$, NDVI$_{CCD}$ and NDVI$_{OLI}$ at the sample plot level are 35.3%,

31.6% and 29.1%, respectively. Their yield per unit area estimations for grassland AGB in the experimental area indicate that the exponential model based on $NDVI_{OLI}$ yielded values closest to the ground-measured value; its estimation error for yield per unit area is the smallest (30.7%). The estimation error for yield per unit area for the experimental area with the optimum AGB inversion model based on $NDVI_{OLI}$ decrease by eight and two percentage points, respectively, compared to the optimum inversion models based on $NDVI_{MOD}$ and $NDVI_{CCD}$.

(2) The filtering and de-noising processing of MOD13Q1 NDVI are key for reducing the AGB inversion error of alpine meadow grassland based on MODIS data. At the sample plot level, the estimation errors of the AGB estimation models based on $NDVI_{SG}$, $NDVI_{LO}$ and $NDVI_{GA}$ decreased by 1.40, 1.14 and 1.13 percentage points, respectively, compared to the AGB estimation model based on $NDVI_{MOD}$. On the study area scale (161.36 ha), the estimation errors for the yield per unit area of grassland AGB based on $NDVI_{SG}$, $NDVI_{LO}$ and $NDVI_{GA}$ decreased by 4.48, 0.95 and 0.22, respectively, compared to that based on $NDVI_{MOD}$.

(3) The feasibility study on previous models (I and II, III and IV and V) developed (on MODIS indices) at broad scales to apply to our small study area suggests that the estimation error of these models is higher than that of the $NDVI_{MOD}$ model constructed in this study by 11.9%–36.4% at the sample plot scale and 5.3%–29.6% at the study area scale. Models V, IV and III based on Xiahe County and Gannan Prefecture do not show considerable difference on the estimation error of AGB, ranging from 47.2%–47.8% at the sample plot level and 44.6%–48.0% of the yield per unit area at the study area level. However, Models I and II based on the Tibetan Plateau scale show much larger estimation error, up to 71.7% and 58.6%, respectively, at the sample plot scale and 68.9% and 48.3% of the yield per unit area at the study area scale.

Acknowledgments: This study was supported by the National Natural Science Foundation of China (31672484, 31372367, 41101337 and 41401472), the Project of the Ministry of Agriculture (201203006), the Program for Changjiang Scholars and Innovative Research Team in University (IRT13019) and the Climate Change Special Foundation of China Meteorological Administration (CCSF201603). We also appreciate the editor's and reviewers' constructive suggestions to greatly improve the paper.

Author Contributions: All authors contributed significantly to this manuscript. To be specific, Baoping Meng and Tiangang Liang designed this study. Baoping Meng, Jing Ge, Shuxia Yang, Jinglong Gao, Qisheng Feng, Xia Cui and Xiaodong Huang were responsible for the data processing, analysis and paper writing. Hongjie Xie made valuable revisions and editing of the paper.

Conflicts of Interest: The authors declare no conflict of interest.

References

1. Adams, J.M.; Faure, H.; Faure-Denard, L.; McGlade, J.M.; Woodward, F.L. Increases in terrestrial carbon storage from the last glacial maximum to the present. *Nature* **1990**, *348*, 711–714. [CrossRef]
2. White, R.; Murray, S.; Rohweder, M. *Pilote Analysis of Global Ecosystems: Grassland Ecosystems*; World Resources Institute: Washington, DC, USA, November 2000; Available online: http://earthtrends.wri.org/text/forests-grasslands-drylands/map-229.htm.
3. Scurlock, J.M.O.; Hall, D.O. The global carbon sink: A grassland perspective. *Glob. Chang. Biol.* **1998**, *4*, 229–233. [CrossRef]
4. Feng, X.M.; Zhao, Y.S. Grazing intensity monitoring in Northern China steppe: Integrating CENTURY model and MODIS data. *Ecol. Indic.* **2011**, *11*, 175–182. [CrossRef]
5. Ma, W.H.; Fang, J.Y.; Yang, Y.H.; Mohammat, A. Biomass carbon stocks and their changes in northern China's grasslands during 1982–2006. *Sci. China Life Sci.* **2010**, *53*, 841–850. [CrossRef] [PubMed]
6. Lauenroth, W.K.; Hunt, H.W.; Swift, D.M.; Singh, J.S. Estimating aboveground net primary production in grasslands—A simulation approach. *Ecol. Model.* **1986**, *33*, 297–314. [CrossRef]
7. Jobbagy, E.G.; Sala, O.E. Controls of Grass and Shrub Aboveground Production in the Patagonian Steppe. *Ecol. Appl.* **2000**, *10*, 541–549. [CrossRef]
8. Soussana, J.F.; Loieau, P.; Vjichard, N.; Ceachia, E.; Balesdent, J.; Chevallier, T.; Arrouays, D. Carbon Cycling and Sequestration Opportunities in Temperate Grasslands. *Soil Use Manag.* **2004**, *20*, 219–230. [CrossRef]

9.   Hopkins, A. Relevance and functionality of semi-natural grassland in Europe–status quo and future prospective. In Proceedings of the International Workshop of the Salvere, Raumberg-Gumpenstein, Austria, 26–27 May 2009; pp. 9–14.

10. Moreau, S.; Bosseno, R.; Gu, X.F.; Baret, F. Assessing the biomass dynamics of Andean bofedal and totora highprotein wetland grasses from NOAA/AVHRR. *Remote Sens. Environ.* **2003**, *85*, 516–529. [CrossRef]

11. Nordberg, M.L.; Evertson, J. Vegetation index differencing and linear regression for change detection in a Swedish mountain range using Landsat TM and ETMt imagery. *Land Degrad. Dev.* **2004**, *16*, 139–149. [CrossRef]

12. Ali, I.; Cawkwell, F.; Dwyer, E.; Barrett, B.; Green, S. Satellite remote sensing of grasslands: From observation to management—A review. *J. Plant Ecol.* **2016**, *9*, 649–671. [CrossRef]

13. Rouse, J.W.; Haas, R.H.; Schell, J.A.; Deering, D.W. Monitoring vegetation systems in the Great Plains with ERTS. In Proceedings of the Third ERTS-1 Symposium, Washington DC, USA, 10–14 December 1973; Fraden, S.C., Marcanti, E.P., Becker, M.A., Eds.; NASA SP-351: Washington, DC, USA, 1973; pp. 309–317.

14. Tucker, C.J. Red and Photographic Infrared Linear Combinations for Monitoring Vegetation. *Remote Sens. Environ.* **1979**, *8*, 127–150. [CrossRef]

15. Tucker, C.J.; Justice, C.O.; Prince, S.D. Monitoring the grasslands of the Sahel 1984–1985. *Int. J. Remote Sens.* **1986**, *7*, 1571–1581. [CrossRef]

16. Ullah, S.; Si, Y.; Schlerf, M.; Skidmore, A.K.; Shafique, M.; Iqbal, I.A. Estimation of grassland biomass and nitrogen using MERIS data. *Int. J. Appl. Earth Obs. Geoinf.* **2012**, *1*, 196–204. [CrossRef]

17. Li, F.; Zeng, Y.; Li, X.S.; Zhao, Q.J.; Wu, B.F. Remote sensing based monitoring of interannual variations in vegetation activity in China from 1982 to 2009. *Sci. China Earth Sci.* **2014**, *57*, 1800–1806. [CrossRef]

18. Yang, F.; Sun, J.L.; Fang, H.L.; Yao, Z.F.; Zhang, J.H.; Zhu, Y.Q.; Song, K.S.; Wang, Z.M.; Hu, M.G. Comparison of different methods for corn LAI estimation over northeastern China. *Int. J. Appl. Earth Obs. Geoinf.* **2012**, *18*, 462–471.

19. Xu, B.; Yang, X.; Tao, W.; Qin, Z.H.; Liu, H.Q.; Miao, J.M. Remote sensing monitoring upon the grass production in China. *Acta Ecol. Sin.* **2007**, *27*, 405–413. [CrossRef]

20. Li, F.; Zeng, Y.; Luo, J.H.; Ma, R.H.; Wu, B.F. Modeling grassland aboveground biomass using a pure vegetation index. *Ecol. Indic.* **2016**, *62*, 279–288. [CrossRef]

21. Yang, Y.H.; Fang, J.Y.; Pan, Y.D.; Ji, C.J. Aboveground biomass in Tibetan grasslands. *J. Arid Environ.* **2009**, *73*, 91–95. [CrossRef]

22. Xie, Y.; Sha, Z.Y.; Yu, M.; Bai, Y.F.; Zhang, L. A comparison of two models with Landsat data for estimating above ground grassland biomass in Inner Mongolia, China. *Ecol. Model.* **2009**, *220*, 1810–1818. [CrossRef]

23. Chen, J.; Jönsson, P.; Tamura, M.; Gu, Z.H.; Matsushita, B.; Eklundh, L. A simple method for reconstructing a high-quality NDVI time-series data set based on the Savitzky–Golay filter. *Remote Sens. Environ.* **2004**, *91*, 332–344. [CrossRef]

24. Dusseux, P.; Hubert-Moy, L.; Corpetti, T.; Vertès, F. Evaluation of SPOT imagery for the estimation of grassland biomass. *Int. J. Appl. Earth Obs. Geoinf.* **2015**, *38*, 72–77. [CrossRef]

25. Wang, X.P.; Guo, N.; Zhang, K.; Wang, J. Hyperspectral Remote Sensing Estimation Models of Aboveground Biomass in Gannan Rangelands. *Procedia Environ. Sci.* **2011**, *10*, 697–702.

26. Liu, B.K.; Du, Y.E.; Liang, T.G.; Feng, Q.S. Outburst Flooding of the Moraine-Dammed Zhuonai Lake on Tibetan Plateau: Causes and Impacts. *IEEE Geosci. Remote Sens. Lett.* **2016**, *13*, 570–574. [CrossRef]

27. Jia, W.X.; Liu, M.; Yang, Y.H.; He, H.L.; Zhu, X.D.; Yang, F.; Yin, C.; Xiang, W.N. Estimation and uncertainty analyses of grassland biomass in Northern China: Comparison of multiple remote sensing data sources and modeling approaches. *Ecol. Indic.* **2016**, *60*, 1031–1040. [CrossRef]

28. Irons, J.R.; Dwyer, J.L.; Barsi, J.A. The next Landsat satellite: The Landsat Data Continuity Mission. *Remote Sens. Environ.* **2012**, *122*, 11–21. [CrossRef]

29. Geng, L.L.; Ma, M.G.; Wang, X.F.; Yu, W.P.; Jia, S.Z.; Wang, H.B. Comparison of Eight Techniques for Reconstructing Multi-Satellite Sensor Time-Series NDVI Data Sets in the Heihe River Basin, China. *Remote Sens.* **2014**, *6*, 2024–2049. [CrossRef]

30. Feng, Q.S.; Gao, X.H.; Huang, X.D.; Yu, H.; Liang, T.G. Remote sensing dynamic monitoring of grass growth in Qinghai-Tibet plateau from 2001 to 2010. *J. Lanzhou Univ.* **2011**, *47*, 75–90.

31. Cui, X.; Guo, Z.G.; Liang, T.G.; Shen, Y.Y.; Liu, X.Y.; Liu, Y. Classification management for grassland using MODIS data: A case study in the Gannan region, China. *Int. J. Remote Sens.* **2012**, *33*, 3156–3175. [CrossRef]

32. Wang, Y.; Xia, W.T.; Liang, T.G.; Wang, C. Spatial and temporal dynamic changes of net primary product based on MODIS vegetation index in Gannan grassland. *Acta Pratacult. Sin.* **2010**, *19*, 201–210.

33. Bao, H.M. Dynamic Monitoring and Prediction of Aboveground Biomass of Natural Grassland—A Case Study in Xiahe County of Gansu Province. Master's Thesis, Gansu Agricultural University, Lanzhou, China, 2010.

34. Porter, T.F.; Chen, C.C.; Long, J.A.; Lawrence, R.L.; Sowell, B.F. Estimating biomass on CRP pastureland: A comparison of remote sensing techniques. *Biomass Energy* **2014**, *66*, 268–274. [CrossRef]

35. Reddersen, B.; Fricke, T.; Wachendorf, M. A multi-sensor approach for predicting biomass of extensively managed Grassland. *Comput. Electron. Agric.* **2014**, *109*, 247–260. [CrossRef]

36. Gao, T.; Yang, X.Y.; Jin, Y.X.; Ma, H.L.; Li, J.Y.; Yu, Q.Y.; Zheng, X. Spatio-Temporal Variation in Vegetation Biomass and Its Relationships with Climate Factors in the Xilingol Grasslands, Northern China. *PLoS ONE* **2013**, *8*, e83824. [CrossRef] [PubMed]

37. Ahamed, T.; Tian, L.; Zhang, Y.; Ting, K.C. A review of remote sensing methods for biomass feedstock production. *Biomass Bioenergy* **2011**, *35*, 2455–2469. [CrossRef]

38. Gitelson, A.A. Wide Dynamic Range Vegetation Index for remote quantification of biophysical characteristics of vegetation. *J. Plant Physiol.* **2004**, *161*, 165–173. [CrossRef] [PubMed]

39. Viña, A.; Henebry, G.M.; Gitelson, A.A. Satellite monitoring of vegetation dynamics: Sensitivity enhancement by the wide dynamic range vegetation index. *Geophys. Res. Lett.* **2004**, *31*, 373–394. [CrossRef]

40. Viña, A.; Gitelson, A.A. New developments in the remote estimation of the fraction of absorbed photosynthetically active radiation in crops. *Geophys. Res. Lett.* **2005**, *32*, 195–221. [CrossRef]

41. Nagol, J.R.; Sexton, J.O.; Kim, D.H.; Anand, A.; Morton, D.; Vermote, E.; Townshend, J.R. Bidirectional effects in Landsat reflectance estimates: Is there a problem to solve? *ISPRS J. Photogramm. Remote Sens.* **2015**, *103*, 129–135. [CrossRef]

42. Zhang, H.K.; Roy, D.P. Landsat 5 Thematic Mapper reflectance and NDVI 27-year time series inconsistencies due to satellite orbit change. *Remote Sens. Environ.* **2016**, *186*, 217–233. [CrossRef]

43. Gao, F.; Jin, Y.; Schaaf, C.B.; Strahler, A.H. Bidirectional NDVI and atmospherically resistant BRDF inversion for vegetation canopy. *IEEE Trans. Geosci. Remote Sens.* **2002**, *40*, 1269–1278.

44. Li, A.; Wang, Q.; Bian, J.; Lei, G. An Improved Physics-Based Model for Topographic Correction of Landsat TM Images. *Remote Sens.* **2015**, *7*, 6296–6319. [CrossRef]

45. Roy, D.P.; Zhang, H.K.; Ju, J.; Gomez-Dansd, J.L.; Lewisd, P.E.; Schaaf, C.B.; Sun, Q.; Li, J.; Huang, H.; Kovalskyy, V. A general method to normalize Landsat reflectance data to nadir BRDF adjusted reflectance. *Remote Sens. Environ.* **2016**, *176*, 255–271. [CrossRef]

46. Jacques, D.C.; Kergoat, L.; Hiernaux, P.; Mougin, E.; Defourny, P. Monitoring dry vegetation masses in semi-arid areas with MODIS SWIR bands. *Remote Sens. Environ.* **2014**, *153*, 40–49. [CrossRef]

47. Liu, M.; Liu, G.H.; Gong, L.; Wang, D.B.; Sun, J. Relationships of biomass with environmental factors in the grassland area of Hulunbuir, China. *PLoS ONE* **2014**, *9*, e102344. [CrossRef] [PubMed]

48. Gao, T.; Xu, B.; Yang, X.C.; Jin, Y.X.; Ma, H.L.; Li, J.Y.; Yu, H.D. Using MODIS time series data to estimate aboveground biomass and its spatio-temporal variation in Inner Mongolia's grassland between 2001 and 2011. *Int. J. Remote Sens.* **2013**, *34*, 7796–7810. [CrossRef]

49. Li, F.; Jiang, L.; Wang, X.F.; Zhang, X.Q.; Zheng, J.J.; Zhao, Q.J. Estimating grassland aboveground biomass using multitemporal MODIS data in the West Songnen Plain, China. *J. Appl. Remote Sens.* **2013**, *7*, 124–131. [CrossRef]

50. Liang, T.G.; Yang, S.X.; Feng, Q.S.; Liu, B.K.; Zhang, R.P.; Huang, X.D.; Xie, H.J. Multi-factor modeling of above-ground biomass in alpine grassland: A case study in the Three-River Headwaters Region, China. *Remote Sens. Environ.* **2016**, *186*, 164–172. [CrossRef]

51. Diouf, A.A.; Hiernaux, P.; Brandt, M.; Faye, G.; Djaby, B.; Diop, M.B.; Ndione, J.A.; Tvchon, B. Do Agrometeorological Data Improve Optical Satellite-Based Estimations of the Herbaceous Yield in Sahelian Semi-Arid Ecosystems? *Remote Sens.* **2016**, *8*, 668. [CrossRef]

remote sensing

MDPI

Article

Modeling Biomass Production in Seasonal Wetlands Using MODIS NDVI Land Surface Phenology

Maria Lumbierres [1,2]**, Pablo F. Méndez** [2,]*****, Javier Bustamante** [1,2]**, Ramón Soriguer** [3] **and Luis Santamaría** [2,]*****

[1] Remote Sensing and GIS Lab (LAST-EBD), Estación Biológica de Doñana (CSIC), C/Américo Vespucio 26, Isla de la Cartuja, Sevilla 41092, Spain; maria.lumbierres@gmail.com (M.L.); jbustamante@ebd.csic.es (J.B.)

[2] Department of Wetland Ecology, Estación Biológica de Doñana (CSIC), C/Américo Vespucio 26, Isla de la Cartuja, Sevilla 41092, Spain

[3] Department of Ethology & Biodiversity Conservation, Estación Biológica de Doñana (CSIC), C/Américo Vespucio 26, Isla de la Cartuja, Sevilla 41092, Spain; soriguer@ebd.csic.es

***** Correspondence: pfmendez@ebd.csic.es (P.F.M.); luis.santamaria@ebd.csic.es (L.S.); Tel.: +34-954-23-23-40 (L.S.)

Academic Editors: Lalit Kumar, Onisimo Mutanga, Xiaofeng Li and Prasad S. Thenkabail
Received: 3 March 2017; Accepted: 17 April 2017; Published: 21 April 2017

Abstract: Plant primary production is a key driver of several ecosystem functions in seasonal marshes, such as water purification and secondary production by wildlife and domestic animals. Knowledge of the spatio-temporal dynamics of biomass production is therefore essential for the management of resources—particularly in seasonal wetlands with variable flooding regimes. We propose a method to estimate standing aboveground plant biomass using NDVI Land Surface Phenology (LSP) derived from MODIS, which we calibrate and validate in the Doñana National Park's marsh vegetation. Out of the different estimators tested, the Land Surface Phenology maximum NDVI (LSP-Maximum-NDVI) correlated best with ground-truth data of biomass production at five locations from 2001–2015 used to calibrate the models ($R^2 = 0.65$). Estimators based on a single MODIS NDVI image performed worse ($R^2 \leq 0.41$). The LSP-Maximum-NDVI estimator was robust to environmental variation in precipitation and hydroperiod, and to spatial variation in the productivity and composition of the plant community. The determination of plant biomass using remote-sensing techniques, adequately supported by ground-truth data, may represent a key tool for the long-term monitoring and management of seasonal marsh ecosystems.

Keywords: Land Surface Phenology; wetlands; above ground biomass; NDVI; MODIS time series

---

1. Introduction

Plant primary production is a key driver of ecosystem dynamics and can thus influence several ecosystem functions, such as water purification capacity and secondary production by animals. Knowledge of the spatio-temporal dynamics of plant biomass production is essential to inform the management of natural resources, in conservation areas and in agro-pastoral systems [1–4]—particularly in the Mediterranean and semiarid regions, where inter-annual changes in precipitation often result in large variations in plant production [5].

Traditional methods for plant biomass estimation are based on in-situ observations. They can be highly accurate but often involve intensive field work and destructive methods, which makes them costly and inapplicable to inaccessible or sensitive areas, or when involving endangered species [4,6,7]. Remote sensing constitutes an increasingly used alternative [8], based on the relationship between satellite-derived metrics and primary production [7]. Remote sensing may allow for the non-destructive, high-resolution coverage of large, remote, and/or inaccessible areas, such as mountains [9], deserts [10], or wetlands [4,11–13]. Remote sensing allows for the reconstruction of historical trends as well, using satellite image time series: for example, the reconstruction of the hydroperiod in Doñana marsh from 1974–2014 [14], or the assessment of rangeland conditions in semiarid regions [15]. The most widely used methods to monitor vegetation are based on the use of vegetation indexes, such as the Normalized Difference Vegetation Index (NDVI) or the Enhanced Vegetation Index (EVI), as proxies of aboveground biomass [7,16,17]. However, the use of these indexes is also subjected to limitations and criticism; for example, they have been shown to saturate asymptotically at high biomass values [12,18].

In general, the assessment of plant production can be based on the analysis of single (i.e., one-date) images, bi-temporal change detection, or temporal trajectory analysis, followed by the interpretation of results over time [19]. For vegetation, a traditional group of methods relies on quantifications of the differences in statistical metrics of the vegetation-index time series like, for example, the beginning and end of the growing season, the maximum and minimum values, the annual mean, or the variance [9]. In regions with strongly seasonal climates, production is typically assessed by searching for anomalies in the current NDVI against the average of the whole time series, or against reference values from the same period of the year, which informs about the current status of vegetation as compared to other seasons, or to an average condition [20]. This is made at predefined fixed dates, which works well when seasonal cycles are regular, but is often problematic when they vary across years due to climatic or environmental variability. In such cases, observed anomalies in NDVI data may simply constitute a temporal shift of the growth season—i.e., an early (positive NDVI anomaly) or delayed (negative NDVI anomaly) start of the growth season [10]. Other key problems may be related to the lack of consistency and reliability of the NDVI images used for analysis due to noise or errors, especially when a single image per year is used. Examples include variation in viewing and illumination geometry, resolution and calibration, digital quantization errors, ground and atmospheric conditions, as well as orbital and sensor degradation [7,21].

To overcome these limitations, the use of smoothed NDVI time series including a number of consecutive growing seasons (instead of a single image per growing season) is being proposed. Such time series analyses make use of all the information accumulated at the end of the growing season to estimate the parameters describing vegetation phenology (e.g., [21,22]). Indeed, the study of vegetation phenology has become very relevant in several realms, such as productivity and the carbon cycle (e.g., [23,24]), climate change and its impacts on ecosystems [25–27], as well as crop and pasture monitoring [10]. During the last decade, on-the-ground phenological studies have been complemented by studies focusing on large-scale remote sensing [28], technically referred to as Land Surface Phenology (LSP, [12]). LSP can be defined as the timing of recurring changes in the reflectance of electromagnetic radiation from the land surface due to concurrent life-cycle changes of vegetation [29]. It is generally measured by deriving either vegetation parameters (e.g., leaf area index (LAI), fraction of absorbed photosynthetically active radiation (FAPAR) or vegetation indexes (e.g., NDVI, EVI) from remote-sensing data [9,30–32]. These vegetation indexes are used to maximize the extraction of variability assigned to certain plant features (e.g., leaf area, canopy cover, photosynthetic activity) while minimizing other unwanted effects (e.g., geometric, soil color, or atmospheric effects), thus enhancing the information contained in spectral reflectance data [12,33]. LSP is then characterized using different mathematical procedures such as the identification of global/local thresholds and points of maximum increase/decrease, curve fitting and the subsequent extraction of inflection points or thresholds, and harmonic analysis [20].

In this article, we present a method for estimating plant biomass production in seasonal wetlands based on the NDVI from the Moderate-resolution Imaging Spectroradiometer (MODIS). Method development included the comparison of the two different approaches discussed above, namely the use of single images versus the characterization of LSP using the whole time-series; as well as the use of different estimators within each of these two approaches to estimate biomass production across the whole study area for the 16-year series.

We developed and applied this method in a particularly challenging study area: the semiarid marsh and wetlands of the Guadalquivir river estuary (Doñana National Park, SW Spain; 'Doñana marsh' hereafter). As in many arid and semiarid regions, the determination of biomass production is particularly challenging due to the flooding regime, the color influence of soils, and the spatial variation in vegetation communities and species composition [34–37]. The Doñana marsh consists of a diverse and complex array of ecosystems affected by a highly dynamic interplay among vegetation, soil and water [38], whose prolonged land-use history fostered a mix of natural and semi-natural vegetation [39]. Its vegetation provides habitat and food for a highly diverse fauna, making the area a biodiversity hotspot; but grazing by wild and domestic herbivores largely determines plant standing crop and may result in overgrazing, particularly during dry years [40,41]. Studying the primary production of the marsh vegetation is therefore essential for the management and conservation of the Doñana National Park; while its huge size and accessibility problems during most of the flooding period makes this a particularly challenging task using solely on-the-ground approaches.

2. Materials and Methods

2.1. Study Area

The study focuses on the helophyte community of the Doñana marsh, an iconic wetland included in the Doñana National Park (SW Spain, Figure 1, 37°01′ N, 6°26′ W). Doñana has a sub-humid Mediterranean climate characterized by mild winters and hot summers, and rainfall concentrated in autumn (October–December) and spring (March–May). The Doñana marsh is a seasonal floodplain with a flooding regime that depends on rainfall [14,42]. The helophyte community is strongly synchronized with flooding, starting to grow after the water level reaches a peak (February–March), then growing rapidly to create a vegetation layer of approximately 1 m height, and becoming senescent by August when the marsh is dry [42]. More specifically, we focused on nearly-monospecific stands of saltmarsh bulrush (*Bolboschoenus maritimus*) belonging to the phytosociological association *Bolboschoenetum maritimi*. The saltmarsh bulrush represent one of the key primary producers of the marsh and thus sustains many elements of its food chain, including wintering waterfowl that consumes its tubers and seeds (e.g., greylag geese *Anser anser*; [43,44]). Domestic (cattle and horse) and wild (red deer, fallow deer and wild boar) ungulates also make use of the study area, grazing on saltmarsh bulrush and other plants [40].

Vegetation species composition fluctuates interannually as a consequence of climatic variability. To define the limits of the study area in the Doñana marsh, we selected an area as homogenous as possible, by performing an unsupervised classification in ENVI 5.4 using five classes and 10 iterations. We used the descriptive statistics (mean, median, standard deviation, maximum value and minimum value) of a time series of NDVI images from the Landsat satellite TM and ETM+ sensors from 1984 to 2015 [45]. The resulting class that spatially corresponded better with saltmarsh bulrush dominance for the study period defined the study area limits.

The study area includes two estates with different ownership regimes (public and communal land, respectively). During the study period (2001–2015, see below), there were no major changes in the area apart from natural variation in rainfall (typical in a Mediterranean climate), small changes in the hydrology of the marsh, and moderate shifts in the stocking rates allowed at communal land (which had decreased during the 1990s, but remained relatively stable during the 2000s and tended to increase after 2010).

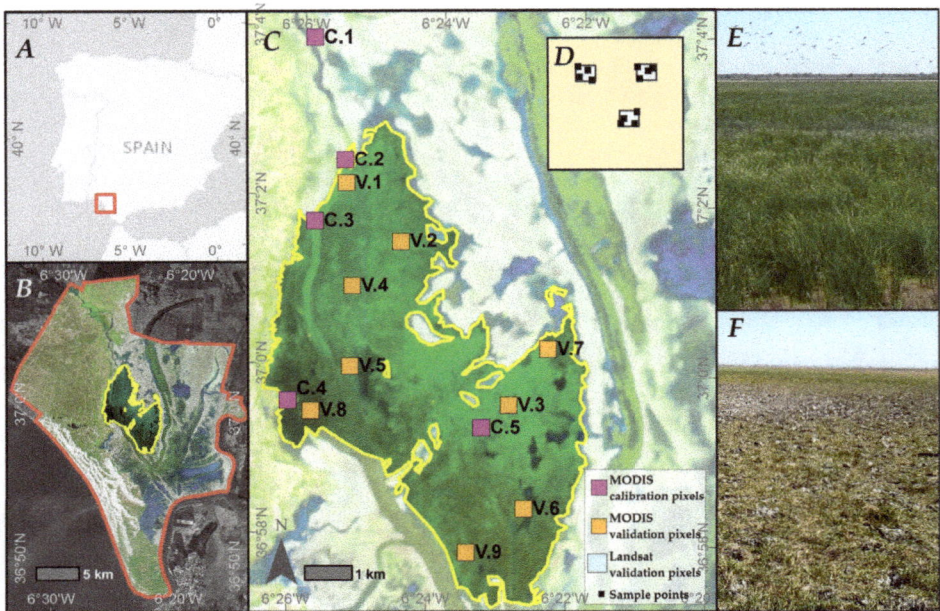

Figure 1. (**A**) Location of the Doñana National Park in the southwest of Spain. (**B**) Location of the study area inside the Doñana National Park marsh. (**C**) Ground-truth biomass plots inside the study area. (**D**) Zoom to a MODIS validation pixel that exemplifies the sample design stratification. (**E**) Picture of the helophyte community (May 2016). (**F**) Picture of an area heavily grazed by cattle (June 2016).

2.2. Satellite Data

Remote sensing data consisted of satellite images from the MODIS (Moderate Resolution Imaging Spectroradiometer) sensor onboard the TERRA satellite (NASA). In particular, we used a raw time series of 387 images provided by the data service platform from the University of Natural Resources and Life Sciences of Vienna (BOKU; [46]). This platform offers a modification of Global MOD13Q1 data, which is the NDVI vegetation index product provided by NASA Land Processes Distributed Active Archive Center, in smooth and raw images (from 2000 to the present) every 16 days as the product of an algorithm that calculates the Maximum Value Composite [47].

2.3. Biomass Data

Plant biomass data consisted of two datasets: one used to calibrate and select among alternative remote sensing models, and another used to validate the best model. The calibration dataset belongs to a long-term study on the impact of ungulate grazing initiated in 1982 by Ramón Soriguer (Doñana Biological Station). Currently, it is the only long-term data set available on above-ground biomass for the Doñana marsh. It was designed neither for this study nor to be a ground-truthing set for data extracted from satellite images; hence, it presents some limitations such as being restricted to localities that remain accessible in high-flood years [40]. Calibration data consisted of regular annual harvests of aboveground biomass production (standing crop, in kg dw/ha) in five fixed locations within the Doñana marsh (Figure 1) from 2001 to 2015, amounting to 75 samples [40]. These five locations were selected due to their accessibility and representativeness of the saltmarsh bulrush community. One of the calibration pixels (C.1 in Figure 1) was placed outside the study area, but it showed a similar vegetation community (dominated by saltmarsh bulrush) and ecological characteristics. For each calibration location and date, biomass production values are based on 5 samples spaced 40 meters along a 200 m transect. Each sample consist of the aboveground biomass present in a randomly oriented rectangle 10 cm wide by 100 cm long, clipped at ground level using an electric grass shear. Samples were transported to the lab, sorted by species discarding dry biomass from the previous year, dried in paper envelopes (72 h at 60 °C) and weighted (accuracy: ±0.01 g). After two years of monthly sampling to assess the phenological cycle, harvests were taken twice a year: (i) one in May–August to estimate peak biomass production (date adjusted to flooding levels, which determine plant phenology), and (ii) another in September-October to estimate biomass "leftovers" (biomass production not consumed by herbivores after the end of the growth season) [40]. We used as ground-truth the annual maximum biomass estimate sampled at each location regardless of the month in which it was collected.

Validation data were collected in August 2016 following a new stratified random sampling scheme. In order to ensure that the sampling areas covered adequately the complete productivity range, we first classified the study area in three biomass-production categories using the averaged value of maximum NDVI from 2001 to 2015: low (<0.5 NDVI), medium (0.5–0.6 NDVI) and high (>0.6 NDVI). These thresholds divided the area in three sub-areas of similar size. Then, among all the MODIS pixels we randomly selected 3 pixels within each NDVI category (250 × 250 m, 9 MODIS pixels in total). Within each MODIS pixel we randomly selected 3 Landsat pixels (30 × 30 m). Within each Landsat pixel we randomly selected four sampling points (1 × 1 m; Figure 1). This design resulted in a total of 108 sampling points (12 biomass estimates per MODIS pixel). The design was chosen to adequately sample the spatial variability in biomass in the study area, to obtain more precise estimates of mean MODIS pixel biomass. The sampling design was aimed at also providing information on the spatial scale at which variation in biomass production occurs and how it relates to the resolution provided by MODIS and Landsat images. Aboveground biomass samples were collected using 100 × 100 cm squares with the electric grass shear. Samples were transported to the lab, sorted by species discarding dry biomass from the previous year, dried in paper envelopes (72 h at 80 °C), and weighted (accuracy: ±0.1 g).

2.4. Environmental Data

Environmental data were used to evaluate model robustness and analyze model results. They consisted of two variables: (1) hydroperiod, i.e., the number of days that each pixel remained flooded in each annual cycle estimated by Díaz-Delgado et al. [14]; and (2) precipitation data, in particular the cumulative precipitation in the hydrometeorological year from 1 September to 31 August (data provided by Doñana's Long-Term Monitoring program, ESPN at ICTS-RBD).

2.5. Biomass Production Models

We used three approaches to model annual biomass production based on NDVI data and selected the best-performing model using the calibration dataset (see above). The calibration dataset was also used to select among alternative model parameters (see model 3, below) and to obtain a relationship between NDVI-based estimates and biomass production. The three model types were based on the following information:

1. The maximum NDVI value observed in each given year and MODIS pixel ('Maximum-NDVI' hereafter).
2. The NDVI value at the time at which peak biomass occurs in an average year (8 May), for each given year and MODIS pixel ("May-NDVI" hereafter). The average time of the biomass peak was calculated as the mean date of the maximum NDVI values observed at the five MODIS pixels included in the calibration dataset from 2001 to 2015.
3. The maximum and small integral NDVI values derived from phenological models fitted using the Land Surface Phenology (LSP) techniques available in the software package TIMESAT [48] ("LSP-Maximum-NDVI" and "LSP-Accumulated-NDVI" hereafter). The model was fit to the

complete series of observed NDVI data (2001–2015) and then compared to the calibration data. The calibration procedure was also used to inform the choice of three settings that must be decided by the user before fitting the curves to NDVI data [49], namely: (1) The baseline value of the phenological curve, a parameter that discards all the values below a specific NDVI value from the growth season under analysis. (2) The criterion that defines the beginning and end of the growth season. We evaluated two options: a fixed threshold value and a fixed proportion of the seasonal amplitude observed during each growth season. (3) The fitting method used to filter noise in the data: Savitzky-Golay filter, Asymmetric Gaussian and Double Logistic. For all other settings, we used the default values in TIMESAT, namely: no spike method, one season per year, no adaptation to the upper envelope of the curve, and normal adaptation strength.

All subsequent analyses were done using packages "car" [50] and "lmodel2" [51] in R [52]. We fitted linear regression models using the annual ground-truth values of biomass production (calibration dataset) at each location and year as response variables and the NDVI-based estimates at a MODIS pixel and year as predictor variables. Models were fitted to untransformed and transformed variables (linear, exponential, logarithmic, power and log-log regressions), to test for an improved fit. We selected among alternative models using the proportion of explained variance as estimated by the R^2 [53], the root mean square error (RMSE) and the percentage of the normalized RMSE (calculated dividing for the mean value).

In addition, we used the calibration data and the best model to evaluate the robustness of model predictions, i.e., to test whether the relationship between NDVI-based estimates and observed biomass production was influenced by (i) changes in two key environmental variables (precipitation and hydroperiod), and (ii) spatial variation in productivity (i.e., among-site variation in soil fertility). For the first purpose, we used multiple regression of observed biomass production on NDVI-based estimates and either precipitation or hydroperiod, and compared them to the univariate regression (only NDVI-based estimates) using F-tests [53], and adjusted R^2 values. For the second purpose, we compared the relationship between biomass production and NDVI-based estimates among the five calibration locations (which showed consistent differences in biomass-production range across the whole data series; Figure 2). We used an Analysis of Covariance (ANCOVA, [53]) to assess the differences in the mean slope of biomass in relation to the sampling locations used for calibration. We included "location" as a fixed, categorical factor. A significant effect of the interaction between the continuous (NDVI-based estimates) and fixed (location) factors would indicate that the slopes of the relationship between NDVI and biomass production varies significantly among localities, thus a common calibration line should not be used across the whole study area.

2.6. Model Validation

The best model was validated with the new validation (2016) dataset. For validation, we regressed with a major axis regression the biomass field data on the predictions of the best NDVI-based model, evaluating whether the slope differed significantly from 1 (i.e., whether model predictions significantly over- or under-estimated observed values), and estimating model performance with the RMSE and the percentage of the normalized RMSE. In addition, since field data indicated a high variability in species composition in the validation dataset, we evaluated whether model performance was affected by such variability (i.e., if model predictions were better in areas dominated by *B. maritimus*). For this purpose, we performed a multiple regression with NDVI-based estimates and plant composition (the proportion of *B. maritimus* biomass present in each sample) as independent variables, following arcsine (square root) transformation of the latter variable to ensure residuals' normality and homoscedasticity.

2.7. Trend Analysis

After model validation, we evaluated the spatial and temporal patterns of biomass production in the Doñana marsh by generating model predictions for the study area and time range (2001–2016). We used these estimates to analyze the spatial and temporal variability in biomass production. Change in spatial variability over time was analyzed using IDRISI Earth Trends Modeler [54] to calculate the Theil-Sen slope estimator. This is a temporal trend estimator more robust than the least-squares slope because it is much less sensitive to outliers and skewed data. In our analysis, it was used to identify pixels where biomass increased or decreased, considering a significance level of ($\alpha = 0.05$). The main driver of temporal variability in biomass production was analyzed by regressing biomass production (averaged across the whole study area) on cumulative precipitation, calculated over five different time intervals (September–March, September–April, September–May, September–June and September–July) to identify the period over which precipitation was most influential (i.e., the one providing the best fit).

3. Results

3.1. Biomass Production Models

3.1.1. Model Parametrization

Estimators based on a single image (Maximum-NDVI and May-NDVI) were obtained directly from the MODIS images. For the two NDVI estimators modeled using TIMESAT (LSP-Maximum-NDVI and LSP-Accumulated-NDVI), the first step was to choose the three model settings that resulted in the best calibration:

- Baseline value: The best results were obtained with a baseline value of 0.27, which corresponds to the average value of NDVI in September across the whole study area—i.e., the NDVI value of senescent *B. maritimus* vegetation on dry marsh soil. This baseline value resulted in a much better regression fit than using no fixed baseline value ($R^2 = 0.63$ vs. $R^2 = 0.22$, in the best-performing model and filter: LSP-Accumulated-NDVI with Savitzky-Golay, see below). Other baseline values, like the NDVI value of open water (NDVI = 0.31), resulted in the failure to recognize the growing season—probably because it results in large variations in baseline values between early- and late-flooding years, which is unrelated with plant primary production.
- Beginning and end of the growth season: The criterion based on a proportion of the seasonal amplitude performed better than the one based on a fixed threshold value, which resulted in TIMESAT failing to recognize the growth season for most of the years, due to their strong inter-annual variability. Among the different threshold-amplitude values tested, a value of 10% performed best ($R^2 = 0.65$, as compared to $R^2 = 0.63$ for 3% and $R^2 = 0.61$ for 5%, in the best-performing model and filter: LSP-Maximum-NDVI with Savitzky-Golay, see below), allowing for the recognition of the growth season of all years and succeeding with the filtering of the noise.
- Fitting method: The metrics derived from the Savitzky-Golay filter performed slightly better than those obtained with the other two methods (Table 1). Hence, we solely use and report this method hereafter.

Table 1. Comparison among TIMESAT curve-fitting methods to predict *B. maritimus* biomass using R^2.

	Asymmetrical Gaussian	Double Logistic	Savitzky-Golay
LSP-Maximum-NDVI	0.60	0.62	0.63
LSP-Accumulated-NDVI	0.53	0.54	0.61

3.1.2. Model Calibration

The results showed that there was a statistically significant relationship between each of the four NDVI biomass estimators (Maximum-NDVI, May-NDVI, LSP-Maximum-NDVI, LSP-Accumulated-NDVI) and biomass production. The best results were obtained with a log transformation of the response variable (ground-truth biomass production), (Table 2; Appendix A Table A1). The two estimators based on Land Surface Phenology models performed considerably better, with LSP-Maximum-NDVI providing the best fit. Parameter values from this calibration fit (Table 2, Figure 2) are used hereafter to estimate biomass production from LSP-Maximum-NDVI values.

Table 2. Results of model calibration. Relationship between each of the four NDVI estimators tested and biomass production. Best fits were obtained with ln (y) = a * x + b transformation. The other two transformations, y = a * x + b and ln (y) = a * ln (x) + b, are included in Appendix A. SE = Standard Error. RMSE = Root mean square error. %RMSE = Percentage of RMSE.

Predictor	Intercept ± SE	Slope ± SE	F-Test	DF	p-Value	R^2	RMSE	%RMSE
Maximum-NDVI	4.75 ± 0.39	4.51 ± 0.63	51	1, 73	5.71×10^{-10}	0.41	0.96	12.9
May-NDVI	5.00 ± 0.70	4.46 ± 0.65	47.3	1, 73	1.76×10^{-9}	0.39	0.97	13.1
LSP-Maximum-NDVI	3.77 ± 0.34	6.71 ± 0.59	128	1, 69	$< 2.2 \times 10^{-16}$	0.65	0.74	10.1
LSP-Accumulated-NDVI	5.88 ± 0.19	0.75 ± 0.08	97	1, 69	8.1×10^{-15}	0.59	0.81	11.0

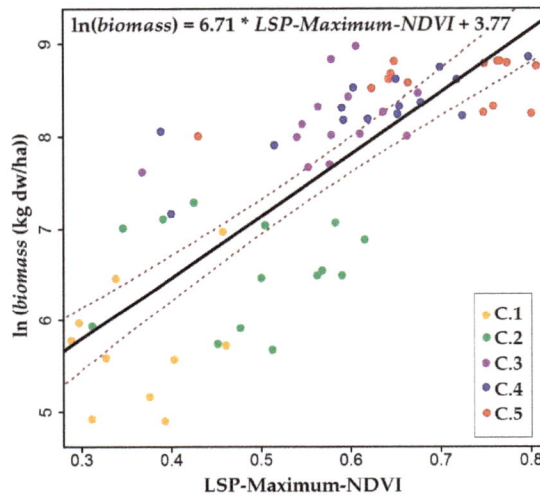

Figure 2. Model calibration. Relationship between the best NDVI estimator tested (LSP-Maximum-NDVI) and the logarithm of biomass production (kg dw/ha). Continuous line: regression line. Dotted lines: 95% confidence intervals. The dot colors represent the five different locations of the calibration biomass plots.

The use of multiple regression models including as a second predictor a key environmental variable (either precipitation or hydroperiod) did not improve the fit of the best model with a single predictor (LSP-Maximum-NDVI, Table 3). Therefore, the relationship between LSP-Maximum-NDVI and biomass production was apparently not influenced by either of these two environmental variables.

The ANCOVA including the five calibration locations as a categorical factor indicated that the slopes were not heterogeneous (i.e., the interaction location * LSP-Maximum-NDVI was not significant: F (9, 61) = 0.36, P = 0.83). However, the model with location without interaction showed a better fit (adjusted R^2 = 0.85) than the model without location (R^2 = 0.64, Table 3). This indicates that the localities differ significantly in average biomass production across the years, but the slope between LSP-Maximum-NDVI and biomass production is not affected by such variation.

Table 3. Model robustness to environmental and spatial variation. Results of multiple regression models of biomass production on LSP-Maximum-NDVI plus a key environmental variable (either precipitation or hydroperiod, as continuous variables) or spatial location (as a categorical variable). Adj.R^2 = adjusted R^2. DV = dummy variable. SE = Standard Error. RMSE = Root mean square error. %RMSE = Percentage of RMSE.

Predictors		Estimates ± SE	T-test	p-Values	Whole-model Parameters					
					F-test	DF	p-Value	Adj. R^2	RMSE	%RMSE
Intercept		3.69 ± 0.41	9.01	2.4×10^{-16}						
LSP-Maximum-NDVI		6.61 ± 0.63	10.4	9.9×10^{-16}	63.6	2, 68	2.7×10^{-16}	0.64	0.75	10.1
Precipitation		2.8 ± 6.3 × 10^{-4}	0.45	0.66						
Intercept		3.73 ± 0.34	10.8	5.0×10^{-16}						
LSP-Maximum-NDVI		6.73 ± 0.64	10.5	1.7×10^{-15}	63.0	2, 63	9.3×10^{-16}	0.66	0.74	10.0
Hydroperiod [1]		6.3 ± 14 × 10^{-4}	0.46	0.65						
Intercept		6.98 ± 0.45	15.2	$< 2.2 \times 10^{-16}$						
Location	DV1	−1.59 ± 0.22	−6.69	1.98×10^{-9}						
	DV2	−0.11 ± 0.19	−0.60	0.55	4.1	9, 61	$< 2.2 \times 10^{-16}$	0.85	0.46	6.2
	DV3	−2.17 ± 0.29	−7.47	2.56×10^{-10}						
	DV4	−0.09 ± 0.18	−5.51	0.61						
LSP-Maximum-NDVI		2.26 ± 0.63	3.61	5.91×10^{-4}						

[1] Model based on fewer observations (N = 66), due to missing hydroperiod data for 2015.

3.2. Model Validation

Biomass production varied considerably among validation plots, ranging from less than 500 kg dw/ha to almost 3000 kg dw/ha (mean = 1595 kg dw/ha, median = 1524 kg dw/ha, standard deviation = 678 kg dw/ha) (Figure 3). Species composition showed an unexpectedly high variation among sampling localities. *B. maritimus* represented 47% of the biomass production, followed by *Eleocharis palustris* (28%) and *Scirpus lacustris* (10%). *B. maritimus* was dominant in 5 of the 9 MODIS pixels, while the other 4 pixels were dominated by *E. palustris* (3 pixels) and *S. lacustris* (1 pixel).

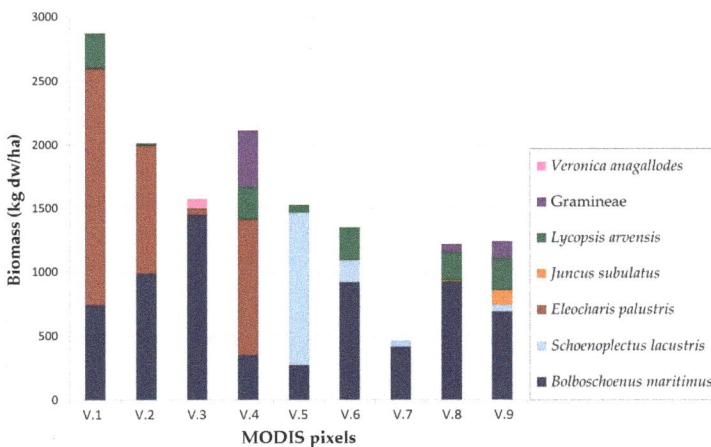

Figure 3. Results of the validation survey. Aboveground biomass production per species at each of the nine MODIS pixels sampled. N = 12 sample plots per pixel.

The results of the validation exercise showed that the model based on LSP-Maximum-NDVI could explain a reasonable percentage of the variance of the biomass ($R^2 = 0.70$; Figure 4), particularly regarding the high spatial and temporal variability present in the study area. The prediction error estimate, based on RMSE, was 354 kg dw/ha and the %RMSE was 22%. The slope of the predicted-observed relationship did not differ significantly from 1 (95% CI = 0.30; 1.07), indicating that model predictions neither under-nor over-estimate observed biomass production.

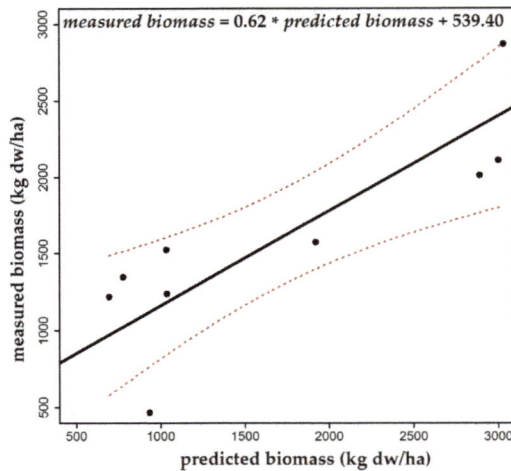

Figure 4. Model validation. Major axis regression between measured and predicted biomass. Continuous line: regression line. Dotted lines: 95% confidence intervals.

Variability in species composition did not significantly influence the relationship between measured and predicted biomass production. Results of the multiple regression model including LSP-Maximum-NDVI and the proportion of *B. maritimus* showed that the latter did not significantly influence biomass yield (Adj. R^2 = 0.61 in contrast to R^2 = 0.70; significance of proportion of *B. maritimus* t-test = −0.39, p-value = 0.71).

3.3. Trend Analysis

Based on our LSP-Maximum-NDVI model, we produced 16 maps representing biomass production per pixel (in kg dw/ha) for each growth season between 2001 and 2016 (Figure 5). The average value per pixel (across all years) was 3869 ± 1781 kg dw/ha. The maps reveal a high spatial variation in biomass production, resulting from a combination of high spatial heterogeneity and high inter-annual variation. In some years, such as 2010, there were extensive areas with high biomass production (up to 10,000 kg dw/ha); while in other years, such as 2005, biomass production was one order of magnitude lower (i.e., it did not reach 1000 kg dw/ha at any pixel across the study area). However, the areas with high and low biomass production were not stationary, but strongly varied among years. For example, in 2001 biomass production peaked at the southern part of the study area, while in 2010 it peaked at its northernmost part.

Values of the Theil-Sen slope estimator showed that there is a general trend towards diminishing biomass production over the last 16 years—i.e., there were more areas where biomass production decreased than areas where it increased (Figure 6A). Biomass production tended to decrease in the central part of the study area, whereas it tended to increase in its periphery. A comparison with the spatial distribution of the average biomass production from 2001 to 2016 revealed that biomass production tended to decrease in areas with high productivity (high average biomass production) and to increase in areas with low productivity (low average biomass production) (Figure 6A,B) (R = −0.24, t-test (725) = −6.51, p-value = 1.38 × 10^{-10}). Disentangling the causes behind this pattern probably deserves further analyses.

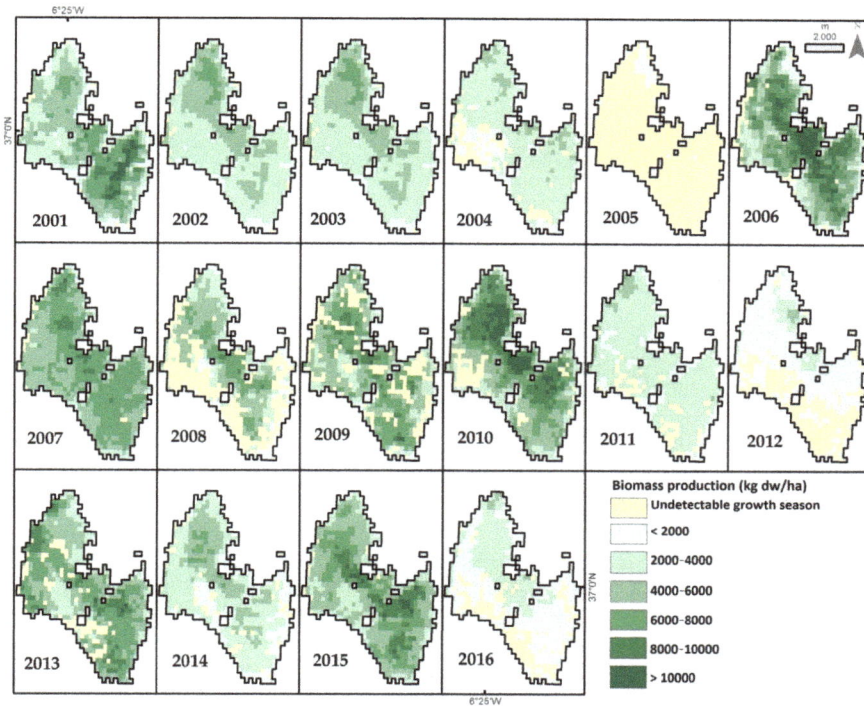

Figure 5. Model predictions. Estimated biomass production (in kg dw/ha) per pixel across the study area.

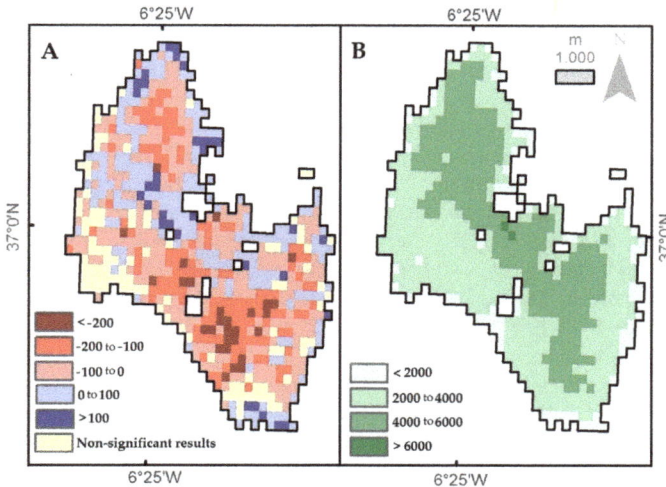

Figure 6. Trend analysis. (**A**) Changes in biomass production from 2001 to 2016, based on the Theil-Sen slope estimator. Positive values (blue colors): increase. Negative values (red colors): decrease. (**B**) Average biomass production (kg dw/ha) from 2001 to 2016. All categories except the one for "non-significant results" indicate Theil-Sen slope estimator values significantly different from zero.

Inter-annual variability in biomass production (summed across the whole study area) was strongly influenced by annual precipitation—with cumulative precipitation from September to April exerting the strongest influence of all periods tested. The best fit between both variables was obtained using a log-log transformation, which indicates that the effect is stronger at low precipitation values and saturates when precipitation is very high ($R^2 = 0.69$, $F(1, 14) = 30.8$, p-value $= 7.15 \times 10^{-5}$; Figure 7).

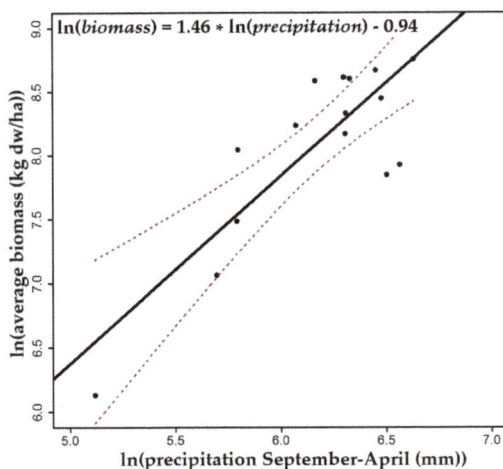

Figure 7. Effect of cumulative precipitation (from September to April) on biomass production (average across the study area). Continuous line: regression fit. Dotted line: 95% confidence intervals. Note the log-transformation in both axes.

4. Discussion

We have shown that MODIS Global MOD13Q1 NDVI data provides a good source of information for estimating biomass production in a challenging situation—a seasonal marsh characterized by high spatio-temporal variation in precipitation and hydroperiod [55]. While the use of a single image per growth season provided estimates of reasonable quality (39–41% of variance explained in the calibration dataset), modeling the phenological cycle using Land Surface Phenology (LSP) techniques considerably improved the quality and robustness of such estimates (65% and 70% of variance explained using LSP-Maximum-NDVI, in the calibration and the validation datasets, respectively; see also [56]). Furthermore, biomass production estimates derived from the best-performing model for the whole study area and time period indicate a strong role of a key climatic driver, the inter-annual variation in precipitation; and a pattern of spatio-temporal change (decreasing yields in the most productive areas) that could be consistent either with changes in vegetation community composition due to marsh siltation and changes in hydroperiod [14] or with the impact of a key biotic driver, overgrazing by domestic and wild herbivores.

The modeling process was particularly challenging because marshes are highly dynamic and heterogeneous wetland ecosystems where the reflectance signal can change rapidly, sometimes within hours or days [6,38]. Despite these challenges, the four different, NDVI-based estimators predicted biomass production with reasonable quality (39–65% of variance explained during calibration). However, the two NDVI biomass estimators derived from TIMESAT models of LSP performed significantly better than those based on a single image per year only—reinforcing previous suggestions that LSP may improve biomass determination in complex ecosystems [20,57]. The improved performance of LSP estimators is probably caused by the higher sensitivity of single-image estimators to several sources of error and noise, such as sensor resolution and calibration, digital quantization errors, ground and atmospheric conditions, or orbital and sensor degradation [7]; and to the rapid changes in the NDVI signal in heterogeneous ecosystems—which may bias such estimators, for example, if an image is taken after a rainfall episode [38]. LSP makes use of the information gathered across the complete growth season to produce a smooth NDVI curve that integrates the whole vegetation cycle, thus reducing noise and errors [12,21,58,59]. On the one hand, the difference among fitting methods was marginal for the best-performing predictor (LSP-Maximum-NDVI; Table 1), suggesting that the smoothing provided by all fitting procedures sufficed to remove noise and ensure predictor quality—in contrast with works reporting that the over-smoothing introduced by the Asymmetric Gaussian and Double Logistic methods affected the accuracy of parameter estimates [21]. On the other hand, the use of a baseline criterion that removed the influence of water removed the strong bias introduced on NDVI-based estimates by early-flooding years—which caused a drop in NDVI values, unrelated to plant productivity.

Besides their statistical properties, the choice of estimator may influence its potential use by managers or policy makers. Management applications that rely on an early prediction of the season's standing crop, for example to adjust the stocking rates of domestic herbivores (cattle and horses), will be best served by those based on single images taken at early dates—such as the May-NDVI, chosen to coincide with the average NDVI maximum without requiring the uptake of ulterior images to identify the exact time of the season's maximum. Similarly, one of the two indicators based on LSP can be calculated at a much earlier point than the other—since LSP-Maximum-NDVI only requires the maximum value to be reached, while LSP-Accumulated-NDVI can only be calculated at the end of the growth season. Under such circumstances, it might be more useful to use a statistically weaker estimator that can be estimated earlier, as long as the associated decrease in accuracy is acceptable. Unfortunately, single-image estimators such as May-NDVI had a much lower accuracy than LSP-based estimators (39–41% vs. 65–70% of variance explained). We therefore recommend the use of LSP-Maximum-NDVI, which provides the best estimates at a relatively early date.

Estimators based on NDVI have been shown to saturate asymptotically at high biomass values [12,18]. While the relationship between NDVI and biomass production was multiplicative (i.e., the slope decreased with increasing NDVI, following a logarithmic relationship), the best-performing estimator LSP-Maximum-NDVI was far from reaching a plateau at the highest biomass production values we measured. As a consequence, estimates based on LSP-Maximum-NDVI performed reasonably well in the validation exercise. We cannot rule out, however, a saturation of these estimators in situations (years or localities) with higher biomass production—which would result in a disproportionate increase in prediction errors. We decided to build our models using NDVI because it is the most frequently used vegetation index, but as it is prone to saturation, and to noise caused by soil color and water, it would be interesting to test whether models can be improved using EVI, a vegetation index less prone to these problems [60].

Testing the robustness and validating the performances of the best estimator with independent data was particularly relevant given the high heterogeneity, complexity and unpredictability of the Doñana marsh ecosystems [14,61]. Validation yielded satisfactory levels of predictive ability, particularly given the characteristics of the study system and the high variation in species composition detected. More importantly, the estimator also proved to be robust to the influence of environmental variables (precipitation and hydroperiod), spatial variation in baseline productivity, and species composition—suggesting that it can be safely used under the variety of situations present in the Doñana marshes, as well as in similar systems.

The analysis of the spatial and temporal variation of biomass production in the Doñana marsh confirmed that production is both highly variable and highly heterogeneous. Based on previous studies we expected precipitation, which determines the flooding regime, to account for a large percentage of the variation in biomass production [62]. Indeed, precipitation explained 69% of the temporal variation in biomass production (summed across the study area). The relationship between precipitation and biomass production was however non-linear, indicating that biomass production is strongly dependent on precipitation in dry years but it tends to saturate in very wet years (similar to what Coe et al. [63] report). Whether this saturation results from self-thinning effects (intra- and inter-specific competition) and/or from the negative effect of prolonged inundation on plant development remains a topic for future studies.

The effect of herbivores on the marsh vegetation is another important source of variability. Specifically, changes in plant consumption caused by variation in the number and distribution of domestic (cattle and horses) and wild (fallow deer *Dama dama*, red deer *Cervus elaphus*, wild boar *Sus scrofa*) herbivores have been shown to determine the abundance and distribution of plant biomass, reducing it severely in dry years [40]. The spatio-temporal trends detected using the Theil-Sen slope estimator suggest that biomass production has decreased, during the last 16 years, precisely in the areas where this production was more abundant. This pattern could be consistent with changes in vegetation community composition due to temporal trends in mean hydroperiod [14] probably due to marsh siltation. However, they could also be reflecting the effect of overgrazing by herbivores, which may be expected to concentrate their grazing (thus consuming more biomass) in the areas with higher biomass yield—particularly in dry years with low biomass production. Indeed, herbivores do not distribute uniformly in the marsh; they move tracking food and water availability, and avoiding heavily flooded areas. Mapping the biomass is an important first step to monitor and manage the effects of herbivores [5,39,64]. It can support management programs that rationalize the number of domestic animals and find a dynamic balance between cattle and vegetation [65,66], helping to prevent land degradation, soil erosion and biodiversity loss [67]. In this regard the study of the vegetation patterns could be improved by correlating the changes in vegetation biomass with hydroperiod trends, and with the spatial distribution and movements of domestic and wild herbivores. The modeling process in a heterogeneous ecosystem such as the Doñana marsh could also benefit from increasing the spatial resolution using other sensors such as Landsat.

5. Conclusions

We show that by using Land Surface Phenology (LSP) techniques and relatively simple statistical models, it is possible to provide accurate estimations of plant biomass production in a large seasonal wetland, the Doñana marsh. Estimators based on LSP models provided substantially better predictions than those based on a single image, and were robust to environmental variation and spatial heterogeneity. Model predictions indicate that the marsh areas with highest productivity coincide with those in which productivity has been declining during the last 16 years—suggesting changes in vegetation communities or the potential effect of overgrazing by wild and domestic herbivores. The estimation of plant biomass using remote-sensing techniques, adequately supported by ground-truth data, may represent a key tool for the long-term monitoring and management of ecosystems, especially in protected areas where the natural world and human activities coexist.

Acknowledgments: David Aragonés, Isabel Afán, Ricardo Díaz-Delgado and Diego García Díaz (EBD-LAST) provided support for remote-sensing and LSP analyses. Alfredo Chico, José Luis del Valle and Rocío Fernández Zamudio (ESPN, ICTS-RBD) provided logistic support and taxonomic expertise during the field work (validation dataset). Ernesto García and Cristina Pérez assisted with biomass harvesting and processing (calibration dataset). Gerrit Heil provided support in the project design. This study received funding from Ministerio de Medio Ambiente-Parque Nacional de Doñana, Consejeria de Medio Ambiente, Junta de Andalucia (1999–2000): RNM118 Junta de Andalucia (2003); the European Union's Horizon 2020 Research and Innovation Program under grant agreement No. 641762 to ECOPOTENTIAL project; and the Spanish Ministry of Economy, Plan Estatal de I+D+i 2013–2016, under grant agreement CGL2016-81086-R to GRAZE project.

Author Contributions: All authors are intellectually responsible for the conducted research. M.L., L.S. and J.B. designed the work. M.L processed and analyzed the satellite time series, ran the statistical analyses, and designed the validation sampling and collected the ground-truth data in 2016. R.S. designed and collected the ground-truth data for the calibration. M.L. and P.F.M. drafted the first version of the manuscript. All authors participated in writing the final version of the manuscript, and reviewed and approved it.

Conflicts of Interest: The authors declare no conflict of interest.

Appendix A

Table A1. Results of model calibration. Relationship between each of the four NDVI estimators tested and biomass production with different transformations: y = a * x + b and ln(y) = a * ln(x) + b. SE = Standard Error. RMSE = Root mean square error. %RMSE = Percentage of RMSE.

	Predictor	Intercept ± SE	Slope ± SE	F-Test	DF	*p*-Value	R^2	RMSE	%RMSE
y = a * x + b	Maximum-NDVI	−1270 ± 783	7023 ± 1267	30.7	1, 73	4.52×10^{-7}	0.29	1919	66.5
	May-NDVI	−1104 ± 718	7348 ± 1259	34.7	1, 73	1.36×10^{-7}	0.32	1888	65.5
	LSP-Maximum-NDVI	−3641 ± 629	12 085 ± 1110	118.5	1, 69	$< 2.2 \times 10^{-16}$	0.63	1400	48.6
	LSP-Accumulated-NDVI	21 ± 327	1400 ± 133	110.7	1, 69	5.49×10^{-16}	0.61	1430	49.6
ln (y) = a * ln (x) + b	Maximum-NDVI	8.26 ± 0.18	1.36 ± 0.21	39.6	1, 73	2.09×10^{-8}	0.35	1.01	13.5
	May-NDVI	8.85 ± 0.22	2.11 ± 0.28	55.8	1, 73	1.40×10^{-10}	0.43	0.94	12.7
	LSP-Maximum-NDVI	9.60 ± 0.21	3.31 ± 0.29	133.5	1, 68	$< 2.2 \times 10^{-16}$	0.64	0.74	10.0
	LSP-Accumulated-NDVI	6.89 ± 0.11	1.18 ± 0.11	109	1, 69	7.68×10^{-16}	0.61	0.79	10.6

References

1. Cho, M.A.; Skidmore, A.; Corsi, F.; van Wieren, S.E.; Sobhan, I. Estimation of green grass/herb biomass from airborne hyperspectral imagery using spectral indices and partial least squares regression. *Int. J. Appl. Earth Obs. Geoinf.* **2007**, *9*, 414–424. [CrossRef]

2. Fava, F.; Colombo, R.; Bocchi, S.; Meroni, M.; Sitzia, M.; Fois, N.; Zucca, C. Identification of hyperspectral vegetation indices for Mediterranean pasture characterization. *Int. J. Appl. Earth Obs. Geoinf.* **2009**, *11*, 233–243. [CrossRef]

3. Xiaoping, W.; Ni, G.; Kai, Z.; Jing, W. Hyperspectral remote sensing estimation models of aboveground biomass in Gannan rangelands. *Procedia Environ. Sci.* **2011**, *10*, 697–702. [CrossRef]

4. Moeckel, T.; Safari, H.; Reddersen, B.; Fricke, T.; Wachendorf, M. Fusion of ultrasonic and spectral sensor data for improving the estimation of biomass in grasslands with heterogeneous sward structure. *Remote Sens.* **2017**, *9*, 98. [CrossRef]

5. Carmona, C.P.; Röder, A.; Azcárate, F.M.; Peco, B. Grazing management or physiography? Factors controlling vegetation recovery in Mediterranean grasslands. *Ecol. Model.* **2013**, *251*, 73–84. [CrossRef]

6. Adam, E.; Mutanga, O.; Rugege, D. Multispectral and hyperspectral remote sensing for identification and mapping of wetland vegetation: A review. *Wetl. Ecol. Manag.* **2010**, *18*, 281–296. [CrossRef]

7. Pettorelli, N.; Vik, J.O.; Mysterud, A.; Gaillard, J.M.; Tucker, C.J.; Stenseth, N.C. Using the satellite-derived NDVI to assess ecological responses to environmental change. *Trends Ecol. Evol.* **2005**, *20*, 503–510. [CrossRef] [PubMed]

8. Tiner, R.W.; Lang, M.W.; Klemas, V.V. *Remote Sensing of Wetlands: Applications and Advances*; CRC Press: Abingdon, UK, 2015.

9. Yan, E.; Wang, G.; Lin, H.; Xia, C.; Sun, H. Phenology-based classification of vegetation cover types in Northeast China using MODIS NDVI and EVI time series. *Int. J. Remote Sens.* **2015**, *36*, 489–512. [CrossRef]

10. Meroni, M.; Verstraete, M.M.; Rembold, F.; Urbano, F.; Kayitakire, F. A phenology-based method to derive biomass production anomalies for food security monitoring in the Horn of Africa. *Int. J. Remote Sens.* **2014**, *35*, 2472–2492. [CrossRef]

11. Diaz-Delgado, R.; Aragonés, D.; Ameztoy, I.; Bustamante, J. Monitoring marsh dynamics through remote sensing. In *Conservation Monitoring in Freshwater Habitats: A Practical Guide and Case Studies*; Springer: Berlin, Germany, 2010; pp. 1–415.

12. Reed, B.C.; Schwartz, M.D.; Xiao, X. Remote sensing phenology: Status and the way forward. In *Phenology of Ecosystems Processes*; Springer: Berlin, Germany, 2009; pp. 231–246.

13. Engman, E.; Gurney, R. *Remote Sensing in Hydrology*; Springer: Berlin, Germany, 1991.

14. Díaz-Delgado, R.; Aragonés, D.; Afán, I.; Bustamante, J. Long-term monitoring of the flooding regime and hydroperiod of Doñana marshes with Landsat time series (1974–2014). *Remote Sens.* **2016**, *8*, 775. [CrossRef]

15. Vanderpost, C.; Ringrose, S.; Matheson, W.; Arntzen, J. Satellite based long-term assessment of rangeland condition in semi-arid areas: An example from Botswana. *J. Arid Environ.* **2011**, *75*, 383–389. [CrossRef]

16. Yengoh, G.T.; Dent, D.; Olsson, L.; Tengberg, A.E.; Tucker, C.J., III. *Use of the Normalized Difference Vegetation Index (NDVI) to Assess Land Degradation at Multiple Scales: Current Status, Future Trends, and Practical Considerations*; Springer: Berlin, Germany, 2015.

17. Santin-Janin, H.; Garel, M.; Chapuis, J.L.; Pontier, D. Assessing the performance of NDVI as a proxy for plant biomass using non-linear models: A case study on the kerguelen archipelago. *Polar Biol.* **2009**, *32*, 861–871. [CrossRef]

18. Huete, A.; Didan, K.; Miura, T.; Rodriguez, E.P.; Gao, X.; Ferreira, L.G. Overview of the radiometric and biophysical performance of the MODIS vegetation indices. *Remote Sens. Environ.* **2002**, *83*, 195–213. [CrossRef]

19. Lhermitte, S.; Verbesselt, J.; Verstraeten, W.W.; Coppin, P. A comparison of time series similarity measures for classification and change detection of ecosystem dynamics. *Remote Sens. Environ.* **2011**, *115*, 3129–3152. [CrossRef]

20. Meroni, M.; Rembold, F.; Verstraete, M.; Gommes, R.; Schucknecht, A.; Beye, G. Investigating the relationship between the inter-annual variability of satellite-derived vegetation phenology and a proxy of biomass production in the Sahel. *Remote Sens.* **2014**, *6*, 5868–5884. [CrossRef]

21. Lara, B.; Gandini, M. Assessing the performance of smoothing functions to estimate land surface phenology on temperate grassland. *Int. J. Remote Sens.* **2016**, *37*, 1801–1813. [CrossRef]

22. Rojas, O.; Vrieling, A.; Rembold, F. Assessing drought probability for agricultural areas in Africa with coarse resolution remote sensing imagery. *Remote Sens. Environ.* **2011**, *115*, 343–352. [CrossRef]

23. Richardson, A.D.; Black, T.A.; Ciais, P.; Delbart, N.; Friedl, M.A.; Gobron, N.; Hollinger, D.Y.; Kutsch, W.L.; Longdoz, B.; Luyssaert, S.; et al. Influence of spring and autumn phenological transitions on forest ecosystem productivity. *Philos. Trans. R. Soc. B Biol. Sci.* **2010**, *365*, 3227–3246. [CrossRef] [PubMed]

24. Higgins, S.I.; Scheiter, S. Atmospheric CO_2 forces abrupt vegetation shifts locally, but not globally. *Nature* **2012**, *488*, 209–212. [CrossRef] [PubMed]

25. Wright, S.J.; van Schaik, C.P. Light and the phenology of tropical trees. *Am. Nat.* **1994**, *143*, 192–199. [CrossRef]

26. Vanschaik, C.P.; Terborgh, J.W.; Wright, S.J. The phenology of tropical forests—Adaptive significance and consequences for primary consumers. *Annu. Rev. Ecol. Syst.* **1993**, *24*, 353–377. [CrossRef]

27. Wolkovich, E.M.; Cook, B.I.; Allen, J.M.; Crimmins, T.M.; Betancourt, J.L.; Travers, S.E.; Pau, S.; Regetz, J.; Davies, T.J.; Kraft, N.J.B.; et al. Warming experiments underpredict plant phenological responses to climate change. *Nature* **2012**, *485*, 18–21. [CrossRef] [PubMed]

28. Adole, T.; Dash, J.; Atkinson, P.M. A systematic review of vegetation phenology in Africa. *Ecol. Inform.* **2016**, *34*, 117–128. [CrossRef]

29. Hanes, J.M. *Biophysical Applications of Satellite Remote Sensing*; Springer: Berlin, Germany, 2014.

30. Huete, A.; Didan, K.; van Leeuwen, W.; Miura, T.; Glenn, E. MODIS vegetation indices. *Remote Sens. Digit. Image Process.* **2011**, *11*, 579–602.

31. Boyd, D.S.; Almond, S.; Dash, J.; Curran, P.J.; Hill, R.A.; Hill, R.A. Phenology of vegetation in southern England from Envisat MERIS terrestrial chlorophyll index (MTCI) data. *Int. J. Remote Sens.* **2011**, *3223*, 8421–8447. [CrossRef]

32. Zucca, C.; Wu, W.; Dessena, L.; Mulas, M. Assessing the effectiveness of land restoration interventions in dry lands by multitemporal remote sensing—A case study in Ouled DLIM (Marrakech, Morocco). *Land Degrad. Dev.* **2015**, *26*, 80–91. [CrossRef]

33. Moulin, S.; Guérif, M. Impacts of model parameter uncertainties on crop reflectance estimates: A regional case study on wheat. *Int. J. Remote Sens.* **1999**, *20*, 213–218. [CrossRef]

34. Huete, A.R. A soil-adjusted vegetation index (SAVI). *Remote Sens. Environ.* **1988**, *25*, 295–309. [CrossRef]

35. Box, E.O.; Holben, B.N.; Kalb, V. Accuracy of the AVHRR vegetation index as a predictor of biomass, primary productivity and net CO_2 flux. *Vegetatio* **1989**, *80*, 71–89. [CrossRef]

36. Eisfelder, C.; Kuenzer, C.; Dech, S. Derivation of biomass information for semi-arid areas using remote-sensing data. *Int. J. Remote Sens.* **2012**, *33*, 2937–2984. [CrossRef]

37. Mbow, C.; Fensholt, R.; Rasmussen, K.; Diop, D. Can vegetation productivity be derived from greenness in a semi-arid environment? Evidence from ground-based measurements. *J. Arid Environ.* **2013**, *97*, 56–65. [CrossRef]

38. Gallant, A.L. The challenges of remote monitoring of wetlands. *Remote Sens.* **2015**, *7*, 10938–10950. [CrossRef]

39. Röder, A.; Udelhoven, T.; Hill, J.; del Barrio, G.; Tsiourlis, G. Trend analysis of Landsat-TM and -ETM+ imagery to monitor grazing impact in a rangeland ecosystem in Northern Greece. *Remote Sens. Environ.* **2008**, *112*, 2863–2875. [CrossRef]

40. Soriguer, R.C.; Rodriguez Sierra, A.; Domínquez Nevado, L. *Análisis de la Incidencia de los Grandes Herbívoros en la Marisma y Vera del Parque Nacional de Doñana*; Organismo Autónomo de Parques Nacionales: Segovia, Spain, 2001.

41. Soriguer, R.C. Consideraciones sobre el efecto de los conejos y los grandes herbívoros en los pastizales de la Vera de Doñana. *Doñana Acta Vertebr.* **1983**, *10*, 155–168.

42. Clemente, L.; García, L.-V.; Espinar, J.L.; Cara, J.S.; Moreno, A. Las marismas del Parque Nacional de Doñana. *Investig. Cienc.* **2004**, *332*, 72–83.

43. Castroviejo, J. *Memoria Mapa del Parque Nacional de Doñana*; Consejo Superior de Investigaciones Científicas: Madrid, Spain, 1993.

44. Figuerola, J.; Green, A.J.; Santamaría, L. Passive internal transport of aquatic organisms by waterfowl in Doñana, south-west Spain. *Glob. Ecol. Biogeogr.* **2003**, *12*, 427–436. [CrossRef]

45. Aragonés, D.; Díaz-Delgado, R.; Bustamante, J. Estudio de la dinámica de inundación histórica de las marismas de Doñana a partir de una serie temporal larga de imágenes Landsat. In Procceedings of XI Congreso Nacional de Teledetección, Pto. de la Cruz, Spain, 14 September 2005.

46. Vuolo, F.; Mattiuzzi, M.; Klisch, A.; Atzberger, C. Data service platform for MODIS Vegetation Indices time series processing at BOKU Vienna: Current status and future perspectives. In Procceedings of SPIE Remote Sensing, Edinburgh, UK, 24–27 September 2012.

47. Huete, A.R.; Didan, K.; Huete, A.; Didan, K.; Van Leeuwen, W.; Jacobson, A.; Solanos, R.; Laing, T. *MODIS Vegetation Index (MOD 13) Algorithm Theoretical Basis Document*; Vegetation Index and Phenology Lab: Tucson, AZ, USA, 1999.

48. Jonsson, P.; Eklundh, L. TIMESAT—A program for analyzing time-series of satellite sensor data. *Comput. Geosci.* **2004**, *30*, 833–845. [CrossRef]

49. Eklundh, L.; Jönsson, P. *TIMESAT 3.2 with Parallel Processing Software Manual*; Lund University: Lund, Sweden, 2015.

50. Fox, J.; Weisberg, S. *An R Companion to Applied Regression*; SAGE Publication: Thousand Oaks, CA, USA, 2011.

51. Legendre, P. *Model II Regression User's Guide, R Edition*; Département de sciences biologiques, Univ. de Montreal: Quebec, QC, Canada, 2014.

52. R Core Team. *R: A Language and Environment for Statistical Computing*; R Foundation for Statistical Computing: Vienna, Austria, 2013.

53. Crawley, M.J. *The R Book*; John Wiley & Sons Inc.: Hoboken, NJ, USA, 2012.

54. Eastman, J.R. *Earth Trends Modeler in Terrset*; Clark Labs: Worcester, MA, USA, 2009.

55. Bustamante, J.; Aragonés, D.; Afán, I. Effect of protection level in the hydroperiod of water bodies on Doñana's aeolian sands. *Remote Sens.* **2016**, *8*, 867. [CrossRef]

56. Moreau, S.; Bosseno, R.; Gu, X.F.; Baret, F. Assessing the biomass dynamics of Andean bofedal and totora high-protein wetland grasses from NOAA/AVHRR. *Remote Sens. Environ.* **2003**, *85*, 516–529. [CrossRef]

57. Kang, X.; Hao, Y.; Cui, X.; Chen, H.; Huang, S.; Du, Y.; Li, W.; Kardol, P.; Xiao, X.; Cui, L. Variability and changes in climate, phenology, and gross primary production of an alpine wetland ecosystem. *Remote Sens.* **2016**, *8*, 391. [CrossRef]

58. Hird, J.N.; McDermid, G.J. Noise reduction of NDVI time series: An empirical comparison of selected techniques. *Remote Sens. Environ.* **2009**, *113*, 248–258. [CrossRef]

59. Klemas, V. Remote sensing of coastal and ocean currents: An overview. *J. Coast. Res.* **2012**, *282*, 576–586. [CrossRef]

60. Villa, P.; Mousivand, A.; Bresciani, M. Aquatic vegetation indices assessment through radiative transfer modeling and linear mixture simulation. *Int. J. Appl. Earth Obs. Geoinf.* **2014**, *30*, 113–127. [CrossRef]

61. Stephens, P.A.; Pettorelli, N.; Barlow, J.; Whittingham, M.J.; Cadotte, M.W. Management by proxy? The use of indices in applied ecology. *J. Appl. Ecol.* **2015**, *52*, 1–6. [CrossRef]

62. García-Murillo, P.; Bazo, E.; Fernández-Zamudio, R. Las plantas de la marisma del Parque Nacional de Doñana (España): Elemento clave para la conservación de un humedal europeo paradigmático. *CienciaUAT* **2014**, *9*, 60–75.

63. Coe, M.J.; Cumming, D.H.; Phillipson, G. Biomass and production of large african herbivores in relation to rainfall and primary production. *Oecologia* **1976**, *22*, 341–354. [CrossRef] [PubMed]

64. Li, Z.; Huffman, T.; McConkey, B.; Townley-Smith, L. Monitoring and modeling spatial and temporal patterns of grassland dynamics using time-series MODIS NDVI with climate and stocking data. *Remote Sens. Environ.* **2013**, *138*, 232–244. [CrossRef]

65. Archer, E.R.M. Beyond the "climate versus grazing" impasse: Using remote sensing to investigate the effects of grazing system choice on vegetation cover in the eastern Karoo. *J. Arid Environ.* **2004**, *57*, 381–408. [CrossRef]

66. Li, A.; Wu, J.; Huang, J. Distinguishing between human-induced and climate-driven vegetation changes: A critical application of RESTREND in inner Mongolia. *Landsc. Ecol.* **2012**, *27*, 969–982. [CrossRef]

67. Hooke, J.; Sandercock, P. Use of vegetation to combat desertification and land degradation: Recommendations and guidelines for spatial strategies in Mediterranean lands. *Landsc. Urban Plan.* **2012**, *107*, 389–400. [CrossRef]

remote sensing

MDPI

Article

Phenology-Based Biomass Estimation to Support Rangeland Management in Semi-Arid Environments

Anne Schucknecht [1,2,*], Michele Meroni [1], Francois Kayitakire [1] and Amadou Boureima [3]

[1] European Commission, Joint Research Centre, Directorate of Sustainable Resources, Ispra 21027, Italy; michele.meroni@ec.europa.eu (M.M.); francois.kayitakire@ec.europa.eu (F.K.)

[2] Karlsruhe Institute of Technology, Institute of Meteorology and Climate Research—Atmospheric Environmental Research, Garmisch-Partenkirchen 82467, Germany

[3] Ministry of Livestock, General Directorate of Production and Animal Industries, Niamey 23220, Niger; aboureimamadou@gmail.com

* Correspondence: anne.schucknecht@kit.edu; Tel.: +49-8821-183226

Academic Editors: Jose Moreno, Lalit Kumar and Prasad S. Thenkabail
Received: 24 February 2017; Accepted: 2 May 2017; Published: 10 May 2017

Abstract: Livestock plays an important economic role in Niger, especially in the semi-arid regions, while being highly vulnerable as a result of the large inter-annual variability of precipitation and, hence, rangeland production. This study aims to support effective rangeland management by developing an approach for mapping rangeland biomass production. The observed spatiotemporal variability of biomass production is utilised to build a model based on ground and remote sensing data for the period 2001 to 2015. Once established, the model can also be used to estimate herbaceous biomass for the current year at the end of the season without the need for new ground data. The phenology-based seasonal cumulative Normalised Difference Vegetation Index (cNDVI), computed from 10-day image composites of the Moderate-resolution Imaging Spectroradiometer (MODIS) NDVI data, was used as proxy for biomass production. A linear regression model was fitted with multi-annual field measurements of herbaceous biomass at the end of the growing season. In addition to a general model utilising all available sites for calibration, different aggregation schemes (i.e., grouping of sites into calibration units) of the study area with a varying number of calibration units and different biophysical meaning were tested. The sampling sites belonging to a specific calibration unit of a selected scheme were aggregated to compute the regression. The different aggregation schemes were evaluated with respect to their predictive power. The results gathered at the different aggregation levels were subjected to cross-validation (cv), applying a jackknife technique (leaving out one year at a time). In general, the model performance increased with increasing model parameterization, indicating the importance of additional unobserved and spatially heterogeneous agro-ecological effects (which might relate to grazing, species composition, optical soil properties, etc.) in modifying the relationship between cNDVI and herbaceous biomass at the end of the season. The biophysical aggregation scheme, the calibration units for which were derived from an unsupervised ISODATA classification utilising 10-day NDVI images taken between January 2001 and December 2015, showed the best performance in respect to the predictive power (R^2_{cv} = 0.47) and the cross-validated root-mean-square error (398 kg·ha^{-1}) values, although it was not the model with the highest number of calibration units. The proposed approach can be applied for the timely production of maps of estimated biomass at the end of the growing season before field measurements are made available. These maps can be used for the improved management of rangeland resources, for decisions on fire prevention and aid allocation, and for the planning of more in-depth field missions.

Keywords: food security; Sahel; Niger; rangeland productivity; livestock; MODIS; NDVI

1. Introduction

The livestock sector is economically important in Niger (for a map see Figure 1), contributing on average 10% to the gross domestic product (GDP) of Niger during the period 2009–2013 [1]. The agriculture and livestock census of 2005/2007 [2] estimated the total number of livestock to be around 31 million, composed mainly of cattle, sheep, and goats. The geographical distribution of livestock is not homogeneous in the country, and the largest numbers of livestock are in the regions of Zinder, Tahoua, Maradi, and Tillabery. Three different livestock systems exist in Niger, which are adapted to the agro-ecological conditions in the different zones of the country. A sedentary livestock system (accounting for 66% of the livestock) is practised together with the cultivation of crops in farms in the agricultural zone of the south (see Figure S1). Low-distance transhumance (i.e., seasonal movement of livestock) to pastoral enclaves during the rainy season is performed to avoid crop damage. Nomadic herding (18% of the livestock) is the main or only activity of herders in the pastoral zone or in the transition between the agricultural zone and the pastoral zone to its north. Livestock movements depend on the availability of water and pasture. The transhumance system (involving 16% of total livestock) is practised by herdsmen belonging to the Fulani ethnic group and is characterised by the seasonal movement of large herds between two distinct areas through well-defined corridors. Generally, the movement from north to south occurs in late winter, and transboundary transhumance to Benin, Nigeria, or Burkina Faso is common [2].

The rangelands of Niger are mainly located in the Sahel zone, a semi-arid region between the Sahara desert in the north and the Sudanian savannah in the south and are highly vulnerable as a result of the high inter-annual precipitation variability [3,4] and, hence, rangeland production variability. In addition to the inter-annual weather variability, the Sahel region of Niger shows great spatial precipitation heterogeneity, making rangeland production highly variable in space in a single year. Taking this variability into account would allow better management of rangeland resources and animal movement. The availability of a biomass production map at the end of the growing season represents a key tool to enhance sustainable and efficient rangeland management and improve food security in this region as a large number of households depend on livestock to sustain their livelihoods. Together with the number of livestock estimated by administrative units, this map could be used to compile a forage balance, identifying areas potentially affected by forage surplus or deficit that create a risk for fires or livestock mortality, respectively.

There is a long tradition of remote sensing (RS)-based herbaceous biomass estimation in Niger. The first studies in the late 1980s [5–8] applied linear regression between maximum standing biomass and normalised difference vegetation index (NDVI)-based indices (time-integrated and maximum NDVI) derived from the National Oceanic and Atmospheric Administration (NOAA) Advanced Very High Resolution Radiometer (AVHRR) imagery. The Ministry of Livestock in Niger currently uses a similar method based on NDVI data from the Satellite Pour l'Observation de la Terre—Vegetation (SPOT-VGT) and, since 2014, data from the Meteorological Operational satellite programme (MetOp) AVHRR. The ministry fits a linear regression model between ground measurements of the current season and a NDVI-based metric (maximum NDVI, mean NDVI, or cumulative NDVI are tested, and the best-performing metric is selected) to create a map of rangeland biomass. In 2014, Nutini et al. [9] estimated the extent of rangeland biomass at three sites in Niger with a radiation use efficiency model utilising cumulative dry matter productivity derived from SPOT-VGT [10] over a fixed time period, corrected with the evaporative fraction and derived from Moderate-resolution Imaging Spectroradiometer (MODIS) albedo and thermal measurements.

In addition to Niger, aboveground biomass has been estimated in several other semi-arid regions using RS data, including optical, radar, and combined multi-sensor imagery and modelling approaches (see [11] for a review). The majority of studies reviewed by Eisfelder et al. [11] applied medium- and low-resolution optical or radar data to derive a RS indicator that was used in an empirical relationship with field biomass measurements. The review by Ali et al. [12] on satellite RS of grasslands in general (not restricted to semi-arid regions) includes methodologies to retrieve grassland biophysical parameters. The authors concluded that regression models based on vegetation indices were the predominant approach for biomass estimation, while machine learning algorithms and the fusion of multisource data in biophysical simulation models still require further research to better utilise their potential [12].

Recent studies in the Sahel zone have used linear regression to estimate aboveground biomass from satellite-derived phenological metrics such as maximum value, the start-of-season and end-of-season values, and the cumulative value of the RS index during the growing season. These studies used time series of different RS indices such as the NDVI [13], the fraction of absorbed photosynthetically active radiation (FAPAR) [14,15], and the vegetation optical depth (VOD) [16]. In addition to simple linear regression, recent studies have also applied multi-linear models of phenological metrics [14] and machine-learning models combining FAPAR seasonal metrics and agro-meteorological data [17] to estimate annual (herbaceous) biomass production in Sahelian ecosystems of Senegal.

High spatial heterogeneity due to topography, cover, or relatively small area like in the Mediterranean rangelands challenge RS-based vegetation mapping and monitoring [18]. This heterogeneity can be addressed with higher spatial resolution imagery (for a review see [11]) or spectral unmixing modelling as a substitute for higher spatial resolution [18]. Landsat and very high-resolution data (e.g., IKONOS, QuickBird, HyMap) have been used for biomass estimation in Botswana, the Sahel and Sudan, West Africa, South Africa, and Zimbabwe [11]. However, biomass estimation has been mostly limited to a single year or very few years and relatively small target areas. An example of the application of an unmixing method for the retrieval of herbaceous biomass is the study by Svoray and Shoshany [19] that merged synthetic aperture radar (SAR) images with unmixed Landsat TM images.

This study aims to support existing rangeland management activities in Niger by developing an RS-based approach for mapping rangeland biomass. In contrast to the current method used by the Niger Ministry of Livestock, which is calibrated with the field observation of the current year, we intended to develop a method that does not require field measurements of the current season to estimate the herbaceous biomass at the end of the season. The proposed approach uses the phenological timings of the start and end of season extracted from the RS time series to retrieve a proxy of rangeland biomass at the end of the growing season; namely, the cumulative value of NDVI during the growing season (cNDVI). A linear regression model is calibrated with multi-annual field measurements of herbaceous biomass. The calibration of the regression model using multi-annual and multi-site observations can be performed with different aggregation levels, ranging from the site level (i.e., site by site) to the global level (i.e., all sites pooled together). The site-level calibration can be used to gain insights into the robustness of the relationship by analysing the slope and intercept across sites [9]. However, some level of aggregation is needed to map the biomass. Pooling all observations to derive a single regression model may represent an optimal use of the data if the relationship between measured biomass and the RS variable is constant in space. However, location-specific factors such as herbaceous species type and background reflectance may generate differences in the relationship between the RS variable and biomass. This spatial heterogeneity in the relationship can be modelled by grouping the field observations into more homogeneous sets. Different spatial aggregation schemes are proposed and tested to select the design providing the highest prediction power in a similar way to that suggested by Meroni et al. [20]. The method that has been developed can also be used in a predictive fashion to estimate the biomass at the end of the current season before field survey results are made available. The model can also serve as a backup solution in the event that field surveys are not carried out in a specific year or a specific region.

2. Materials and Methods

2.1. Study Area

The West African state of Niger features a pronounced climatic gradient. According to the Köppen-Geiger climate classification [21], the northern part is characterised by a hot arid desert and the southern part by a hot arid steppe climate. Mean annual precipitation (Figure 1) ranges from <100 mm in the north to 600–700 mm in the south of the country (calculated from TAMSAT rainfall estimates from the TARCAT v2.0 dataset for the period 1986–2015 [22,23]). Most of the precipitation falls during the rainy season from June to September and is associated with the West African monsoon season [3]. The mean annual surface temperature ranges from about 23 °C in the northernmost part of the country to 27–29 °C in the central and southern parts, with the highest mean monthly values occurring during the summer (e.g., range across the country in June: 31–35 °C) and the lowest during the winter (e.g., January: 13–25 °C) during the period 1981–2010 [24,25].

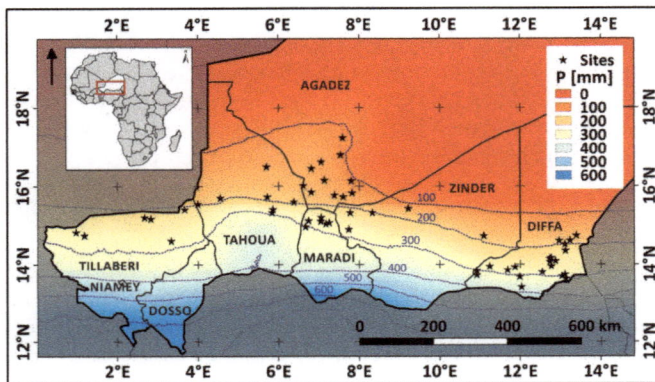

Figure 1. Mean annual precipitation totals (1986–2015), location of sample sites used for the regression model, and administrative regions of Niger (dark grey). Precipitation totals were computed from 10-day TARCAT data [22,23]. Dark blue dotted lines refer to 100 mm isohyets.

Following the precipitation gradient, the vegetation also changes from north to south. The main ecoregions in Niger include the Sahara desert, the south Saharan steppe and woodlands, the Sahelian acacia savannah, and the west Sudanian savannah [26]. The focus of this study is the pastoral zone of Niger, which is mainly located in the area with a mean annual precipitation of 100–300 mm (see Figure 1) and generally belongs to the Sahelian acacia savannah. As a precise delineation of the pastoral area is not available, the following limits were adopted: the 380 mm mean annual precipitation isohyet marks the southern limit, while the extent of vegetated area (as defined in Section 2.3.1) marks the northern limit (the final outline is shown in Section 3). In this way, we excluded the more agricultural landscape in the south while including all field measurement sites.

The mean annual precipitation sum of the sampling sites ranges from about 107 mm to 372 mm (calculated from TARCAT data [22,23]), while the mean annual temperature ranges from about 25.6 to 29.3 °C (calculated from ECMWF ERA INTERIM model data [27]). The dominant species in the herbaceous layer at the sample sites are *Cenchrus biflorus*, *Aristida* spp., *Schoenefeldia gracilis*, *Dactyloctenium aegyptium*, and *Alysicarpus ovalifolius*. In general, the species composition does not vary significantly annually at any particular site unless there is a bush fire. In contrast, the species composition varies between sites and is mainly linked to the soil type (A. Boureima, unpublished data).

2.2. Data

2.2.1. Biomass Data

In this study we used a dataset of measured aboveground herbaceous biomass (B) at the end of the growing season for the period 2001–2015, provided by the Ministry of Livestock of Niger. The dataset included 90 sites within the pastoral zone, with varying levels of information available for the study period, and comprised 926 records of dry matter production (kg·ha^{-1}) in total. The fresh and dry leaf matter are considered and are here assumed to represent the total aboveground biomass of the concluded season. The date of sampling varied but was usually between mid-September and the end of October. Each sampling site represents a (relatively) homogeneous area of 3 km × 3 km. The sampling of herbaceous biomass was performed along a transect, applying a double-sampling approach combining weighed biomass and visual estimates that finally yields estimates of dry matter production. The sampling method is described in more detail in Maidagi et al. [5] and Wylie et al. [8].

Biomass measurements should be made in the initial stages of the senescence phase, after the time of maximum vegetation development, typically between mid-August and mid-September. In this way, the measured biomass should not be affected by the natural decay of leaf matter and possible grazing that may take place during senescence. In addition to the tendency to underestimate the total biomass, late biomass measurements are less correlated to the RS-derived biomass proxies, which are dominated by the signal recorded during the period of maximum vegetation cover. To exclude measurements that are poorly representative of the seasonal biomass production we thus discarded biomass measurements from sampling carried out after 25 October (date proposed by local experts).

The quality of data from the database is known to be variable and field operator-dependent; therefore, we analysed the dataset and screened it for outliers. We inspected the site-level relationship between biomass and the RS indicator (see Section 2.3.1). As a relatively good site-specific relation (i.e., positive correlation) between the biomass and the RS indicator is expected, we opted to exclude from further analysis those sites with a coefficient of determination (R^2) of less than 0.2 for the linear regression between the two time series (number of sites excluded = 13). Sampling sites with fewer than four observations in the study period were discarded from the dataset (number of sites excluded = 21). Finally, 56 sites with a total of 616 records were retained for model parameterisation.

The biomass data at the end of the growing season of the rangelands in Niger showed high variability in space and time during the period 2001–2015. The multi-annual average biomass (dry matter production) of single sites ranged from 331 kg·ha^{-1} to 1778 kg·ha^{-1}, with a mean value of 760 kg·ha^{-1}. The annual spatial average of all sites varied over the study period from 246 kg·ha^{-1} (2004) to 1228 kg·ha^{-1} (2007), with an overall mean value of 732 kg·ha^{-1}. This high spatial and temporal variability of end-of-season biomass is mainly due to the high precipitation variability e.g., [28,29] and is in accordance with observations in other parts of the Sahel region e.g., [14,30,31].

2.2.2. Remote Sensing Data

This study used the eMODIS NDVI product, based on MODIS data acquired by the Terra satellite and provided by the United States Geological Survey (USGS; data portal: http://earlywarning.usgs. gov/fews/). The eMODIS product is a 10-day maximum value NDVI composite [32], temporally smoothed with the Smets algorithm [10]. Composite images are produced every five days; hence six temporally overlapping composites are generated per month. Here we only used the composite images for days 1–10, days 11–20, and days 21 to the last day of each month for the period January 2001 to December 2015. More information on the eMODIS NDVI product can be retrieved from the product documentation [33].

2.3. Methods

2.3.1. RS Proxy for Biomass Production

The seasonal cumulative NDVI (cNDVI) is defined as the integral of NDVI during the growing season subtracted from the area under the baseline (i.e., the NDVI minimum level before the start of the growing season), as proposed by Meroni et al. [34] for the seasonal cumulative FAPAR. The time interval for integration is defined by the start of season (SOS) and the end of season (EOS), estimated for each pixel and season. The model-fit approach of Meroni et al. [35], with further modification as in Vrieling et al. [36], was used to calculate the phenology parameters (SOS, EOS, maximum value of NDVI). SOS was estimated as the point at which the fitted NDVI model for the season exceeded 20% of the local growing amplitude (i.e., between minimum NDVI before green-up and maximum NDVI of that season), and EOS was estimated as the point at which the model falls below 80% of the decay amplitude (i.e., between maximum NDVI of the season and the following minimum NDVI after decay). Pixels are considered non-vegetated when the variability of the NDVI time series, as measured by the difference between the 95th and 5th percentiles, is below 0.05 NDVI units. Other methodological details can be found in Vrieling et al. [36]. The value of cNDVI is thus controlled by the integration limits, the baseline value, and the amplitude and shape of the NDVI seasonal trajectory.

2.3.2. Linear Regression Model and Spatial Aggregation Levels

Linear regression models are built by matching the field measurements at a given site and time with the corresponding cNDVI. The cNDVI values for the site were calculated as the mean of a kernel of 11 × 11 pixels centred on the site coordinates to roughly match the 3 km × 3 km area for which the field measurements are representative. A smaller kernel size (2750 m) was set to avoid border effects.

Relationships between cumulative vegetation indices and dry biomass are empirical and have only local values [37], meaning that a regression established in a given agro-ecological context may not be successfully extrapolated to a different region. One approach to tackle this problem would be to define smaller geographical units with similar physical characteristics and to treat them separately. However, improved performance is not guaranteed as the total number of observations available is constant, and the sample size (over which the single regression is calibrated) is reduced by subsampling. Thus, a trade-off between model specificity and data availability exists. We explored this trade-off empirically to find the modelling solution providing the best performance in prediction.

Different levels of spatial data aggregation (and thus model parameterisation) were considered for model calibration. In the simplest considered model, the whole set of measurements was pooled, and a single linear regression model was established (hereafter referred to as the 'global' model). Such a model assumes a constant relationship between biomass and cNDVI in both space and time. However, this relation might not hold in all circumstances, in particular with respect to spatial variation. The relation between the two variables may, in fact, vary spatially as a result of different species composition or background reflectance affecting NDVI estimates. Such spatial heterogeneities may be accounted for by limiting the spatial domain at which the model is calibrated. The stratification of the study area into smaller and more homogeneous units is thus expected to reduce the estimation error in the temporal domain. We therefore attempted a number of stratification strategies (Figure 2) with the aim of selecting the one providing the best performance in prediction. This selection was done empirically on the basis of a cross-validation (cv), applying a jackknife technique (leaving out one year at a time) to assess the predictability of biomass based on cNDVI. For each year left out, the linear regression coefficients were estimated based on the remaining dataset and used to predict the biomass values that were not used in the calibration. Then, all single jackknifed predictions were compared with the actual measurements to compute the cross-validated R^2 (R^2_{cv}). Based on the predicted biomass values from the jackknifing, a cross-validated root-mean-square error ($RMSE_{cv}$) was calculated as follows:

$$RMSE_{cv} = \sqrt{\frac{\sum\limits_{i=1}^{n} (\hat{B}_i - B_i)^2}{n}},$$

(1)

where $\hat{B}_i$ is the predicted biomass derived from the jackknifing, B_i is the measured biomass, and n is the number of samples. The $RMSE_{cv}$ gives an indication of the magnitude of the errors in prediction.

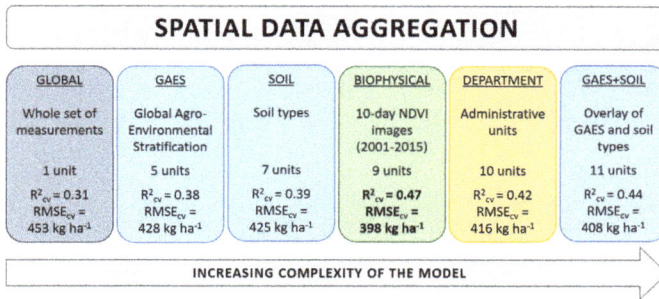

Figure 2. Overview of applied spatial aggregations for model calibration indicating the name of the model, the input data for the delineation of units, the number of spatial calibration units, and the cross-validated R^2 (R^2_{cv}) and root-mean-square error (RMSE$_{cv}$). The colour code is related to the biophysical meaning of single units: yellow = no biophysical meaning; blue = biophysical meaning exists but units have been taken from global or continental maps; green = biophysical meaning exists. GAES, Global Agro-Environmental Stratification.

An initial attempt to include local effects in the regression was the use of spatial proximity for stratification of available field measurements. For that we used administrative units (departments) (referred to as the 'department' model) taken from the Global Administrative Unit Layers (GAUL) level 2 [38], as administrative units are often used as stratification layers in official statistics. Note that the department with just one sample site (Goure) was assigned to the closest department, leading to a final total of 10 units. However, the departments are administrative units with arbitrary boundaries that do not guarantee any internal affinity in a biophysical sense. That said, the department model can still serve as a benchmark, a kind of 'arbitrary reference', to evaluate its performance relative to more biologically meaningful models. For a better parameterisation of the relationship between biomass and cNDVI, different stratifications with a more biophysical rationale were tested.

First, we used the Global Agro-Environmental Stratification (GAES [39]). The GAES is derived from the segmentation of 13 input layers with 1 km spatial resolution, aiming to stratify agricultural production zones according to the region's agro-environmental characteristics, including climate, altitude, irrigation, production, phenology, growing cycles, crop type, and field size parameters. The GAES is produced at four hierarchical levels (level 4 being the most detailed) using eCognition software on a continental basis for the segmentation. For our study, we used GAES level 4 to stratify the sample sites (referred to as the "GAES" model), which are represented by six GAES strata. The GAES stratum represented by just one site (ID 177) was assigned to another GAES stratum based on spatial proximity and agro-environmental type description, resulting in a total of five GAES units of the model.

Second, soil types from the African soil map [40] were used for stratification (referred to as the 'soil' model). The basic information for this harmonised continental scale soil map was derived from the Harmonized World Soil Database [41], and the naming of the soil types followed the World Reference Base for Soil (WRB) classification and correlation system [42]. The sample sites were characterised by nine different soil types. Soil types represented by just one sample site were assigned to other soil types based on spatial proximity and similar characteristics, i.e., eutric fluvisol was merged with haplic vertisols (both soil types are influenced by water) and vertic cambisols were merged with eutric cambisols (both are cambisols), resulting in a total of seven soil units.

Third, we created a stratification by intersecting GAES and soil type units (referred to as the "GAES + soil" model) to further refine the segmentation of the study area. This intersection resulted in a total of 13 combinations. Combinations represented by just one sample site were merged with another GAES + soil combination with the same GAES unit, i.e., "GAES unit 1 + soil unit 2" was assigned to "GAES unit 1 + soil unit 1" and "GAES unit 5 + soil unit 2" was merged with "GAES unit 5 + soil unit 3", resulting in a total of 11 GAES + soil units.

Fourth, we generated a stratification layer based on the unsupervised classification of NDVI images from eMODIS (see Section 2.2.2), applying an unsupervised clustering algorithm, namely the Iterative Self-Organizing Data Analysis Technique (ISODATA), similar to the approach of de Bie et al. [43]. The aim was to obtain a biophysical stratification purely relying on RS data. All 10-day NDVI images from January 2001 to December 2015 (n = 540; spatial subset defined by the mask of the study area) were used as input for the ISODATA clustering. We ran the ISODATA clustering with a fixed number of clusters, testing all options from five to 15. We then calibrated the regression model for each of the obtained 11 stratifications and selected the one with the highest R^2_{cv} as the 'biophysical model' for comparison with the other aggregation schemes. Table S1 and Figure S2 show the model results for the different ISODATA clustering-based stratifications. Note that not all the clusters obtained were represented by sample sites. This does not affect the calibration of the model (there will just be no regression for the uncovered clusters), but it does need to be addressed in the creation of the biomass maps (see Section 2.3.3).

The biophysical model that performed best in prediction was obtained with 11 clusters (clusters 1 and 11 were not represented by samples; therefore, nine calibration units were used) and is shown in Figure 3. In the northern part of the study area, south of the Sahara desert, there occur some well-defined bands of clusters that are linked to the precipitation gradient (see isohyets in Figure 3). In the south, the spatial pattern of clusters is more complex, especially in the southwest of Niger. Here, a small-scale heterogeneous pattern occurs, attributed to higher variability of morphological features (e.g., rivers, depressions) leading to different plant growth conditions and different land uses (e.g., presence of agriculture). Figure S3 shows the ISODATA clusters compared with the elevation (derived from Shuttle Radar Topography Mission data [44]) of Niger, highlighting that some spatial patterns of the clusters are clearly linked to elevation differences.

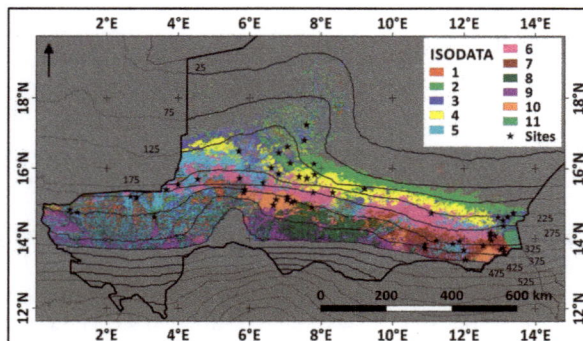

Figure 3. Results of the best-performing ISODATA clustering (11 clusters, of which nine were represented by sample sites). Dark grey lines represent mean annual isohyets (in mm) computed from 10-daily TARCAT data from 1986–2015 [22,23]. Areas outside the study area (as defined in 2.1) are masked out in grey.

In summary, a set of different stratification schemes with varying spatial detail was tested for the model calibration, ranging from the more detailed models at the GAES + soil (number of calibration units = 11), department (*n* = 10), and biophysical (*n* = 9) levels to those at the soil (*n* = 7), GAES (*n* = 5), and, finally, global levels (*n* = 1). Given the linear model used, the total number of parameters to be estimated (with the same amount of data) is 2*n* (*n* being the number of stratification units). The final selection of the most suitable model is made in cross-validation to address the trade-off between the increased accuracy resulting from the increased model parameterisation and the decreased robustness and predictive power related to a reduced sample size on which each model is calibrated through an increased number of calibration units.

2.3.3. Map of Estimated Biomass

The estimated biomass (B_e) at the end of the season is then mapped using the developed regression model:

$$B_e = MAX(0, a \times cNDVI + b) \tag{2}$$

where *a* and *b* are the model coefficients and cNDVI is the cumulative NDVI over the growing season of the year of interest. The maximum value between zero and the linear regression estimate is taken to discard negative biomass estimations that may be originated by models with negative intercept when the cNDVI is close to or equal to zero. The same coefficients are applied for all pixels when using the global regression model. In case of a stratified model, unit-specific coefficients are applied for each pixel within a certain unit.

In the case of the biophysical model, two clusters were not represented by sampling sites (i.e., cluster 1 and cluster 11). To obtain an estimation of biomass for those clusters, the coefficients of the next closest ISODATA clusters were used (cluster 2 parameters for cluster 1 and cluster 10 parameters for cluster 11).

3. Results and Discussion

Spatial patterns of the mean and coefficient of variation of the cNDVI over the period 2001–2015, show a rough north–south gradient (Figure 4), following the annual precipitation gradient (see Figure 1). The north of the study area (at the border to the Sahara) is characterised by a low mean cNDVI and high coefficient of variation values, the latter indicating high inter-annual variability. The south shows a higher mean cNDVI and lower inter-annual variability.

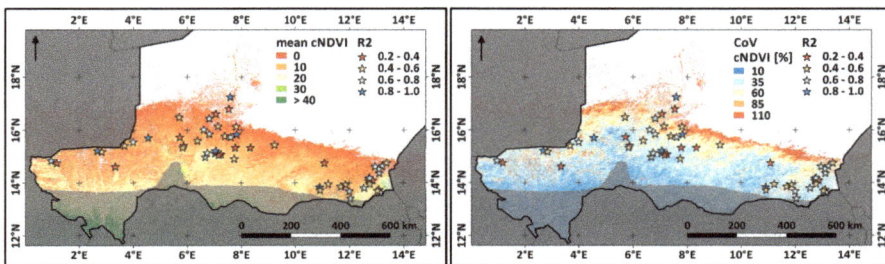

Figure 4. Mean (**left**) and coefficient of variation (CoV) (**right**) values of cumulative normalised difference vegetation index (cNDVI) over the period 2001–2015 and R^2 values of the site specific regression between biomass and cNDVI. Stars show the location of the field measurement sites while the R^2 has been colour coded from red to blue. Non-vegetated areas (as defined in Section 2.3.1) are in white.

3.1. Regression Models

The R^2 value between measured biomass and cNDVI at each site can serve as a primary indication of the strength of the relationship between these two parameters. R^2 varies notably among the sites (ranging from 0.25 to 0.99), as indicated in Figure 4, without showing any clear spatial pattern. This variability might be due to data quality issues and/or other processes affecting the measured biomass (e.g., grazing) that are not considered in the simple linear regression with cNDVI. Mbow et al. [45] showed in their study in the Sahel of Senegal that species composition significantly affects the relationship between NDVI and biomass, concluding that temporal and spatial variation in species dominance could add noise to this relationship. In their study, the effect was less pronounced when applying the seasonal integral of NDVI instead of the peak NDVI value. Olsen et al. [13] demonstrated on the basis of a long-term grazing trial in north Senegal that grazing-induced variations in the composition of annual plants changed the relationship between the seasonal integral of NDVI and the end of the season biomass.

Considering the regression models with the different aggregation levels, all stratified models show a notably better performance than the global model (Table 1). There is a tendency for increasing R^2 in fitting values with increasing complexity of the model (increasing number of coefficients to be adjusted). While the global single-unit model has an R^2 of 0.33, the models with five to seven units have R^2 values in the range 0.42–0.44, and the models with 10 or 11 units show R^2 values in the range 0.50–0.51. However, it is not strictly the case that an increase in calibration units always causes an increase in R^2. The biophysical model is characterised by the highest R^2 without having the highest number of calibration units (Figure 5). This indicates that the type of the stratification plays an important role in determining the performances of the model.

Table 1. Results of the regression model for different stratifications.

Stratification	No. of Units	R^2	R^2_{cv}	$RMSE_{cv}$ (kg·ha^{-1})	R^2 of Elementary Model Units	R^2_{cv} of Elementary Model Units
Global	1	0.33	0.31	453		
GAES	5	0.42	0.38	428	0.29–0.75	0.25–0.66
Soil	7	0.44	0.39	425	0.29–0.57	0.08–0.51
Biophysical	9	0.52	0.47	398	0.34–0.73	0.25–0.67
Department	10	0.51	0.42	416	0.29–0.67	0.22–0.52
GAES + soil	11	0.50	0.44	408	0.20–0.75	0.08–0.66

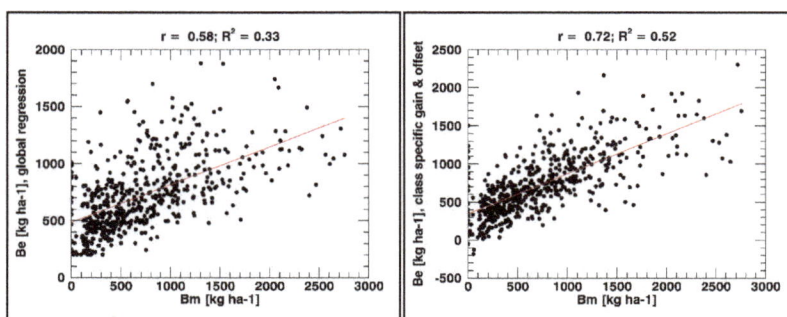

Figure 5. Scatterplot of measured biomass (B_m) versus estimated biomass (B_e) for the global model (**left**) and for the biophysical model (**right**). Negative B_e values in the biophysical model can occur when the linear regression of a certain class has a negative offset and when B_m is very low (i.e., less than this offset).

The performance of the different models in prediction is indicated by the R^2_{cv} and the $RMSE_{cv}$. The global model has by far the lowest R^2_{cv} and the highest $RMSE_{cv}$. The models building on existing thematic stratifications show increasing R^2_{cv} and decreasing $RMSE_{cv}$, with increasing numbers of calibration units. The biophysical model deviates from these relationships and shows the highest R^2_{cv} (0.47) and the lowest $RMSE_{cv}$ (393 kg·ha^{-1}), while having nine calibration units. The second highest R^2 value for the department model appears to be due to overfitting as this model shows the largest drop when changing to the prediction mode (R^2 compared with R^2_{cv}).

Note that the regressions over elementary units of the models vary in performance. For example, the biophysical model shows a range of R^2 values from 0.34 to 0.73, and, in cross-validation, the values range from 0.25 to 0.67. Elementary model units with a very low R^2 indicate that it is difficult to establish a good relationship between measured biomass and the RS-based biomass proxy in this area. This is not necessarily related to the small number of sites/samples in these units (e.g., the elementary model units of the biophysical model with the lowest R^2 are represented by eight and 10 sites, respectively). These low R^2 values could be the result of unsuitable stratification or the reasons mentioned for the single sites (data quality issues of certain samples, grazing, changing species composition).

In summary, the stratification of the study area in several calibration units increases the performance of the global regression model. An improved model performance for the estimation of biomass based on seasonal RS-derived indicators and in situ measurements through the subdivision of the study area into biophysically meaningful units was also shown for the Sahel in Senegal. Diouf et al. [14] stratified their study area by ecoregions and increased the R^2 from 0.68 to 0.77, applying a multi-linear model of three phenology-related parameters.

The biophysical model appears to be the most suitable among the tested stratifications for the estimation of end-of-the-season biomass in Niger as it shows the highest R^2_{cv} and the lowest $RMSE_{cv}$. Furthermore, this model presents a biophysically meaningful stratification of the study area that can be easily retrieved from the same RS data that are used to build the model and does not require ancillary data.

Compared with field surveys, our approach provides continuous biomass estimates across the study region and not just point information like other RS-based methods. The advantage of the proposed method compared with the current RS-based approach of the Niger Ministry of Livestock is that it can be applied before the field measurements are made available. This translates to a time-gain of two to four weeks that the field trips normally last. Therefore, the produced maps of estimated biomass could be used for the planning of more in-depth field missions and for the better management of rangeland resources. Additionally, our model can also serve as a backup solution in the event that field surveys are not carried out in a specific year or a specific region. For example, the number of available measurement sites per year varied between 20 and 54 during the study period (see Figure S4) and the frequency of measurements at a single site between four and 14. However, we strongly recommend to continue the field measurements as they provide additional information like species composition and allow the recalibration of the model, yielding to an improved statistical reliability of the model.

Three ways to improve the model can be tested: first, incorporating more seasonal metrics in a multi-linear model as proposed by Diouf et al. [14]; second, utilising a seasonal metric or a combination of parameters that can be retrieved earlier in the season and therefore allow a biomass estimation before the end of the season; and, third, using a combination of NDVI- or FAPAR-derived seasonal metrics with meteorological data [17].

3.2. Estimated Biomass Map

Sample maps of estimated biomass at the end of the growing season derived from the biophysical model are presented in Figure 6 for the years 2004 and 2007. As shown by field measurements (see Section 2.2.2), 2004 was characterised by a very low spatial average of measured biomass, while 2007 showed very high measured biomass values. The biomass maps highlight the high inter-annual but also spatial variability of biomass production in the region, underlining the importance of flexible and production-adapted rangeland management.

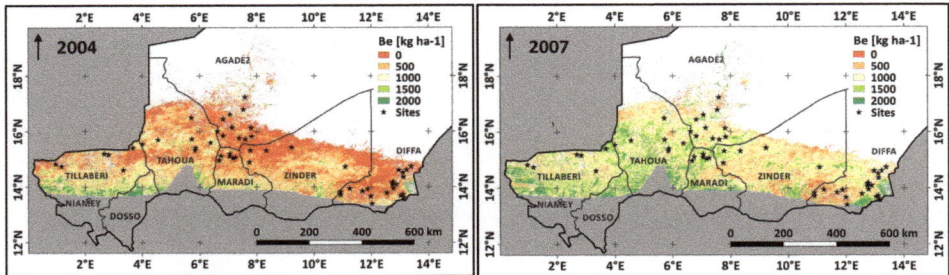

Figure 6. Maps of estimated biomass levels based on the biophysical model for the years 2004 (**left**) and 2007 (**right**).

Figure 7 shows the estimated biomass map for 2015 as the most recent example, indicating one possible use of such biomass maps. In addition to the biomass for 2015, the biomass anomaly for this year was also calculated as the difference between the estimated biomass of 2015 and the long-term average (in our example, the mean of the years 2001–2014). The anomaly map highlights areas with above or below average biomass production. With this information at hand, the relevant authorities could make certain rangeland management decisions. In the 2015 example, there is a large negative anomaly in the Eastern Tahoua and Southern Agadez regions, while there is a significant positive anomaly in the Eastern Zinder and Western Diffa regions (circled in Figure 7). Together with information about the availability of water for the animals and the location and size of herds, the authorities could then assess if it is necessary to reduce the livestock stocking rates or to provide additional feed (e.g., enriched alfalfa, wheat bran, cotton seed cake) in areas with large negative anomalies. In areas with pronounced positive anomalies authorities could evaluate if it would be appropriate to channel additional animals to these areas (simultaneously strengthening the livestock safety in view of the high concentration of animals) or if some fire protection measures are required (e.g., firewalls and awareness-raising campaigns). The report on the pastoral campaign from the Ministry of Livestock [46] from November 2015 reported variable forage production in the pastoral areas of Niger, ranging from good to mediocre production. Large pockets with low watering resulting in low biomass production were observed in South Ingall, Aderbissenat, North-West Abalak, Tchintabaraden, and Bermo, which are areas of large negative anomalies in Figure 7. Having considered the biomass production and the number of livestock (resident livestock and exceptional cases of livestock from Malian refugees), the report concluded with a note of concern for the 2015/2016 dry season as all regions were experiencing an overall forage deficit.

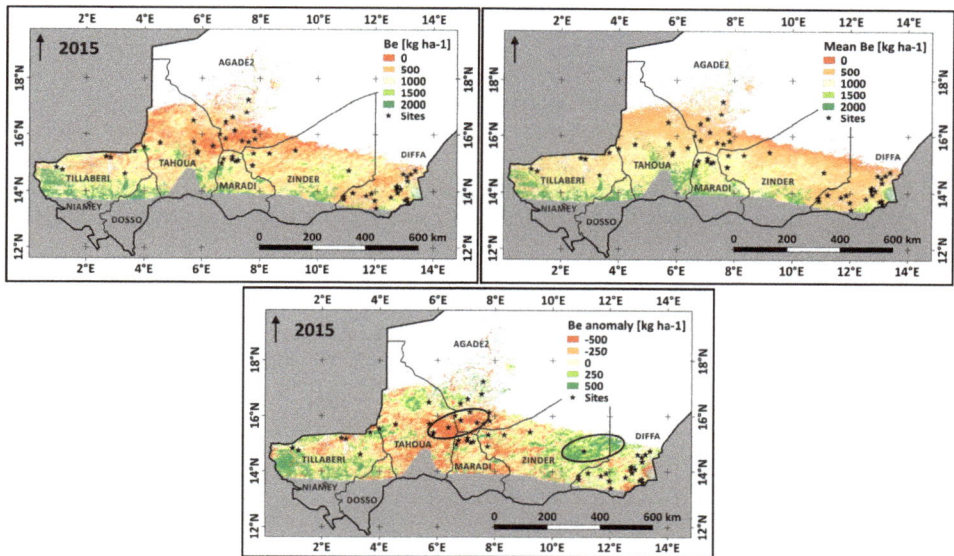

Figure 7. Estimated biomass (B$_e$) for 2015 based on the biophysical model (**upper left**), mean B$_e$ for the years 2001–2014 (**upper right**), and B$_e$ anomaly for 2015 (**bottom**), calculated as the difference between B$_e$ for 2015 and mean B$_e$ for the years 2001–2014.

4. Conclusions

In this study a phenology-based predictive model to estimate rangeland biomass in Niger was developed. The cNDVI during the growing season, derived from 10-day MODIS NDVI data applying a model-fit phenology retrieval method, was used as a remote sensing proxy for biomass production. The relationship between this variable and the measured herbaceous biomass at the end of the growing season was modelled by linear regression. Different spatial aggregation levels for the model calibration were tested to find the most suitable one for biomass prediction.

A general, but not strict, tendency for increased estimation performance with increased model complexity (in terms of number of parameters to be calibrated) was observed. This indicates the importance of additional and spatially heterogeneous agro-ecological unobserved effects (which might relate to grazing, species composition, optical soil properties, etc.) in modifying the relationship between cNDVI and herbaceous biomass at the end of the season. The biophysical model showed the best performance with respect to the predictive power (R$^2_{cv}$ = 0.47) and RMSE$_{cv}$ value (398 kg·ha^{-1}), without having the highest number of parameters to be calibrated. All models that were tested using several calibration units outperformed the simple model, using a unique linear relationship for the entire study area.

The presented approach can be applied for the timely production of estimated biomass production maps at the end of the growing season and before the field measurements are made available. This would mean a time gain of two to four weeks, which is the length of time the field trips normally last. Therefore, such maps could be used for the planning of more in-depth field missions, for the better management of rangeland resources, and for timely decisions on aid allocation and fire prevention. Additionally, the approach can serve as a backup solution in the event that field surveys are not carried out in a specific year or a specific region.

In summary, the presented approach should be used as a supplement to and not as a replacement for field measurements. We highly recommend that field measurements are continued for the following reasons:

- They provide additional information such as species composition that currently cannot be retrieved by RS data.
- The presented model to estimate biomass was calibrated with field data obtained over a 15-year time period. However, the full variability in biomass production in the study area may not be captured by this 15-year period, as this variability may also be affected by longer-term climate variability. As longer satellite image time series become available in the coming years, future research should analyse this temporal variability. In general, whenever a new field survey becomes available, the model should be recalibrated as more data improve the statistical reliability of the model.

Supplementary Materials: The following supplementary data are available online at www.mdpi.com/ 2072-4292/9/5/463/s1. Figure S1: Agro-ecological zones of Niger; Figure S2: Performance of different ISODATA-clustering-based stratifications with respect to R^2_{cv} and $RMSE_{cv}$; Figure S3: Results of the best-performing ISODATA clustering compared with elevation from SRTM data; Figure S4: Available number of biomass measurements during the study period; Table S1: Model results for ISODATA clustering with a fixed number of classes.

Acknowledgments: We thank our colleague, Tharcisse Nkunzimana, for his support regarding the communication and exchange with the Ministry of Livestock of Niger. Furthermore, we thank the three anonymous referees for their helpful comments. The source of administrative boundaries in all presented maps is the Global Administrative Unit Layers (GAUL) dataset, implemented by the Food and Agriculture Organization within the CountrySTAT and Agricultural Market Information System (AMIS) projects.

Author Contributions: Anne Schucknecht, Michele Meroni., and Francois Kayitakire conceived and designed the experiments; Anne Schucknecht and Michele Meroni performed the experiments and analysed the data; Amadou Boureima contributed field data and local information; Anne Schucknecht drafted the manuscript; and all authors revised the manuscript and contributed to the discussion of the results.

Conflicts of Interest: The authors declare no conflict of interest.

References

1. INS-Niger—Institut National de la Statistique, Republique du Niger. *Annuaire Statistique 2009–2013;* Institut National de la Statistique: Niamey, Niger, 2014. Available online: http://www.stat-niger.org/statistique/ file/Annuaires_Statistiques/Annuaire%20Statistique%202009-2013.pdf (accessed on 6 May 2016).

2. Republique du Niger. *Recensement General de L'Agriculture et du Cheptel (RGAC 2005/2007)—Vol. II Resultats Definitifs (Volet Cheptel);* Republique du Niger, Ministere du Developpement Agricole, Ministere des Ressources Animales: Niamey, Niger, 2007. Available online: http://harvestchoice.org/sites/default/ files/downloads/publications/Niger_2005--07_Vol2.pdf (accessed on 17 February 2015).

3. Nicholson, S.E. The West African Sahel: A review of recent studies on the rainfall regime and its interannual variability. *ISRN Meteorol.* **2013**, *2013*. [CrossRef]

4. Tarhule, A.; Zume, J.T.; Grijsen, J.; Talbi-Jordan, A.; Guero, A.; Dessouassi, R.Y.; Doffou, H.; Kone, S.; Coulibaly, B.; Harshadeep, N.R. Exploring temporal hydroclimatic variability in the Niger Basin (1901–2006) using observed and gridded data. *Int. J. Climatol.* **2014**, *35*, 520–539. [CrossRef]

5. Maidagi, B.; Denda, I.; Wylle, B.; Harrington, J. Pasture Production in the Central and Eastern Pastoral Zones of Niger. Available online: http://pdf.usaid.gov/pdf_docs/PNABG559.pdf (accessed on 2 March 2015).

6. Wylle, B.; Harrington, J.A.; Pieper, R.; Maman, A.; Denda, I. Pasture Assessment Early Warning System—Research on Satellite-based Pasture Assessment Implementation Techniques. Available online: http://pdf.usaid.gov/pdf_docs/PNABB570.pdf (accessed on 4 March 2015).

7. Wylle, B.K.; Harrington, J.A.; Prince, S.D.; Denda, I. Satellite and ground-based pasture production assessment in Niger: 1986–1988. *Int. J. Remote Sens.* **1991**, *12*, 1281–1300. [CrossRef]

8. Wylle, B.K.; Denda, I.; Pieper, R.D.; Harrington, J.A.; Reed, B.C.; Southward, G.M. Satellite-based herbaceous biomass estimates in the pastoral zone of Niger. *J. Range Manag.* **1995**, *48*, 159–164. [CrossRef]

9. Nutini, F.; Boschetti, M.; Candiani, G.; Bocchi, S.; Brivio, P.A. Evaporative fraction as an indicator of moisture condition and water stress status in semi-arid rangeland ecosystems. *Remote Sens.* **2014**, *6*, 6300–6323. [CrossRef]

10. Smets, B.; Eerens, H.; Jacobs, T.; Royer, A. BioPar Product User Manual—Dry Matter Productivity (DMP), Version 0 from SPOT/VEGETATION data. Available online: http://land.copernicus.eu/global/sites/default/files/products/GIO-GL1_PUM_DMP_I1.00.pdf (accessed on 5 March 2015).

11. Eisfelder, C.; Kuenzer, C.; Dech, S. Derivation of biomass information for semi-arid areas using remote-sensing data. *Int. J. Remote Sens.* **2012**, *33*, 2937–2984. [CrossRef]

12. Ali, I.; Cawkwell, F.; Dwyer, E.; Barrett, B.; Green, S. Satellite remote sensing of grasslands: From observation to management—A review. *J. Plant Ecol.* **2016**, *9*, 649–671. [CrossRef]

13. Olsen, J.L.; Miehe, S.; Ceccato, P.; Fensholt, R. Does EO NDVI seasonal metrics capture variations in species composition and biomass due to grazing in semi-arid grassland savannas? *Biogeosciences* **2015**, *12*, 4407–4410. [CrossRef]

14. Diouf, A.A.; Brandt, M.; Verger, A.; El Jarroudi, M.; Djaby, B.; Fensholt, R.; Ndione, J.A.; Tychon, B. Fodder biomass monitoring in Sahelian rangelands using phenological metrics from FAPAR time series. *Remote Sens.* **2015**, *7*, 9122–9148. [CrossRef]

15. Schucknecht, A.; Meroni, M.; Kayitakire, F.; Rembold, F.; Boureima, A. Biomass estimation to support pasture management in Niger. In Proceedings of the 36th International Symposium on Remote Sensing of Environment, Berlin, Germany, 11–15 May 2015.

16. Tian, F.; Brandt, M.; Liu, Y.Y.; Verger, A.; Tagesson, T.; Diouf, A.A.; Rasmussen, K.; Mbow, C.; Wang, Y.; Fensholt, R. Remote sensing of vegetation dynamics in drylands: Evaluating vegetation optical depth (VOD) using AVHRR NDVI and in situ green biomass data over West African Sahel. *Remote Sens. Environ.* **2016**, *177*, 265–276. [CrossRef]

17. Diouf, A.A.; Hiernaux, P.; Brandt, M.; Faye, G.; Djaby, B.; Diop, M.B.; Ndione, J.A.; Tychon, B. Do agrometeorological data improve optical satellite-based estimations of the herbaceous yield in Sahelian semi-arid ecosystems? *Remote Sens.* **2016**, *8*. [CrossRef]

18. Svoray, T.; Perevolotsky, A.; Atkinson, P.M. Ecologocial sustainability in rangelands: the contribution of remote sensing. *Int. J. Remote Sens.* **2013**, *34*, 6216–6242. [CrossRef]

19. Svoray, T.; Shoshany, M. Herbaceous biomass retrieval in habitats of complex composition: A model merging SAR images with unmixed landsat TM data. *IEEE Trans. Geosci. Remote Sens.* **2003**, *41*, 1592–1601. [CrossRef]

20. Meroni, M.; Marinho, M.; Verstraete, M.; Sghaier, N.; Leo, O. Remote sensing based yield estimation in a stochastic framework—Case study of Tunisia. *Remote Sens.* **2013**, *5*, 539–557. [CrossRef]

21. Kottek, M.; Grieser, J.; Beck, C.; Rudolf, B.; Rubel, F. World map of the Köppen-Geiger climate classification updated. *Meteorol. Z.* **2006**, *15*, 259–263. [CrossRef]

22. Maidment, R.; Grimes, D.; Allan, R.P.; Tarnavsky, E.; Stringer, M.; Hewison, T.; Roebeling, R.; Black, E. The 30 year TAMSAT African rainfall climatology and time series (TARCAT) data set. *J. Geophys. Res. Atmos.* **2014**, *119*. [CrossRef]

23. Tarnavsky, E.; Grimes, D.; Maidment, R.; Black, E.; Allan, R.; Stringer, M.; Chadwick, R.; Kayitakire, F. Extension of the TAMSAT satellite-based rainfall monitoring over Africa and from 1983 to present. *J. Appl. Meteorol. Clim.* **2014**, *53*, 2805–2822. [CrossRef]

24. Kalnay, E.; Kanamitsu, M.; Kistler, R.; Collins, W.; Deaven, D.; Gandin, L.; Iredell, M.; Saha, S.; White, G.; Woollen, J.; et al. The NCEP/NCAR 40-year reanalysis project. *B. Am. Meteorol. Soc.* **1996**, *77*, 437–471. [CrossRef]

25. NOAA—National Oceanic and Atmospheric Administration. Monthly/Seasonal Climate Composites from the NCEP Reanalysis and Other Datasets. Available online: https://www.esrl.noaa.gov/psd/cgi-bin/data/composites/printpage.pl (accessed on 18 August 2016).

26. Olson, D.M.; Dinerstein, E.; Wikramanayake, E.D.; Burgess, N.D.; Powell, G.V.N.; Underwood, E.C.; D'Amico, J.A.; Itoua, I.; Strand, H.E.; Morrison, J.C.; et al. Terrestrial ecoregions of the world: A new map of life on earth. *Bioscience* **2001**, *51*, 933–938. [CrossRef]

27. Dee, D.P.; Uppala, S.M.; Simmons, A.J.; Berrisford, P.; Poli, P.; Kobayashi, S.; Andrae, U.; Balmaseda, M.A.; Balsamo, G.; Bauer, P.; et al. The ERA-Interim reanalysis: Configuration and performance of the data assimilation system. *Q. J. R. Meteorol. Soc.* **2011**, *137*, 553–597. [CrossRef]

28. Herrmann, S.M.; Anayamba, A.; Tucker, C.J. Recent trends in vegetation dynamics in the African Sahel and their relationship to climate. *Glob. Environ. Chang.* **2005**, *15*, 394–404. [CrossRef]

29. Hickler, T.; Eklundh, L.; Seaquist, J.W.; Smith, B.; Ardö, J.; Olsoon, L.; Sykes, M.T.; Sjöström, M. Precipitation controls Sahel greening trend. *Geophys. Res. Lett.* **2005**, *32*. [CrossRef]

30. Brandt, M.; Mbow, C.; Diouf, A.A.; Verger, A.; Samimi, C.; Fensholt, R. Ground- and satellite-based evidence of the biophysical mechanisms behind the greening Sahel. *Glob. Chang. Biol.* **2015**, *21*, 1610–1620. [CrossRef] [PubMed]

31. Dardel, C.; Kergoat, L.; Hiernaux, P.; Mougin, E.; Grippa, M.; Tucker, C.J. Re-greening Sahel: 30 years of remote sensing data and field observations (Mali, Niger). *Remote Sens. Environ.* **2014**, *140*, 350–364. [CrossRef]

32. Jenkerson, C.B.; Maiersperger, T.; Schmidt, G. *eMODIS: A User-Friendly Datasource*; U.S. Geological Survey Open-File Report 2010–1055; U.S. Geological Survey: Reston, VA, USA, 2010.

33. USGS—U.S. Geological Survey. Product Documentation: eMODIS TERRA Normalized Difference Vegetation Index (NDVI). 2013. Available online: https://earlywarning.usgs.gov/fews/product/115 (accessed on 5 May 2017).

34. Meroni, M.; Rembold, F.; Verstraete, M.M.; Gommes, R.; Schucknecht, A.; Beye, G. Investigating the relationship between the inter-annual variability of satellite-derived vegetation phenology and a proxy of biomass production in the Sahel. *Remote Sens.* **2014**, *6*, 5868–5884. [CrossRef]

35. Meroni, M.; Verstraete, M.M.; Rembold, F.; Urbano, F.; Kayitakire, F. A phenology-based method to derive biomass production anomaly for food security monitoring in the Horn of Africa. *Int. J. Remote Sens.* **2014**, *35*, 2471–2492. [CrossRef]

36. Vrieling, A.; Meroni, M.; Mude, A.G.; Chantarat, S.; Ummenhofer, C.C.; de Bie, K. Early assessment of seasonal forage availability for mitigating the impact of drought on East African pastoralists. *Remote Sens. Environ.* **2016**, *174*, 44–55. [CrossRef]

37. Moulin, S.; Bondeau, A.; Delecolle, R. Combining agricultural crop models and satellite observations: From field to regional scales. *Int. J. Remote Sens.* **1998**, *19*, 1021–1036. [CrossRef]

38. FAO—Food and Agriculture Organization of the United Nations. The Global Administrative Unit Layers (GAUL) 2014. Available online: http://www.fao.org/geonetwork/srv/en/metadata.show?currTab=simple&id=12691 (accessed 05 May 2017).

39. Mücher, S.; De Simone, L.; Kramer, H.; De Wit, A.; Roupioz, L.; Hazeu, G.; Boogaard, H.; Schuiling, R.; Fritz, S.; Latham, J.; et al. A New Global Agro-Environmental Stratification (GAES). Available online: http://library.wur.nl/WebQuery/wurpubs/fulltext/400815 (accessed on 24 February 2017).

40. Jones, A.; Breuning-Madsen, H.; Brossard, M.; Dampha, A.; Deckers, J.; Dewitte, O.; Gallali, T.; Hallett, S.; Jones, R.; Kilasara, M.; et al. *Soil Atlas of Africa*; Publications Office of the European Union: Luxembourg, 2013.

41. Dewitte, O.; Jones, A.; Spaargaren, O.; Breuning-Madsen, H.; Brossard, M.; Dampha, A.; Deckers, J.; Gallali, T.; Hallett, S.; Jones, R.; et al. Harmonisation of the soil map of Africa at the continental scale. *Geoderma* **2013**, *211–212*, 138–153. [CrossRef]

42. IUSS Working Group WRB. World Reference Base for Soil Resources 2006. Available online: http://www.fao.org/fileadmin/templates/nr/images/resources/pdf_documents/wrb2007_red.pdf (accessed 5 May 2017).

43. De Bie, C.A.J.M.; Khan, M.R.; Smakhtin, V.U.; Venus, V.; Weir, M.J.C.; Smaling, E.M.A. Analysis of multi-temporal SPOT NDVI images for small-scale land-use mapping. *Int. J. Remote Sens.* **2011**, *32*, 6673–6693. [CrossRef]

44. Jarvis, A.; Reuter, H.I.; Nelson, A.; Guevara, E. Hole-Filled Seamless SRTM Data V4. International Centre for Tropical Agriculture (CIAT). Available online: http://srtm.csi.cgiar.org (accessed on 6 February 2014).

45. Mbow, C.; Fensholt, R.; Rasmussen, K.; Diop, D. Can vegetation productivity be derived from greenness in a semi-arid environment? Evidence from ground-based measurements. *J. Arid. Environ.* **2013**, *97*, 56–65. [CrossRef]

46. Republique du Niger. *Rapport de Synthese des Resultats de la Campagne Pastorale 2015–2016*; Ministere de l'Elevage: Niamey, Niger, 2015.

![remote sensing logo] *remote sensing*

MDPI

Article

Seasonal Timing for Estimating Carbon Mitigation in Revegetation of Abandoned Agricultural Land with High Spatial Resolution Remote Sensing

Ning Liu [1,2,*], Richard J. Harper [1,2], Rebecca N. Handcock [1,3], Bradley Evans [4], Stanley J. Sochacki [1], Bernard Dell [1,2], Lewis L. Walden [1] and Shirong Liu [2,*]

[1] School of Veterinary and Life Sciences, Murdoch University, South Street, Murdoch, WA 6150, Australia; R.Harper@murdoch.edu.au (R.J.H.); rebecca.handcock@ecu.edu.au (R.N.H.); s.sochacki@murdoch.edu.au (S.J.S.); b.dell@murdoch.edu.au (B.D.); L.Walden@murdoch.edu.au (L.L.W.)

[2] Key Laboratory of Forest Ecology and Environment of State Forestry Administration, Institute of Forest Ecology, Environment and Protection, Chinese Academy of Forestry, Beijing 10091, China

[3] Edith Cowan University, Joondalup, WA 6027, Australia

[4] School of Life and Environmental Sciences, The University of Sydney, Sydney, NSW 2006, Australia; bradley.evans@sydney.edu.au

* Correspondence: ln1267@gmail.com or N.Liu@murdoch.edu.au (N.L.); Liusr@caf.ac.cn (S.L.); Tel.: +61-8-9360-2191 (N.L.); +86-10-6288-9311 (S.L.)

Academic Editors: Lalit Kumar, Onisimo Mutanga and Randolph H. Wynne
Received: 7 March 2017; Accepted: 24 May 2017; Published: 1 June 2017

Abstract: Dryland salinity is a major land management issue globally, and results in the abandonment of farmland. Revegetation with halophytic shrub species such as *Atriplex nummularia* for carbon mitigation may be a viable option but to generate carbon credits ongoing monitoring and verification is required. This study investigated the utility of high-resolution airborne images (Digital Multi Spectral Imagery (DMSI)) obtained in two seasons to estimate carbon stocks at the plant- and stand-scale. Pixel-scale vegetation indices, sub-pixel fractional green vegetation cover for individual plants, and estimates of the fractional coverage of the grazing plants within entire plots, were extracted from the high-resolution images. Carbon stocks were correlated with both canopy coverage (R^2: 0.76–0.89) and spectral-based vegetation indices (R^2: 0.77–0.89) with or without the use of the near-infrared spectral band. Indices derived from the dry season image showed a stronger correlation with field measurements of carbon than those derived from the green season image. These results show that in semi-arid environments it is better to estimate saltbush biomass with remote sensing data in the dry season to exclude the effect of pasture, even without the refinement provided by a vegetation classification. The approach of using canopy cover to refine estimates of carbon yield has broader application in shrublands and woodlands.

Keywords: aboveground biomass; *Atriplex nummularia*; carbon mitigation; carbon inventory; forage crops; remote sensing; vegetation index

1. Introduction

Global climate change is resulting from an imbalance in global greenhouse gas emissions [1]. A major strategy to mitigate carbon dioxide emissions is to sequester or remove carbon from the atmosphere through changing land use and increasing storage in plant biomass or soils [2,3]. Indeed, 83% of the mitigation targets or Intended Nationally Determined Contributions (INDCs) published following the 2015 Paris Climate Change Conference included the land sector [4]. However, carbon mitigation activities on farmland can displace food production [5] or affect water supplies [6].

Alternative mitigation approaches have been advocated, such as using low value or otherwise abandoned farmland to avoid competitive effects of vegetation [7].

In 2002, about 20,000 farms and 2 million hectares of agricultural land showed actual signs of salinity [8], and up to 170,000 km^2 of land in Australia is predicted to be affected from salinity by 2050 [9] and up to 4 million km^2 globally [10]. One option for salt-affected land is revegetation with salt-tolerant grazing plants such as *Atriplex* spp. [7]. Revegetation is a specific category of mitigation activity within the United Nations Framework Convention on Climate Change and is defined as the establishment of vegetation that does not meet the definitions of afforestation or reforestation. In Australia, this is defined as plants that do not exceed 2 m in height. Walden et al. [11] found consistent amounts of aboveground carbon stock (0.2–0.6 t C·ha^{-1}·year^{-1}) by *A. nummularia* at six sites across southern Australia, with potential total sequestration of 1.1–3.6 Mt C·year^{-1}, and studies (e.g., Harper et al. [2] and Harper et al. [12]) have described its usefulness as a grazing shrub in animal based farming systems.

To participate in carbon trading schemes, such as the Clean Development Mechanism or the Australian Carbon Farming Initiative [13], it is essential to measure and report amounts of carbon stocks. Ground-based field measurements of biomass are expensive. This study therefore evaluates the suitability of less expensive remote sensing approaches to estimate carbon stocks following farmland revegetation, with a focus on areas where there are low rates of sequestration. The calibration of remote sensing data with in situ measurements of biomass has the potential to be a cost effective means of reporting carbon stocks across landscapes, as well as being a timely source of data originating from direct observation of actual carbon stocks rather than being solely modelled values.

Analysis of high-resolution remotely sensed images can be at the scale of individual pixels (i.e., pixel-based) [14] or use approaches to extract multi-pixel features from the image (i.e., feature extraction, fractional coverage) [15], including the ability to use individual canopy crowns e.g. Bunting and Lucas [16] where the images are of sufficiently fine spatial resolution. Sochacki et al. [7] and Walden et al. [11] both found that aboveground biomass of *A. nummularia* followed a strong linear trend (R^2 = 0.81) in relation to a crown volume index (CVI), calculated from crown width, length, and height. Walden et al. [11] also found that canopy diameter measurements were only slightly less predictive (R^2 = 0.68) compared to CVI, and could be used to estimate aboveground biomass (AGB). The allometric relationship between biomass measurements and carbon estimates has been established [11] and these relationships could therefore be used as a basis for estimating carbon stocks in situations where individual crowns can be delineated from remote sensing images.

Although the use of pixel-based vegetation indices as proxies for estimating vegetation biomass is well established [17,18], including examples such as the normalized difference vegetation index (NDVI), enhanced vegetation index (EVI) [19], and the ratio vegetation index (RVI) on forests [20], grass [21] and woodland [22,23], these techniques have been infrequently used to estimate carbon stocks in shrublands. However, the broader use of remote sensing for calibrating vegetation indices to biomass (e.g., Asner [24]) shows that the relationship between vegetation indices and biomass can differ between species, season, and the scale of the vegetation and pixels.

Estimates of canopy coverage derived from remote sensing images have also been applied as a proxy for calculating individual tree and stand biomass [25,26]. For example, Sousa et al. [27] found that the tree canopy horizontal projection derived from QuickBird satellite images produced highly accurate estimates of AGB of *Quercus rotundifolia* at both individual and plot scales. All of these approaches require remotely sensed images of sufficiently high spatial resolution to resolve the individual plants or stands being monitored.

In this study, we use high-resolution aerial images to explore characteristics of monitoring salt-tolerant grazing plants for carbon stocks in a Mediterranean environment. We derive estimates of canopy coverage from high spatial resolution airborne Digital Multi Spectral Imagery (DMSI) at two times of the year, and determine the utility of these images for estimating aboveground biomass and carbon stocks at both the plant- and stand-scale. Three remote sensing approaches were used: pixel-scale vegetation indices, extraction of the stand crowns from the high-resolution images, and estimation of the fractional coverage of the grazing plants within entire stands (Figure 1).

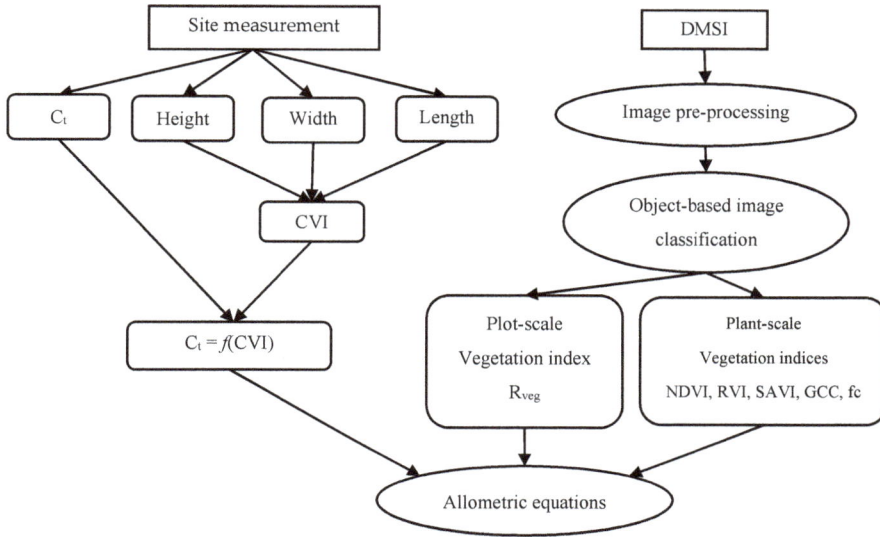

Figure 1. Flow diagram of the methodology followed in this study (where C_t is the carbon stocks (t C·ha^{-1}), DMSI is the Digital Multi Spectral Imagery, and CVI is the crown volume index). For definitions of vegetation indices, see Table 1.

Our specific objective was to determine if it is feasible to develop remote sensing models that can be used economically and efficiently to estimate carbon stocks at scales suitable for project level carbon accounting, and to determine the most suitable timing (e.g., wet season or dry season) for image acquisition for this purpose.

2. Materials and Methods

2.1. Experimental Sites

The study site (32°43′34.33″S, 117°39′55.27″E) was located near Wickepin, Western Australia, and was established to investigate carbon stocks following revegetation of abandoned salinized farmland [7,28]. The region has a semi-arid Mediterranean climate, with a seasonal drought from November to April, and a mean annual rainfall of 357 mm·year^{-1} (2000–2011, Wickepin weather station No. 010654 from the Australian Bureau of Meteorology) and a mean annual pan evaporation of 1789 mm·year^{-1}.

Atriplex nummularia was planted adjacent to a salt scald in 2001, at densities of 0, 500 and 2000 plants ha^{-1}, each with three replicates, in a randomized complete block design, consisting of two blocks (Figure 2). Details of seedling production and planting have been described previously [7]. At the time of field measurement in December 2011 (dry season), the low- and high-density stands had average heights of 2.17 and 1.68 m, and average canopy diameters of 2.66 and 1.52 m, respectively.

Both the control plots (0 plants ha^{-1}) and the areas between the *A. nummularia* plants were comprised of an array of annual volunteer pasture plants including capeweed (*Arctotheca calendula*), geranium (*Erodium* sp.), and various grasses (e.g., *Hordeum glaucum*, *Lolium rigidum*). As a consequence of the Mediterranean environment, the annual pasture plants are only alive in the period April-October, whereas the *A. nummularia* plants bear foliage year-round.

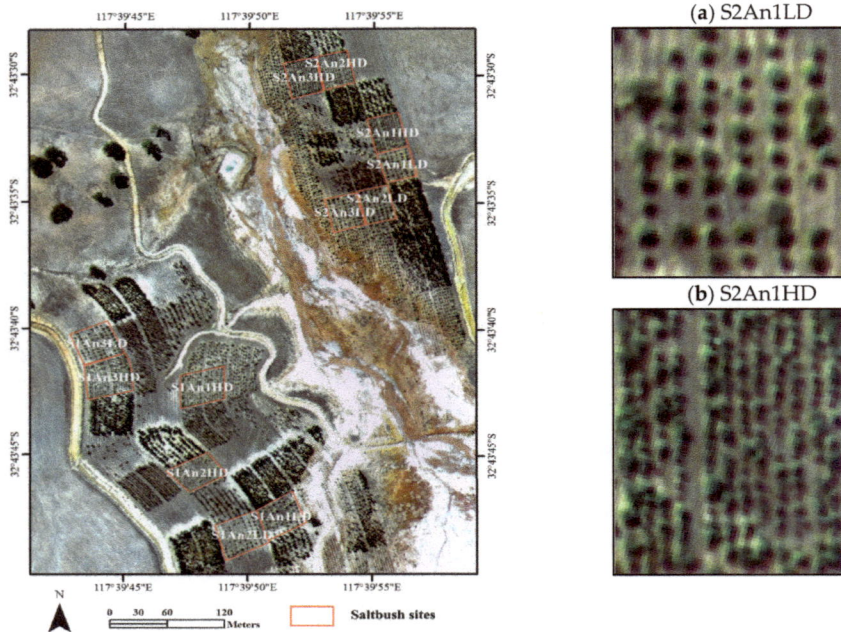

Figure 2. High spatial resolution airborne DMSI image (0.5 m) of the Wickepin experimental site taken on 24 March 2011 (dry season), with examples of: (**a**) low density (500 plants ha^{-1}, plot S2An1LD); and (**b**) high density (2000 plants ha^{-1}, plot S2An1HD). Plots were 40 × 40 m^2, with field imagery measurements taken from an internal 20 × 20 m plot to minimize competitive edge effects. Key to plot name: S1, S2—Block; An—species (*Atriplex nummularia*); 1, 2, 3—Replicate; LD, HD—planting density (500 or 2000 plants ha^{-1}).

2.2. Ground Based Measurements

Permanent measurement plots (20 × 20 m) were established within the main treatment plots to minimise competitive edge effects. Measurement of potential predictor variables of *A. nummularia* shrubs were made of all plots which were applied to allometric relationships for estimates of carbon stocks in above- and belowground biomass.

Shrub crown width was measured on two axes at 90° to each other and used to derive the mean crown diameter (MCD). Shrub height and crown base height were measured to determine the crown height and used to calculate a crown volume index (CVI) (1):

$$CVI = (Ht \times W_1 \times W_2)/3 \tag{1}$$

where Ht is crown height, W_1 is crown width along axis 1, and W_2 is crown width along axis 2 which is 90° to axis 1. All measurements are in meters.

Measurement of treatment plots was made on an annual basis following establishment [7] with the field measurements made on 10 December 2011, which was the closest sampling to the time of aerial digital data capture.

2.3. Biomass Sampling

The destructive harvest method described in Snowdon et al. [29] was used to estimate AGB and belowground biomass (BGB). A total of 54 *A. nummularia* shrubs were sampled for AGB, across the dynamic range of shrub sizes to ensure data were representative, and of these 22 were sampled for BGB.

Sampling the AGB involved the removal of the entire shrub above the soil surface. The shrubs were then weighed in the field to determine total fresh weight and subsamples (0.5–0.7 kg) were taken and dried at 70 °C to constant dry weight and the moisture content (% w/w) determined to calculate the dry above ground mass of the sampled shrubs. Subsamples were further separated into leaf and stem components to determine the proportion of these of the AGB.

Sampling the BGB was achieved by excavating with a backhoe to approximately 0.5 m and collecting all roots with a diameter of approximately $\geq$2 mm. Soil was placed on a sieving table overlaid with 50 mm square mesh and roots were collected as described by Ritson and Sochacki [30]. The roots were washed to remove any adhering soil and then dried to determine the dry root weight.

There was no accumulation of soil organic carbon following *A. nummularia* establishment compared to untreated areas [11], thus it is not considered in this paper.

2.4. Carbon Analysis

Samples were taken from 8 random plants within the plots and analysed for carbon content. These were separated into leaf and stem components then dried at 70 °C to constant dry weight. The determination of carbon content of the leaves and stems was undertaken at a commercial laboratory, using the Leco combustion method [31].

2.5. Allometric Relationships

During Autumn, some *A. nummularia* leaves (L) are removed by livestock grazing, therefore the stable carbon store was considered to consist of the BGB and the stems of the AGB. Leaves represented 14.1% of the total plant biomass at the time of sampling [11]. The carbon store of each plant (C_{pl}) was estimated in Equation (2):

$$C_{pl} = (BGB \times C_i) + ((AGB - L) \times C_{ii}) \tag{2}$$

where C is the carbon content of the *A. nummularia* plants; BGB is belowground biomass; AGB is aboveground biomass; and L is leaves. C_i and C_{ii} are the respective C compositional values of the roots (46%) and stems (49%) from Walden et al. [11].

Plant carbon (C_{pl}) was then regressed against the CVI to develop a predictive allometric Equation (3):

$$C_{pl} = 0.494 + 4.607 \times CVI \tag{3}$$

where C_{pl} was plant carbon and CVI is the crown volume index.

An estimate of total carbon stocks (t C·ha^{-1}) for each measurement plot was made by estimating the carbon content of each plant by applying Equation (3), summing these values for the measurement plot, and converting to a per hectare value:

$$C_t = \sum C_{pl} \times 25 \tag{4}$$

where C_t is the total carbon stocks (t C·ha^{-1}), C_{pl} is the carbon content of each plant and 25 is the value to convert from the 400 m^2 measurement plot to 10,000 m^2 (1 ha).

In this study, the ground measurements of carbon storage of saltbush conducted in December 2011 was the nearest observation to both remotely sensed images. At this site, minimal change in carbon storage in the saltbush planting was observed after 4 years of age [7].

2.6. High Spatial Resolution Remote Sensing Data

The DMSI sensor acquires 12-bit digital number (DN) data simultaneously in four narrow spectral bands (20 nm full width half maximum). The spectral bands are located in the visible and near-infrared (NIR) region of the electromagnetic spectrum using filters centred at 450 nm (blue), 550 nm (green), 675 nm (red), and 780 nm (NIR) [32].

Two high spatial resolution airborne DMSI images with 0.5 m pixels were acquired by SpecTerra Services Proprietary Limited (Perth, WA, Australia) [33] from an altitude of 2000 m. The first DMSI image was acquired on the 28 September 2010 and is designated as the "September-2010-Green" image as this is when the saltbush shrubs are surrounded by green pastures containing photosynthetically active vegetation (PV). The second DMSI image was acquired on the 24 March 2011 and is designated as the "March-2011-Dry" image as this is when the saltbush shrubs are surrounded by dead/senesced pastures comprised predominantly of non-photosynthetically active vegetation (NPV).

The DMSI images were geo-referenced by SpecTerra based on GPS ground control points. Post-flight image processing included a bidirectional reflectance distribution function (BRDF) correction for variations in the sun-sensor-target viewing geometry across each image. The SpecTerra proprietary BRDF correction algorithm preserved the spectral integrity within an image, but produced DN rather than absolute radiance (energy received in $W\,m^{-2}\,sr^{-1}$).

Further radiometric correction of these images was necessary to convert raw DN values to reflectance at ground, which is required for calculating vegetation indices. The atmospheric correction of the DMSI images to reflectance at the ground was made using a QUick Atmospheric Correction (QUAC) [34] applied though the ENVI 5.1 remote sensing package [35]. The QUAC atmospheric correction is applicable where no concurrent atmospheric measurements are available, and can be applied to either raw DN or radiance-at-sensor image values [34]. The resulting QUAC atmospherically corrected images were verified by extracting spectral profiles for "pure" pixels of a variety of materials identified through manual examination of each of the two images. The spectra from these "pure" pixel locations were also validated against laboratory spectra of vegetation, water and soil from the ASTER spectral library [36], to identify any gross differences despite the library materials not representing materials from the Australian environment.

2.7. Vegetation Indices

Five pixel-scale vegetation indices and the sub-pixel fractional green vegetation cover for individual plants were calculated from the high-resolution images in order to highlight the carbon content of saltbush in the high-resolution images. A classification-based index R_{veg}, was also calculated for entire plots each image. These indices are summarized in Table 1.

Table 1. Vegetation indices used in this study.

Vegetation Index	Formula	Reference
Normalized Difference Vegetation Index	NDVI = (NIR − red)/(NIR + red)	[37]
Ratio Vegetation Index	RVI = NIR/red	[38]
Soil Adjusted Vegetation Index	SAVI = 1.5 × (NIR − red)/(NIR + red + 0.5)	[19,39]
Green Chromatic Coordinate	GCC = green/(red + green + blue)	[40]
Fractional green vegetation cover	fc = (NDVI − $NDVI_{soil}$)/($NDVI_{veg}$ − $NDVI_{soil}$)	[41]
R_{veg}	R_{veg} = percentage of vegetation pixels for each plot	[25,42]

Note: NIR is the reflectance of the near-infrared band, red is the reflectance of the red band, green is the reflectance of the green band, and blue is the reflectance of the blue band. The radiometrically corrected DMSI images are used for vegetation index calculations.

The first group of vegetation indices used in this study was pixel-based. The NDVI is one of the most widely used vegetation indices as it provides a measure of absorption of red light by plant chlorophyll as well as the reflection of infrared radiation by water-filled cells [37,43]. The Soil Adjusted Vegetation Index (SAVI) has been found to be robust under variations in soil brightness. The SAVI was selected to reduce the impact of soil in the scene, as the extent of the canopy coverage of saltbush in our study area was relatively small due to the sparse spacing of the shrubs. The RVI was selected to capture the contrast between the red and infrared bands for vegetated pixels [17], and for its use of only two spectral bands. The green chromatic coordinate (GCC) was used in this study to test the performance of general optical bands for estimating saltbush biomass. GCC has been used as an indicator of plant condition and phenology [18,40]. The GCC was chosen as it can be readily calculated from spectral bands found in standard digital cameras, and so has the potential to be available from a variety of sensor platforms.

Two additional indices related to the spatial coverage of vegetation (R_{veg} and fractional green vegetation cover (fc)) were used in this study. Crown horizontal projection, which refers to the vertical projected area of vegetation crown, has been reported as being strongly related to AGB [27], and canopy diameter is also strongly correlated to the AGB of saltbush in the study area [11], suggesting that the spatial coverage of vegetation can be an indicator of saltbush biomass. Calculating vegetation coverage can be complicated when the vegetation does not cover the entire pixel, resulting in mixed pixels on the edge of the saltbush canopy. In the study by Wittich and Hansing [41], the green vegetation fraction within a mixed pixel was shown to be related to the NDVI of the pixel, and also to the NDVI values of pure soil and vegetation in the scene. We therefore calculated fc by first selecting representative soil and vegetation samples manually from each image, and using NDVI to calculate fc for each pixel in the image according to the equation shown in Table 1. R_{veg} is a classification-based index that is calculated at the plot scale. R_{veg} is simply the proportion (%) of vegetation pixels in a plot compared to the total number of pixels in that plot. The use of a classification index such as R_{veg} is appropriate for the sparse nature of the saltbush planting as it focuses just on pixels that have been classified as vegetation, while providing estimates of vegetation fraction at the plot scale. Unlike the fc, the R_{veg} is dependent on the accuracy of the vegetation classification method.

2.8. Object-Based Classification Method

To determine the vegetation classifications for the calculation of R_{veg}, we used an object-based classification method. Object-based classification is suitable for high-resolution images such as DMSI where the relationship between the pixel size and the typical canopy width means that the vegetation is resolved by multiple pixels. Instead of analyzing information in each pixel separately, the object-based classification method takes image objects with a set of similar pixels as the basic unit [16,44]. The aim of object-based classification is to delineate readily usable objects from the background pixels, in order to utilize spectral and contextual information in the image.

For this study, we chose to use the commercially available software eCognition 8.4 [45,46] for object-based image classification as it contains a wide range of tools under a trial version. However, object-based image classification tools are available as both commercial and open-source products [16], which makes it possible to use these methods to develop an inexpensive operational system.

The object-based classification of our images to identify the salt-bush was made as follows. Each of the DMSI images was first segmented to identify individual objects in the scene using a "Bottom-Up" algorithm, "multi-resolution segmentation", in which all bands were used to split the original image into objects according to object shape, size, color, and pixel topology. The second stage of the object classification process is to assign each object identified by the segmentation process to a class based on features and criteria set by the user. We applied the "Assign class" and "Fuzzy membership" algorithms to the segmented objects from step 1, to identify "saltbush" within each image. As part of the input variables for the "Assign class" algorithm, an NDVI > 0.2 was assigned as the "saltbush" class, while NDVI < 0.1 was treated as "Soil" background. As saltbush is very sparse in some plots in the study area, a "Fuzzy membership" algorithm was used to classify "saltbush" and "soil" for NDVI between 0.1 and 0.2. Visual examination of the resulting classification with the high-resolution base images was used to confirm the suitability of the final classification.

The difference in vegetation index values between the interspace pasture and saltbush in the dry season image (March-2011-Dry) was found to be more pronounced than in the green season image (September-2010-Green), as during this period the pasture had died whereas the saltbush (*A. nummularia*) was alive at all times (Figure 2). We therefore used only the dry season image for determining the vegetation classification, although the canopy crowns identified by the classification were applicable to both images.

2.9. Scale and Estimating Carbon Stocks in A. nummularia

Two spatial scales, being the scales of the individual plant and of the whole plot, were used for relating the image data in this study to the carbon stocks of saltbush. "Plant scale" is where the relationship between vegetation indices and carbon stocks is determined only for the pixels in each plot that have been determined to be saltbush in the object-based classification. "Plot scale" is where all pixels within the plot, including vegetated and background pixels, are aggregated by averaging the values of all pixels within the plot.

2.10. Statistical Analys

All statistical analysis was made in the R 3.2.3 statistical software. A two-way analysis of variance (ANOVA) was conducted using the "anova" function in the "stats" package to detect the differences between vegetation type (Pasture and Saltbush) and the season of observation (Green and Dry) for all the image-derived vegetation indices.

The relationship between vegetation indices and carbon stocks was evaluated using a Spearman rank correlation in R 3.2.3 [47]. At the individual plant scale the non-saltbush plant pixels in each plot were treated as null values in all vegetation indices, while at the plot scale all pixels are included in the calculation.

Both non-linear relationships [21,48] and linear relationships [49–51] between vegetation indices and biomass were derived. The best-fit biomass estimation models were selected by comparing several regression models (exponential function, linear function, logarithm function, polynomial function, and power function). Each model was validated using the Leave-one-out cross validation (LOOCV) method. The precision of the estimation models was evaluated by the relative root mean squared error (RMSE, %) and the coefficient of determination (R^2).

3. Results

3.1. Vegetation Classification Based on Difference of Pasture and Saltbush

The values of the vegetation indices for saltbush were generally significantly different from those for pasture ($p < 0.001$) in the vegetation indices for both the green season image (September-2010-Green) and the dry season image (March-2011-Dry). Moreover, the difference in the vegetation indices between saltbush and pasture was larger in the dry season than the green season. For example, the differences of NDVI and RVI were 0.17 and 0.65 in the dry season, and 0.13 and 0.48 in the green season, respectively (Table 2).

NDVI, RVI, SAVI, GCC, and fc all showed significant differences ($p < 0.001$) between pasture and saltbush, which are expected given the different absorption features of the red and NIR spectral bands used in these vegetation indices. The differences of SAVI, fc and GCC between saltbush and pasture in the dry season were 0.08, 0.41 and 0.02, respectively.

Table 2. Comparison of vegetation indices and vegetation cover between pasture and saltbush from DMSI images for the green season (September-2010-Green) and dry season (March-2011-Dry).

Variable		Green Season		Dry Season		ANOVA Two-Way
		Pasture	Saltbush	Pasture	Saltbush	
Vegetation Index	NDVI	0.13	0.26	0.11	0.28	***
	RVI	1.31	1.79	1.25	1.90	***
	SAVI	0.08	0.15	0.07	0.19	***
	GCC	0.35	0.37	0.32	0.34	***
Vegetation coverage	fc	0.14	0.60	0.13	0.54	***
	R_{veg}	0.46	0.54	0.46	0.54	-

Note: The radiometrically corrected DMSI images are used for these vegetation index calculations. *** $p < 0.001$ in a two-way ANOVA test for vegetation type group (Pasture and Saltbush) and seasonal group (green season and dry season).

Figure 3 shows examples of the object-based vegetation classification. The main canopy of saltbush can be clearly recognized, with mean classification stability and best classification results from the eCognition analysis of 0.70 and 0.85, respectively. In addition, RVI, fc, and NDVI also showed significant differences ($p < 0.001$) between pasture and saltbush, which mainly resulted from the different absorption features of red and NIR bands of vegetation. The differences in RVI, fc and NDVI between saltbush and pasture in the dry season were 0.19, 0.26 and 0.12, respectively.

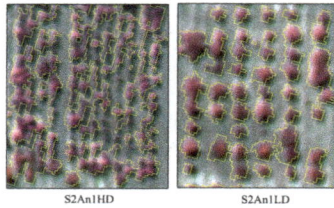

S2An1HD   S2An1LD

Figure 3. Examples of the object-based classification for the high density (S2An1HD) and low density plots (S2An1LD) of *A. nummularia*. The background images are false color composites, and the yellow boundaries are the derived canopy.

3.2. Relationships between Digital Vegetation Indices and Carbon Stocks

Overall, there was a significant relationship between all of the vegetation indices and carbon stocks (Tables 3 and 4, Figures 4, A1 and A2) in the dry season at both individual plant and plot scales, while a strong relationship was only observed at the individual plant scale in the green season. R_{veg} was significantly related to carbon stocks (ϱ of 0.91, $p < 0.001$).

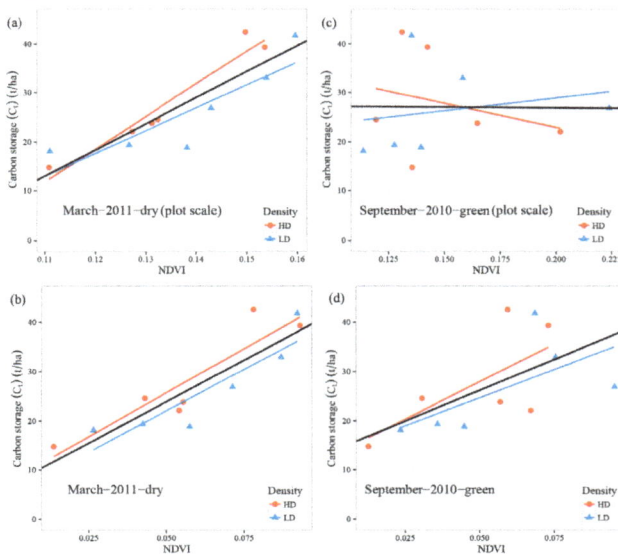

Figure 4. Relationship between vegetation indices and carbon stocks (C_t) for the *A. nummularia* shrubs sampled from the 500 (Δ, LD symbols) and 2000 ($\bullet$, HD symbols) plants ha^{-1} treatments in two seasons (dry season "March-2011-Dry" (**a,b**) and green season "September-2010-Green" (**c,d**)) at the plot scale (**a,c**) and the plant scale (**b,d**) of NDVI. The black line represents the fitted linear model for all plots and the red and blue lines are for high density and low density plots, respectively.

Table 3. Spearman rank correlation tests of vegetation indices against carbon stocks (C_t) in *A. nummularia* at the individual plant and whole plot scales in the green season (September-2010-Green).

Variable		Scale			
		Individual Plant [1]		Plot [2]	
		ϱ	p	ϱ	p
Vegetation Index	NDVI	0.73	0.01	0.16	0.62
	RVI	0.80	0.003	0.23	0.47
	SAVI	0.73	0.01	0.04	0.92
	GCC	0.89	0.001	0.14	0.67
Vegetation coverage	fc	0.55	0.05	0.16	0.62

[1] Pasture pixels were omitted for each plot (only pixels of saltbush canopies were combined for analysis). [2] The indices were the mean values of the whole plot. ϱ is Spearman's correlation coefficient and p is the significance of the Spearman test.

Table 4. Spearman rank correlation tests of vegetation indices against carbon stocks (C_t) in *A. nummularia* at the individual plant and whole plot scales in the dry season (March-2011-Dry).

Variable		Scale			
		Individual Plant [1]		Plot [2]	
		ϱ	p	ϱ	p
Vegetation Index	NDVI	0.86	0.001	0.88	0.001
	RVI	0.895	0.0002	0.88	0.001
	SAVI	0.895	0.001	0.85	0.001
	GCC	0.91	0.0001	0.42	0.18
Vegetation coverage	fc	0.87	0.001	0.88	0.001
	R_{veg}	0.91	0.0001	-	-

[1] Pasture pixels were omitted for each plot (only pixels of saltbush canopies were combined for analysis); [2] The indices were the averaged values of the whole plot. ϱ is Spearman's correlation coefficient and p is the significance of the Spearman test.

At the individual plant scale, RVI and GCC were strongly correlated to carbon storage (ϱ of 0.9, $p < 0.001$) in both seasons. The fc showed a much higher correlation in the dry season (ϱ of 0.87, $p < 0.001$) than in the green season (ϱ of 0.55, $p < 0.05$). The relationship between NDVI and carbon and between SAVI and carbon were not as strong in different seasons (ϱ of 0.7 in September-2010-Green and ϱ of 0.9 in March-2011-Dry).

In contrast, the results at the plot scale were more varied in both seasons. Very weak relationships between vegetation indices and carbon stocks were found in the green season, while a strong relationship, except for GCC, were found in the dry season. Similar strong relationships between NDVI, RVI, SAVI and fc with carbon storage were apparent in the dry season (ϱ of 0.88, $p < 0.001$). However, GCC showed a very weak relationship with carbon in both seasons.

For the data around different plot densities, a similar correlation was observed between vegetation indices and carbon stocks in the dry season, while in the green season significant relationships were only found for GCC and RVI at the individual plant scale (Figure 4). Meanwhile, there was no significant difference between slopes of linear regression lines derived from low and high density plots.

For space limitations, Figure 4 represents the model types of NDVI. All of the fitted model types for each vegetation index are available in Appendix Materials (Table A2 and Figures A1 and A2).

3.3. Comparison of Carbon Estimation Methods for Different Seasons and Scales

In the green season, all vegetation indices performed weakly in estimating carbon storage of saltbush at the plot scale (Figure A1). At the individual plant scale, RVI produced a reasonable result, explaining around 70% ($p < 0.05$) of the variation in carbon storage (Figure A2). GCC was found to be the best index for estimating carbon in the green season (Figure A2), with an R^2 of 0.86 and RMSE of 12.9%. In the dry season, similar results were found at the individual plant scale and the plot scale for each vegetation index except for GCC. GCC was one of the best indicators for carbon estimation at the individual plant scale, with an R^2 of 0.89 (RMSE = 12.4% and LOOCV RMSE = 15.8%) and $p < 0.01$, but it was not suitable at the plot scale ($R^2 = 0.1$, $p > 0.1$). NDVI, RVI and fc all showed similarly good results, explaining around 80% ($p < 0.01$ and RMSE < 16%) of the variation when all plots were included and 85% of the variation ($p < 0.01$, RMSE < 13%) in individual plant scale.

Overall, the relationships for the dry season and at the individual plant scale showed the best results for estimating saltbush biomass from the vegetation indices (Figure A2, Tables 3 and 4). By comparing the measures (R^2, RMSE, and LOOCV RMSE) of each regression model (Table A2), the best-fit regression model for carbon estimation was demonstrated in Table 5. Overall, very similar regressions and correlation coefficients were found from these vegetation indices. When all plots are included, the exponential function model showed almost the same precision as the polynomial function model for all vegetation indices. However, there were differences in the strength of the relationship with planting density. For the lower density plots, the polynomial function model was the best-fit model, explaining 96% of the variation ($p < 0.01$, RMSE < 7%), whereas for the higher density plot, the exponential function model showed a bit better performance than the polynomial function model, which explained 90% ($p < 0.01$, RMSE < 12%) of the variation.

In comparison with fc, the relationship between R_{veg} and C_t was stronger, explaining 88% ($p < 0.001$) of the variation when all plots were considered with a best-fit polynomial function model (Figure 5). Similar to vegetation indices, there were differences in the strength of this relationship with planting density, with the model for the high density plots explaining 87% of the variation ($p < 0.01$, RMSE = 12.9%), whereas that for the lower density plot explained 96% of the variation ($p < 0.01$, RMSE = 6.7%) when the polynomial function model was used for carbon estimation.

Table 5. Models for estimating carbon stocks (C_t) of *A. nummularia* for different planting densities at the individual plant scale in the dry season (March-2010-dry). Model is the best-fit regression model, R^2 is the coefficient of determination, RMSE is the relative root mean square error (%) of carbon estimation.

Variable		Model	R^2	RMSE (%)	Density (Plants ha^{-1})
Vegetation index	NDVI	$y = 12.29e^{13.44x}$	0.89	11.9	2000
		$y = 7916.3x^2 - 618.11x + 29.584$	0.96	6.5	500
		$y = 12.06e^{12.62x}$	0.84	14.6	ALL
	RVI	$y = 12.10e^{1.79x}$	0.9	11.9	2000
		$y = 155.96x^2 - 89.983x + 30.676$	0.96	6.2	500
		$y = 11.61e^{1.78x}$	0.87	12.9	ALL
	SAVI	$y = 12.10e^{21.09x}$	0.88	11.7	2000
		$y = 18485x^2 - 891.27x + 28.446$	0.92	9.6	500
		$y = 11.89e^{20.04x}$	0.84	15	ALL
	GCC	$y = 11.97e^{7.96x}$	0.89	12	2000
		$y = 3573.8x^2 - 457.16x + 32.411$	0.96	6.2	500
		$y = 11.28e^{8.29x}$	0.89	12	ALL
Vegetation coverage	fc	$y = 12.43e^{7.83x}$	0.89	12	2000
		$y = 2525x^2 - 341.96x + 28.97$	0.96	6.7	500
		$y = 12.42e^{7.05x}$	0.81	15.8	ALL
	R_{veg}	$y = 11.97e^{2.72x}$	0.89	12	2000
		$y = 398.08x^2 - 148.79x + 31.764$	0.96	6.5	500
		$y = 178.87x^2 - 22.86x + 16.40$	0.89	11.9	ALL

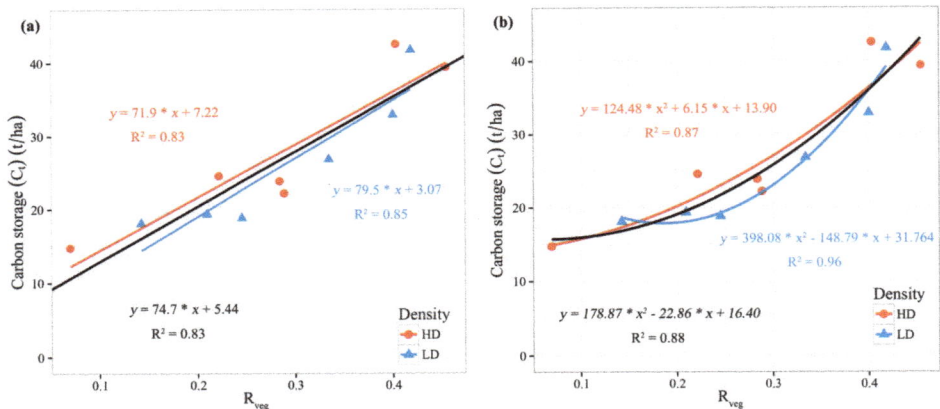

Figure 5. The relationship between the ratio of vegetation pixels (R_{veg}) and carbon stocks (C_t) for the *A. nummularia* shrubs sampled from the 500 (Δ, LD symbols) and 2000 ($\bullet$, HD symbols) plants ha^{-1} treatments with: linear (**a**) and non-linear (**b**) regression models fitted.

4. Discussion

4.1. Characteristics and Dynamics of Vegetation Indices of Saltbush and Annual Pasture

Overall, saltbush showed higher canopy coverage and higher biomass than pasture in both seasons, and the spectral signatures and the derived vegetation indices allowed discrimination between the two vegetation types from the images. These differences were particularly marked in the red band, with differences of the red band between pasture and saltbush of 29.5% in the green season and 34.4% in the dry season. It is thus possible to successfully distinguish pasture and saltbush canopy with an object-based classification method. Due to the senescence of pasture in the dry season, the physical, visual, and spectral differences between saltbush and pasture were much more pronounced in the images than in the green season, thus the dry season is the best time in this Mediterranean climate for vegetation classification. This difference has implications for future design of vegetation monitoring using image data.

Although nearly 90% of the planted saltbush had survived during the study time frame, the values of fc were moderate in both the green season (0.60) and dry season (0.54), which indicates that a greater proportion of the plots were covered by pasture and bare soil than by saltbush. This was confirmed by visual examination of the images. The low vegetation coverage and high salinity (EM38 H ranging from 50 to 300 mS m^{-1}) [7] at the field sites resulted in generally low values of the NIR-based vegetation indices in the study area [52]. For example, the mean NDVI of saltbush was 0.28 in the dry season and 0.26 in the green season, while the mean RVI was 1.9 and 1.8 in the same period.

4.2. Indicators of Carbon Stocks (C_t)

It can be concluded from the Spearman's rank correlation test of vegetation indices against sequestered carbon (C) that vegetation indices calculated from red and NIR bands can accurately reflect the carbon storage for saltbush both at the individual plant and plot scale (Tables 3 and 4) in the dry season (March-2011-Dry image). However, in the green season (September-2010-Green image), carbon storage could only be suitably estimated at the individual plant scale, as the pasture in the green season was still alive (NDVI values of around 0.13), which dramatically changed the estimated total carbon storage for each plot. Meanwhile, only GCC and RVI showed reasonable results for estimating carbon of saltbush in the green season, again indicating that the best time for estimating saltbush biomass with remote sensing data is in the dry season.

GCC was significantly related to carbon storage of saltbush at the individual plant scale in both the dry and green seasons, which suggests that indices derived from bands only in the visible part of the electromagnetic spectrum, and without including a NIR band, can be a useful indicator of saltbush biomass. GCC has also been found to be a good indicator of vegetation health and phenology in other studies [18,40], as GCC best represented differences in healthy vigour and mortality of vegetation. Meanwhile, from the different relationship of GCC between plot and individual plot scales, it can be concluded that the best performance of GCC requires vegetation classification, suggesting that model precision is determined by the accuracy of vegetation classification. In addition, there was only a small difference between the Pasture and Saltbush values of GCC, with these showing a considerable contribution to model performance at plot scale.

Vegetation coverage indices (fc and R_{veg}) were strongly correlated to C_t at both the individual plant and plot scale, suggesting that canopy coverage of saltbush inherently reflects carbon storage. In addition, R_{veg} in this study area produced a better result than fc. The calculation of fc required the NDVI values of pure soil and vegetation, which is a source of additional uncertainty in the index. In this study, mean NDVI values from pure soil and vegetation samples within the image were used for calculating fc. However, there is still a high variation on both vegetation structure and soil properties. In order to increase the accuracy of fc, a spatial interpolation method could be used to predict variations in the spectral characteristics of bare soil and green vegetation across space.

The results for the R_{veg} are consistent with that of Suganuma et al. [25] who used remote sensing derived canopy coverage to estimate stand biomass in forest species (*Acacia aneura* and *Eucalyptus camaldulensis*) in arid Western Australia. Similarly, Sousa et al. [27], working on *Quercus rotundifolia* in southern Portugal, found that AGB as a function of crown horizontal projection had the same trend for individual trees and plots, even though estimation for individual trees produced large individual errors. For our study, the strong relationship between vegetation coverage and C_t can possibly be explained by *A. nummularia* having little variation in height due to the consistency in age and the strong relationship between diameter and biomass reported by Walden et al. [11]. This is in contrast to many forest inventory studies where there is canopy closure and, thus, it is not possible to differentiate between individual trees and height has a large contribution to overall tree mass. For both this study and that of Suganuma et al. [25], the canopies were separated, thus we can suggest that canopy coverage approaches may be applicable to carbon inventory in open woodlands as well as shrubby systems. Similar relationships between canopy coverage and biomass have also been reported in the semiarid savanna of Sudan [53], and in semi-arid Senegal [54]. However, different vegetation types showed significantly different estimation accuracy [26].

For other vegetation indices in our study (i.e., NDVI, SAVI and RVI) that are derived from red and NIR bands, similar strong relationships with carbon storage of saltbush were observed at both individual plant and plot scales. This similarity may be because of their use of the same spectral bands (Table 1). Both NDVI and RVI have been widely used for estimating AGB [17,23,43,55,56].

Linear regression has been widely used to build the relationship between vegetation indices and carbon stocks, which demonstrates a satisfactory performance for carbon estimation. Overall, the linear regression models indicated strong relationships between the vegetation indices and carbon stocks, explaining around 80% of the variation, while the exponential function models explained around 85% of the variation (Table A2). However, in this study, the exponential function and polynomial function models showed much better accuracy than the linear model in the comparison with different densities (Figure 5). Similarly, other studies found close relationships between RVI and AGB with power and exponential functions [57]. A power function model was also found for grassland [21,48]. Furthermore, Santin-Janin et al. [58] developed a generalized non-linear model for the relationship between biomass and NDVI for *Acaena magellanica* and *Taraxacum officinale*. Meanwhile, as for the sensitivity of vegetation indices to planting density, a slightly stronger relationship was found in low density plots (R^2 = 0.96, p ~0.01) than in high density plots (R^2 ~0.90, p ~0.01) at the individual plant scale (Figures 5 and A2), but at the plot scale, a much higher difference occurred (Figure A1).

This difference between the low and high density plots can be ascribed to the effects of a higher ratio of pasture in the low density plots than in high density plots. In addition to the effects of Pasture at plot scale, the accuracy of vegetation classification can be another factor inducing a different relationship between low and high density plots. The difference of the best-fit regression model between high and low density plots also resulted from the different canopy coverage of each plant. Generally, low density plots have a higher canopy coverage (an average of 2.66 × 2.80 m²) than those in high density plots (an average of 1.51 × 1.56 m²) [7], which resulted in the different performance of each vegetation index on carbon estimation. Besides, carbon storage can be different even for the same canopy coverage because of the difference in height of plants between low (an average of 2.1 m) and high density (an average of 1.7 m) plots.

4.3. Limitations and Future Research

The object-based classification method was successfully used to distinguish pasture and saltbush from the high resolution image data. Although the efficacy of the technique was demonstrated here at a single location, the underlying allometric equation between saltbush carbon yield and stand parameters had been calibrated at six sites across southern Australia [11], and this suggests that our results are broadly applicable across other regions.

The mean classification stability and best classification results were 0.7 and 0.85, respectively, but there is still uncertainty related to the identification of the boundary of each saltbush plant. Although the annual pastures had died/senesced by the time the dry season image was acquired, it was still difficult to distinguish the boundary of each saltbush due to its overall low coverage (approximate fc of 0.17 to 0.69). Moreover, compared to the size of the saltbush canopy (an average of 1.79 × 1.85 m²), the pixel size of our image (0.5 m) is still relatively coarse, which makes the pixels in the boundary area to be a mixture of both soil background and saltbush branches, especially in high density plots. Therefore, it is impossible to find a fixed threshold to distinguish saltbush and soil. The spectral response from saltbush in some pixels may be confounded by that from the soil background, and saltbush pixels could therefore be misclassified as pasture during the classification process. Therefore, the potential use of images with finer pixels should enhance the accuracy of remotely sensed data in the future.

The relationships between remote sensing indices and carbon storage will vary in relation to the site-specific properties of soil condition, shadow, different species, and canopy structure. Meanwhile, soil background also has high spatial and seasonal variation. Therefore, the regression quality reported by previous studies varies strongly, R^2 with a range of 0.32 to 0.95 and our relationships may only be suitable for similar climatic and vegetation types as in the study, especially as the assumption of a set root mass to canopy relationship is inherent in our calibration data. However, our findings do demonstrate the capability of this approach to estimate carbon stocks using high-resolution remote sensing images in vegetation with non-overlapping canopies.

The applicability of our results could be further validated in other regions where abandoned farmland is being revegetated to ameliorate negative impacts of agricultural practices. With the advent of unmanned aerial vehicle (UAV) technology, there is the potential to gain significantly higher resolution imagery at a much lower cost (ca. USD$4000 for a DJI Phantom 4 Pro and a NDVI supported camera [59]) and with far greater flexibility of application and hence the rapid and cheap assessment of carbon. Recent examples of sensors mounted on UAV have included pixel resolutions as fine as 0.01 m, which provides sufficiently detailed information for estimating biomass of crops and monitoring forests [60]. The technical specifications of sensors mounted on UAV clearly have the potential to be used for monitoring biomass of vegetation used for carbon stocks, but the design of such a monitoring system has additional requirements, such as determining the best seasonal timing of measurements and assessing the potential for monitoring at the tree- or stand-scale.

With the possibility of finer resolution images for monitoring vegetation, remote sensing methodologies could potentially deliver estimates of biomass with greater precision and accuracy, as more accurate classification results are likely to be achieved for canopy classification with the Objective-based classification method. Finally, as canopy coverage and vegetation indices show high accuracy for estimating carbon stocks at the plot scale, some frequently used sources of image data, for example, Landsat-TM and SPOT, which are of medium spatial resolution and can provide an estimation of canopy coverage, may also be useful for broad scale biomass estimation.

5. Conclusions

This study suggests that there is a potential to use high spatial resolution airborne digital multispectral imagery to rapidly estimate the carbon storage of shrublands resulting from revegetation of abandoned farmland. Carbon stocks were significantly correlated with both canopy coverage and spectrally-based vegetation indices with or without the use of the NIR band. With the comparison of seasonal performances on carbon estimation, we concluded that estimates of saltbush carbon storage could be enhanced by image acquisition during the dry season even without the refinement of using a vegetation classification in the image analysis. This approach will have application in the management of revegetation-based carbon sink projects generally, and particularly in situations where this revegetation is based on discrete shrubs or trees in open woodlands. This is applicable not only to the large areas of land affected by salinity in Australia but also to similar degraded lands in other countries and particularly where these lands form part of the respective countries national carbon mitigation targets (INDCs). Historic aerial photography exists in many areas and the strength of our relationships based on canopy coverage and GCC implies that this photography could be interpreted to produce estimates of long-term carbon dynamics. To extend the present study, further ground-truthing is required to test these models on other *Atriplex* stands in other regions where aboveground biomass estimations are already known.

Acknowledgments: We thank SpecTerra Services Proprietary Limited and Dustin Bridges in particular for their substantial support in providing the ortho-rectified DMSI images utilized for this study. The Martin family is thanked for continuous access to their farm and the experimental site, and John Carter is thanked for comments on the manuscript. Ning Liu is supported by the Chinese Academy of Forestry and a Murdoch University Strategy PhD Scholarship. This study is funded by the Special Research Program for Public-welfare Forestry (Grant No. 201404201).

Author Contributions: Ning Liu, Richard Harper, Bradley Evans, Rebecca N. Handcock and Bernard Dell conceived and designed the experiments; Ning Liu, Richard Harper and Rebecca N. Handcock wrote the paper; Ning Liu analyzed the data; Stanley Sochacki and Lewis Walden contributed data and allometric equations and edited the paper; and Shirong Liu edited the paper.

Conflicts of Interest: The authors declare no conflict of interest.

Appendix A

Table A1 Mean values for vegetation indices and sequestered CO_2 results per plot in 2010 and 2011. Table A2 Models for estimating carbon stocks (C_t) of *A. nummularia* for different planting densities at the individual plant scale in the dry season (March-2010-dry). Figure A1. Relationship between vegetation indices and carbon stocks (C_t) for the *A. nummularia* shrubs sampled from the 500 and 2000 plants ha^{-1} treatments in two seasons (dry season "March-2011-Dry" and green season "September-2010-Green") at plot scale. Figure A2. Relationship between vegetation indices and carbon stocks (C_t) for the *A. nummularia* shrubs sampled from the 500 and 2000 plants ha^{-1} treatments in two seasons (dry season "March-2011-Dry" and green season "September-2010-Green") at individual plant scale.

Table A1. Mean values for vegetation indices and sequestered CO_2 results per plot in 2010 and 2011.

Plot Name	Year	NDVI	RVI	SAVI	GCC	fc	Carbon Stocks (C, t·ha^{-1})
S1An1LD	2010	0.17	1.42	0.10	0.37	0.29	19.3
	2011	0.20	1.52	0.13	0.35	0.35	
S1An2LD	2010	0.16	1.40	0.09	0.37	0.26	18.1
	2011	0.19	1.47	0.12	0.34	0.31	
S1An3LD	2010	0.29	1.82	0.15	0.39	0.69	26.8
	2011	0.21	1.56	0.14	0.34	0.37	
S2An1LD	2010	0.19	1.47	0.11	0.37	0.35	32.9
	2011	0.22	1.57	0.14	0.34	0.38	
S2An2LD	2010	0.16	1.40	0.09	0.37	0.26	41.7
	2011	0.22	1.59	0.14	0.34	0.39	
S2An3LD	2010	0.18	1.46	0.10	0.37	0.33	18.8
	2011	0.23	1.64	0.14	0.34	0.42	
S1An1HD	2010	0.20	1.51	0.12	0.38	0.39	23.8
	2011	0.20	1.49	0.13	0.34	0.33	
S1An2HD	2010	0.18	1.45	0.11	0.38	0.33	14.7
	2011	0.20	1.51	0.13	0.35	0.34	
S1An3HD	2010	0.23	1.62	0.12	0.38	0.50	22.1
	2011	0.19	1.47	0.12	0.34	0.31	
S2An1HD	2010	0.15	1.35	0.08	0.36	0.20	42.5
	2011	0.19	1.49	0.12	0.34	0.33	
S2An2HD	2010	0.14	1.32	0.08	0.37	0.17	24.5
	2011	0.19	1.49	0.13	0.34	0.33	
S2An3HD	2010	0.16	1.39	0.09	0.37	0.25	39.3
	2011	0.20	1.53	0.13	0.34	0.35	

Table A2. Models for estimating carbon stocks (C_t) of *A. nummularia* for different planting densities at the individual plant scale in the dry season (March-2010-dry). Model is the fitted regression models (exponential function model, linear function model, logarithm function model, polynomial function model, and power function model), R^2 is the coefficient of determination, RMSE is the relative root mean square error (%) of estimate, and LOOCV RMSE is the RMSE from leave-one-out cross-validation (LOOCV).

Variable		Model	R^2	RMSE (%)	LOOCV RMSE (%)
Vegetation index	NDVI	$y = 12.06e^{12.62x}$	0.84	14.6	16.8
		$y = 335.87x + 7.09$	0.78	16.1	18.8
		$y = 13.68 \ln(x) + 67.28$	0.63	20.9	27.8
		$y = 3290.91x^2 - 36.87x + 15.64$	0.83	14.4	17.5
		$y = 122.52x^{0.53}$	0.75	17.2	20.4
	RVI	$y = 11.61e^{1.78x}$	0.87	12.9	15.3
		$y = 47.28x + 6.09$	0.81	14.9	17.9
		$y = 14.23 \ln(x) + 40.22$	0.64	20.5	28.6
		$y = 70.47x^2 - 11.85x + 16.24$	0.87	12.5	15.9
		$y = 42.68x^{0.56}$	0.78	16.1	19.3
	SAVI	$y = 11.89e^{20.04x}$	0.84	15	17.6
		$y = 530.30x + 6.83$	0.77	16.3	19.2
		$y = 13.95 \ln(x) + 74.20$	0.63	20.9	27.7
		$y = 7923.11x^2 - 41.81x + 15.24$	0.81	14.8	18.1
		$y = 161.53x^{0.55}$	0.74	17.4	20.7
	GCC	$y = 11.28e^{8.29x}$	0.89	12	15
		$y = 220.24x + 5.35$	0.83	14.4	17.5
		$y = 14.72 \ln(x) + 62.73$	0.65	20.3	29.3
		$y = 1520.43x^2 - 62.34x + 16.19$	0.89	11.5	15.8
		$y = 102.84x^{0.57}$	0.80	15.4	18.6

Table A2. *Cont.*

Variable		Model	R^2	RMSE (%)	LOOCV RMSE (%)
Vegetation coverage	fc	$y = 12.42e^{7.05x}$	0.81	14.9	18
		$y = 187.72x + 7.86$	0.76	16.9	19.5
		$y = 13.28 \ln(x) + 58.98$	0.62	21.2	27.3
		$y = 945.73x^2 + 2.22x + 15.18$	0.79	15.7	18.9
		$y = 88.60x^{0.52}$	0.72	18	21.2
	R_{veg}	$y = 11.32e^{2.81x}$	0.89	11.9	14.6
		$y = 74.71x + 5.44$	0.83	14.3	17.4
		$y = 14.62 \ln(x) + 46.75$	0.65	20.3	29.5
		$y = 178.87x^2 - 22.86x + 16.40$	0.89	11.4	15.4
		$y = 55.11x^{0.57}$	0.80	15.4	18.6

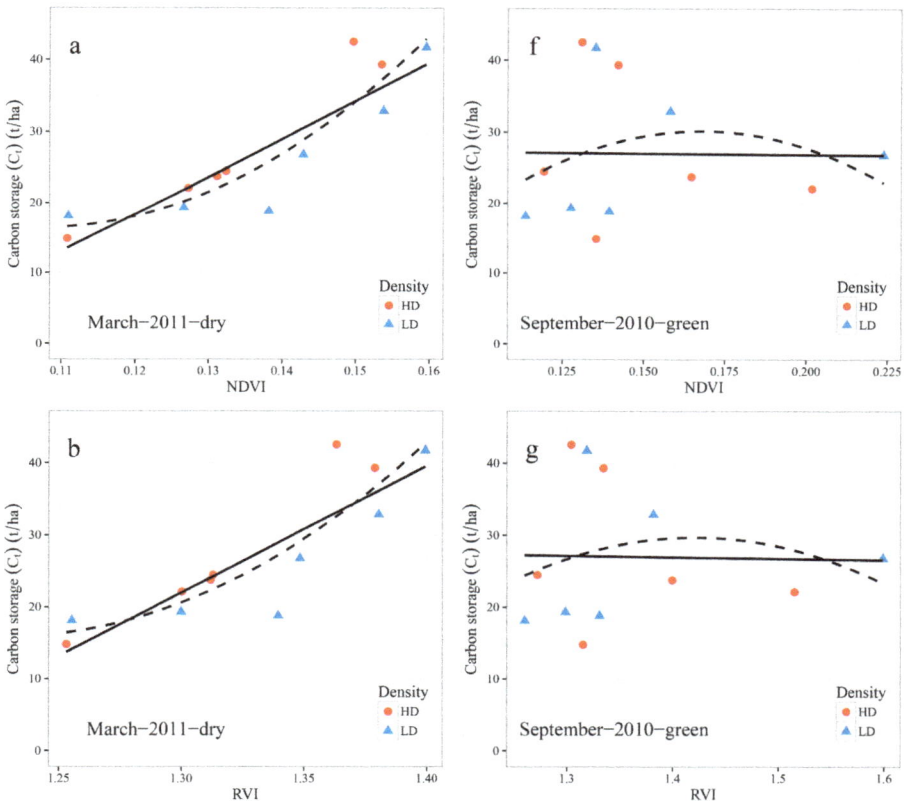

Figure A1. *Cont.*

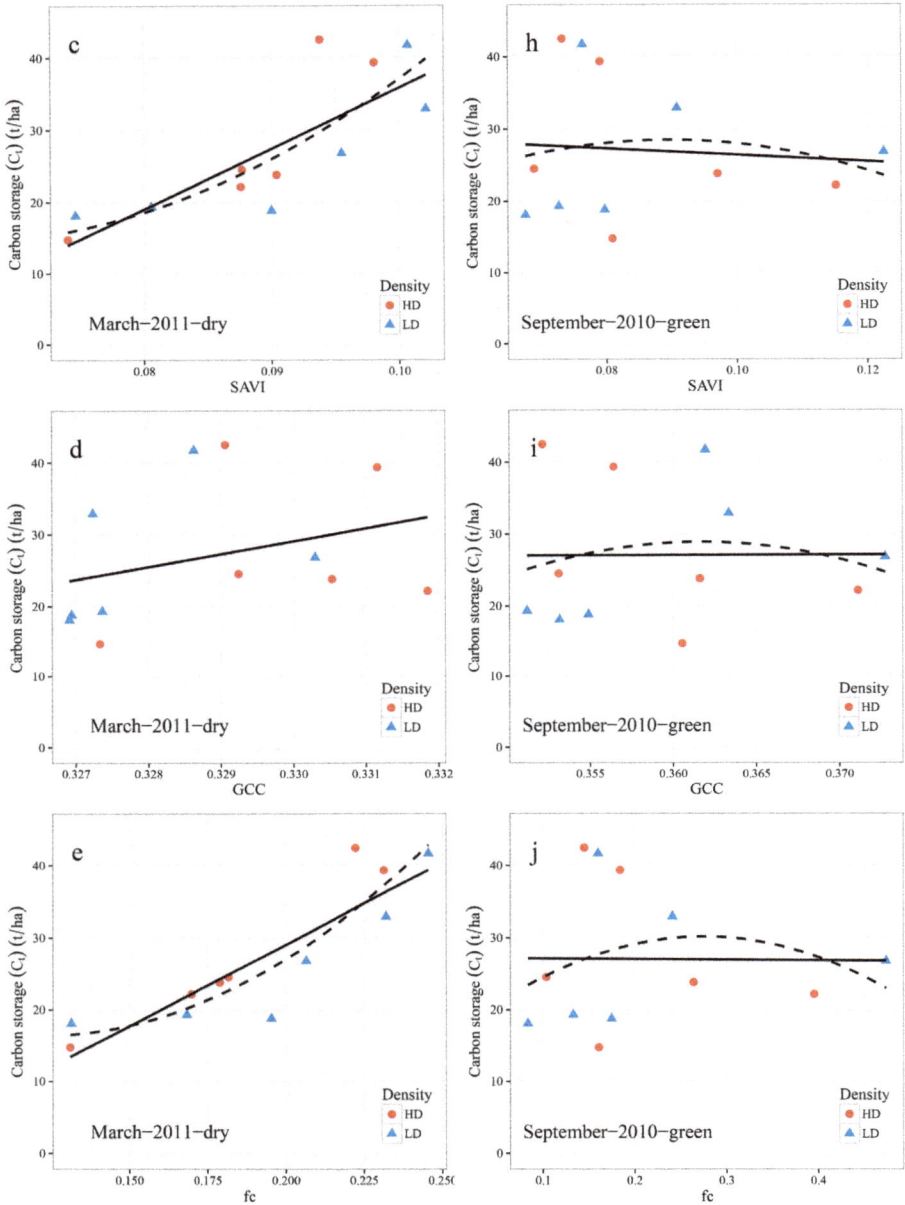

Figure A1. Relationship between vegetation indices and carbon stocks (C$_t$) for the *A. nummularia* shrubs sampled from the 500 (Δ, LD symbols) and 2000 (•, HD symbols) plants ha^{-1} treatments in two seasons (dry season "March-2011-Dry" (**a–e**) and green season "September-2010-Green" (**f–j**)) at plot scale. The straight line is the linear model and dashed line is the non-linear regression model fitted.

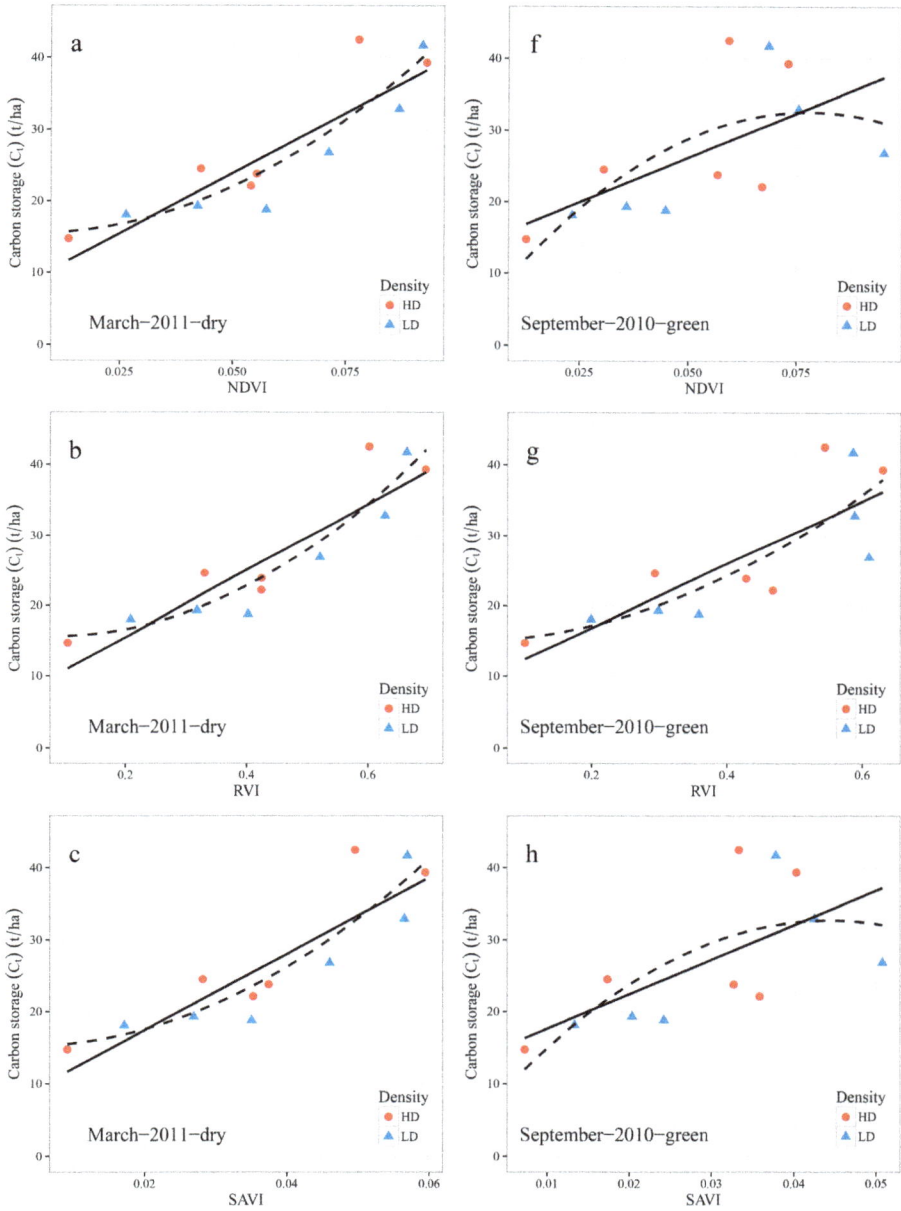

Figure A2. *Cont.*

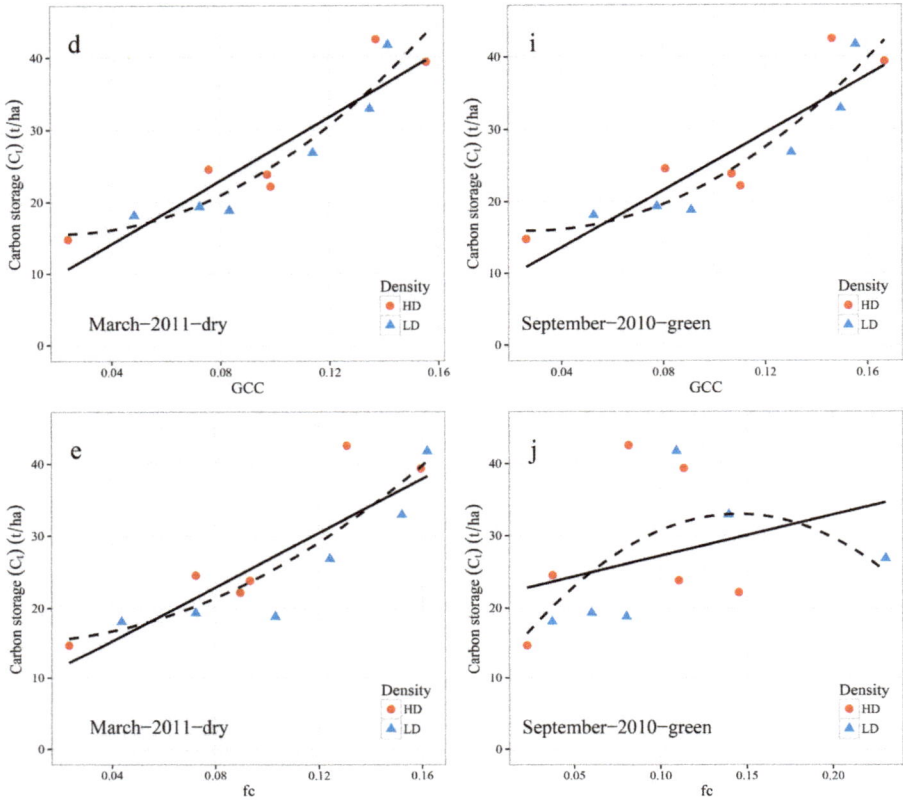

Figure A2. Relationship between vegetation indices and carbon stocks (C_t) for the *A. nummularia* shrubs sampled from the 500 (Δ, LD symbols) and 2000 ($\bullet$, HD symbols) plants ha^{-1} treatments in two seasons (dry season "March-2011-Dry" (**a–e**) and green season "September-2010-Green" (**f–j**)) at individual plant scale. The straight line is the linear model and dashed line is the non-linear regression model fitted.

References

1. Intergovernmental Panel on Climate Change (IPCC). *IPCC, 2013: Climate Change 2013: The Physical Science Basis. Contribution of Working Group I to the Fifth Assessment Report of the Intergovernmental Panel on Climate Change*; Cambridge University Press: Cambridge, UK; New York, NY, USA, 2013; p. 1535.
2. Bustamante, M.; Robledo-Abad, C.; Harper, R.; Mbow, C.; Ravindranat, N.H.; Sperling, F.; Haberl, H.; Pinto Ade, S.; Smith, P. Co-benefits, trade-offs, barriers and policies for greenhouse gas mitigation in the agriculture, forestry and other land use (afolu) sector. *Glob. Chang. Biol.* **2014**, *20*, 3270–3290. [CrossRef] [PubMed]
3. Harper, R.J.; Beck, A.C.; Ritson, P.; Hill, M.J.; Mitchell, C.D.; Barrett, D.J.; Smettem, K.R.J.; Mann, S.S. The potential of greenhouse sinks to underwrite improved land management. *Ecol. Eng.* **2007**, *29*, 329–341. [CrossRef]
4. Food and Agriculture Organization (FAO). *The Agriculture Sectors in the Intended Nationally Determined Contributions: Summary*; Food and Agriculture Organization of the United Nations: Rome, Italy, 2016; pp. 3–4.

5. Smith, P.; Haberl, H.; Popp, A.; Erb, K.H.; Lauk, C.; Harper, R.J.; Tubiello, F.; de Siqueira Pinto, A.; Jafari, M.; Sohi, S.; et al. How much land based greenhouse gas mitigation can be achieved without compromising food security and environmental goals? *Glob. Chang. Biol.* **2013**, *19*, 2285–2302. [CrossRef] [PubMed]

6. Jackson, R.B.; Jobbagy, E.G.; Avissar, R.; Roy, S.B.; Barrett, D.J.; Cook, C.W.; Farley, K.A.; le Maitre, D.C.; McCarl, B.A.; Murray, B.C. Trading water for carbon with biological carbon sequestration. *Science* **2005**, *310*, 1944–1947. [CrossRef] [PubMed]

7. Sochacki, S.J.; Harper, R.J.; Smettem, K.R.J. Bio-mitigation of carbon following afforestation of abandoned salinized farmland. *GCB Bioenergy* **2012**, *4*, 193–201. [CrossRef]

8. Australian Bureau of Statistics. 1370.0-Measures of Australia's Progress. 2010. Available online: http://www.abs.gov.au/ausstats/abs@.nsf/Lookup/by%20Subject/1370.0~2010~Chapter~Salinity%20(6.2.4.4) (accessed on 3 March 2010).

9. National Land and Water Resources Audit. *Australian Dryland Salinity Assessment 2000. Extent, Impacts, Processes, Monitoring and Management Options*; National Land and Water Resources Audit: Canberra, Australia, 2001; p. 129.

10. Food and Agriculture Organization (FAO). More Information on Salt-Affected Soils. Available online: http://www.fao.org/soils-portal/soil-management/management-of-some-problem-soils/salt-affected-soils/more-information-on-salt-affected-soils/en/ (accessed on 15 February 2017).

11. Walden, L.L.; Harper, R.J.; Sochacki, S.J.; Montagu, K.D.; Wocheslander, R.; Clarke, M.; Ritson, P.; Emms, J.; Davoren, C.W.; Mowat, D.; et al. Mitigation of carbon following *Atriplex nummularia* revegetation in southern Australia. *Ecol. Eng.* **2017**, in review.

12. Harper, R.J.; Sochacki, S.J.; Smettem, K.R.J.; Robinson, N. Bioenergy feedstock potential from short-rotation woody crops in a dryland environment. *Energy Fuels* **2010**, *24*, 225–231. [CrossRef]

13. Macintosh, A. The carbon farming initiative: Removing the obstacles to its success. *Carbon Manag.* **2013**, *4*, 185–202. [CrossRef]

14. Phinn, S.; Roelfsema, C.; Dekker, A.; Brando, V.; Anstee, J. Mapping seagrass species, cover and biomass in shallow waters: An assessment of satellite multi-spectral and airborne hyper-spectral imaging systems in moreton bay (Australia). *Remote Sens. Environ.* **2008**, *112*, 3413–3425. [CrossRef]

15. Blaschke, T. Object based image analysis for remote sensing. *ISPRS J. Photogramm. Remote Sens.* **2010**, *65*, 2–16. [CrossRef]

16. Bunting, P.; Lucas, R. The delineation of tree crowns in australian mixed species forests using hyperspectral Compact Airborne Spectrographic Imager (CASI) data. *Remote Sens. Environ.* **2006**, *101*, 230–248. [CrossRef]

17. Silleos, N.G.; Alexandridis, T.K.; Gitas, I.Z.; Perakis, K. Vegetation indices: Advances made in biomass estimation and vegetation monitoring in the last 30 years. *Geocarto Int.* **2006**, *21*, 21–28. [CrossRef]

18. Reid, A.M.; Chapman, W.K.; Prescott, C.E.; Nijland, W. Using excess greenness and green chromatic coordinate colour indices from aerial images to assess lodgepole pine vigour, mortality and disease occurrence. *For. Ecol. Manag.* **2016**, *374*, 146–153. [CrossRef]

19. Huete, A.; Didan, K.; Miura, T.; Rodriguez, E.P.; Gao, X.; Ferreira, L.G. Overview of the radiometric and biophysical performance of the modis vegetation indices. *Remote Sens. Environ.* **2002**, *83*, 195–213. [CrossRef]

20. Ogaya, R.; Barbeta, A.; Başnou, C.; Peñuelas, J. Satellite data as indicators of tree biomass growth and forest dieback in a mediterranean holm oak forest. *Ann. For. Sci.* **2014**, *72*, 135–144. [CrossRef]

21. Jin, Y.X.; Yang, X.C.; Qiu, J.J.; Li, J.Y.; Gao, T.; Wu, Q.; Zhao, F.; Ma, H.L.; Yu, H.D.; Xu, B. Remote sensing-based biomass estimation and its spatio-temporal variations in temperate grassland, northern china. *Remote Sens.* **2014**, *6*, 1496–1513. [CrossRef]

22. Zheng, D.L.; Rademacher, J.; Chen, J.Q.; Crow, T.; Bresee, M.; le Moine, J.; Ryu, S.R. Estimating aboveground biomass using landsat 7 ETM+ data across a managed landscape in northern wisconsin, USA. *Remote Sens. Environ.* **2004**, *93*, 402–411. [CrossRef]

23. Helman, D.; Mussery, A.; Lensky, I.M.; Leu, S. Detecting changes in biomass productivity in a different land management regimes in drylands using satellite-derived vegetation index. *Soil Use Manag.* **2014**, *30*, 32–39. [CrossRef]

24. Asner, G.P. Biophysical and biochemical sources of variability in canopy reflectance. *Remote Sens. Environ.* **1998**, *64*, 234–253. [CrossRef]

25. Suganuma, H.; Abe, Y.; Taniguchi, M.; Tanouchi, H.; Utsugi, H.; Kojima, T.; Yamada, K. Stand biomass estimation method by canopy coverage for application to remote sensing in an arid area of western australia. *For. Ecol. Manag.* **2006**, *222*, 75–87. [CrossRef]

26. Fensham, R.J.; Fairfax, R.J.; Holman, J.E.; Whitehead, P.J. Quantitative assessment of vegetation structural attributes from aerial photography. *Int. J. Remote Sens.* **2002**, *23*, 2293–2317. [CrossRef]

27. Sousa, A.M.O.; Goncalves, A.C.; Mesquita, P.; da Silva, J.R.M. Biomass estimation with high resolution satellite images: A case study of quercus rotundifolia. *ISPRS J. Photogramm. Remote Sens.* **2015**, *101*, 69–79. [CrossRef]

28. Harper, R.J.; Sochacki, S.J.; Smettem, K.R.J.; Robinson, N.; Silberstein, R.P.; Clarke, C.J.; McGrath, J.F.; Crombie, D.S.; Hampton, C.E. *Catchment Scale Evaluation of "Trees, Water and Salt"*; Rural Industries Research and Development Corporation: Kingston, Australia, 2009.

29. Snowdon, P.; Keith, H.; Raison, R.J. *Protocol for Sampling Tree and Stand Biomass*; Australian Greenhouse Office Parkes: Canberra, Australia, 2002.

30. Ritson, P.; Sochacki, S. Measurement and prediction of biomass and carbon content of *Pinus pinaster* trees in farm forestry plantations, south-western Australia. *For. Ecol. Manag.* **2003**, *175*, 103–117. [CrossRef]

31. Rayment, G.; Higginson, F.R. *Australian Laboratory Handbook of Soil and Water Chemical Methods*; Inkata Press Pty Ltd.: Melbourne, Australia, 1992.

32. Evans, B.; Lyons, T.; Barber, P.; Stone, C.; Hardy, G. Enhancing a eucalypt crown condition indicator driven by high spatial and spectral resolution remote sensing imagery. *J. Appl. Remote Sens.* **2012**, *6*, 3605. [CrossRef]

33. SpecTerra Services Pty Ltd.—Global Leaders in Airborne Remote Sensing Technology. Available online: http://www.specterra.com.au (accessed on 29 May 2017).

34. Bernstein, L.S.; Jin, X.; Gregor, B.; Adler-Golden, S.M. Quick atmospheric correction code: Algorithm description and recent upgrades. *Opt. Eng.* **2012**, *51*, 111719. [CrossRef]

35. ENVI Image Analysis Software | ESRI Australia. Available online: https://esriaustralia.com.au/products-specialised-gis-applications-envi (accessed on 29 May 2017).

36. Baldridge, A.M.; Hook, S.J.; Grove, C.I.; Rivera, G. The aster spectral library version 2.0. *Remote Sens. Environ.* **2009**, *113*, 711–715. [CrossRef]

37. Tucker, C.J. Red and photographic infrared linear combinations for monitoring vegetation. *Remote Sens. Environ.* **1979**, *8*, 127–150. [CrossRef]

38. Birth, G.S.; McVey, G. Measuring the color of growing turf with a reflectance spectroradiometer. *Agron. J.* **1968**, *60*, 640–643. [CrossRef]

39. Huete, A.R. A soil-adjusted vegetation index (SAVI). *Remote Sens. Environ.* **1988**, *25*, 295–309. [CrossRef]

40. Nijland, W.; de Jong, R.; de Jong, S.M.; Wulder, M.A.; Bater, C.W.; Coops, N.C. Monitoring plant condition and phenology using infrared sensitive consumer grade digital cameras. *Agric. For. Meteorol.* **2014**, *184*, 98–106. [CrossRef]

41. Wittich, K.-P.; Hansing, O. Area-averaged vegetative cover fraction estimated from satellite data. *Int. J. Biometeorol.* **1995**, *38*, 209–215. [CrossRef]

42. Karlson, M.; Ostwald, M.; Reese, H.; Sanou, J.; Tankoano, B.; Mattsson, E. Mapping tree canopy cover and aboveground biomass in sudano-sahelian woodlands using landsat 8 and random forest. *Remote Sens.* **2015**, *7*, 10017–10041. [CrossRef]

43. Ahamed, T.; Tian, L.; Zhang, Y.; Ting, K.C. A review of remote sensing methods for biomass feedstock production. *Biomass Bioenergy* **2011**, *35*, 2455–2469. [CrossRef]

44. Yu, Q.; Gong, P.; Clinton, N.; Biging, G.; Kelly, M.; Schirokauer, D. Object-based detailed vegetation classification with airborne high spatial resolution remote sensing imagery. *Photogramm. Eng. Remote Sens.* **2006**, *72*, 799–811. [CrossRef]

45. Ding, X.Y. The application of ecognition in land use projects. *Geomat. Spat. Inf. Technol.* **2005**, *28*, 116–120.

46. eCognition | Trimble. Available online: http://www.ecognition.com (accessed on 29 May 2017).

47. Van de Wiel, M.; Di Bucchianico, A. Fast computation of the exact null distribution of spearman's ϱ and page's l statistic for samples with and without ties. *J. Stat. Plan. Inference* **2001**, *92*, 133–145. [CrossRef]

48. Frank, A.B.; Karn, J.F. Vegetation indices, CO_2 flux, and biomass for northern plains grasslands. *J. Rangel. Manag.* **2003**, *56*, 382–387. [CrossRef]

49. Perry, E.M.; Morse-McNabb, E.M.; Nuttall, J.G.; O'Leary, G.J.; Clark, R. Managing wheat from space: Linking modis ndvi and crop models for predicting australian dryland wheat biomass. *IEEE J. STARS* **2014**, *7*, 3724–3731. [CrossRef]

50. Yan, F.; Wu, B.; Wang, Y.J. Estimating spatiotemporal patterns of aboveground biomass using Landsat TM and MODIS images in the Mu US Sandy Land, China. *Agric. For. Meteorol.* **2015**, *200*, 119–128. [CrossRef]

51. Ediriweera, S.; Pathirana, S.; Danaher, T.; Nichols, D. Estimating above-ground biomass by fusion of lidar and multispectral data in subtropical woody plant communities in topographically complex terrain in north-eastern australia. *J. For. Res.* **2014**, *25*, 761–771. [CrossRef]

52. Peñuelas, J.; Isla, R.; Filella, I.; Araus, J.L. Visible and near-infrared reflectance assessment of salinity effects on barley. *Crop Sci.* **1997**, *37*, 198–202. [CrossRef]

53. Olsson, K. Estimating canopy cover in drylands with landsat mss data. *Adv. Space Res.* **1984**, *4*, 161–164. [CrossRef]

54. Woomer, P.L.; Touré, A.; Sall, M. Carbon stocks in senegal's sahel transition zone. *J. Arid Environ.* **2004**, *59*, 499–510. [CrossRef]

55. Roy, P.S.; Ravan, S.A. Biomass estimation using satellite remote sensing data—An investigation on possible approaches for natural forest. *J. Biosci.* **1996**, *21*, 535–561. [CrossRef]

56. Richardson, A.J.; Everitt, J.H.; Gausman, H.W. Radiometric estimation of biomass and nitrogen-content of Alicia grass. *Remote Sens. Environ.* **1983**, *13*, 179–184. [CrossRef]

57. Kaishan, S.; Bai, Z.; Fang, L.; Hongtao, D.; Zongming, W. Correlative analyses of hyperspectral reflectance, soybean lai and aboveground biomass. *Trans. Chin. Soc. Agric. Eng.* **2005**, *1*, 9.

58. Santin-Janin, H.; Garel, M.; Chapuis, J.L.; Pontier, D. Assessing the performance of NDVI as a proxy for plant biomass using non-linear models: A case study on the Kerguelen archipelago. *Polar Biol.* **2009**, *32*, 861–871. [CrossRef]

59. Sentera—Drones + Software to Make Sense of It All. Available online: https://sentera.com/ (accessed on 29 May 2017).

60. Zahawi, R.A.; Dandois, J.P.; Holl, K.D.; Nadwodny, D.; Reid, J.L.; Ellis, E.C. Using lightweight unmanned aerial vehicles to monitor tropical forest recovery. *Biol. Conserv.* **2015**, *186*, 287–295. [CrossRef]

remote sensing

MDPI

Article

Evaluating the Differences in Modeling Biophysical Attributes between Deciduous Broadleaved and Evergreen Conifer Forests Using Low-Density Small-Footprint LiDAR Data

Yoshio Awaya [1],* and Tomoaki Takahashi [2],*

[1]   River Basin Research Center, Gifu University, 1-1 Yanagido, Gifu 501-1193, Japan
[2]   Kyushu Research Center, Forestry and Forest Products Research Institute, 4-11-16 Kurokami, Chuo-ku,
     Kumamoto 860-0862, Japan
*   Correspondence: awaya@green.gifu-u.ac.jp (Y.A.); tomokun@ffpri.affrc.go.jp (T.T.)

Academic Editors: Lalit Kumar, Onisimo Mutanga, Lars T. Waser and Randolph H. Wynne
Received: 15 March 2017; Accepted: 5 June 2017; Published: 7 June 2017

Abstract: Airborne light detection and ranging (LiDAR) has been used for forest biomass estimation for the past three decades. The performance of estimation, in particular, has been of great interest. However, the difference in the performance of estimation between stem volume (SV) and total dry biomass (TDB) estimations has been a priority topic. We compared the performances between SV and TDB estimations for evergreen conifer and deciduous broadleaved forests by correlation and regression analyses and by combining height and no-height variables to identify statistically useful variables. Thirty-eight canopy variables, such as average and standard deviation of the canopy height, as well as the mid-canopy height of the stands, were computed using LiDAR point data. For the case of conifer forests, TDB showed greater correlation than SV; however, the opposite was the case for deciduous broadleaved forests. The average- and mid-canopy height showed the greatest correlation with TDB and SV for conifer and deciduous broadleaved forests, respectively. Setting the best variable as the first and no-height variables as the second variable, a stepwise multiple regression analysis was performed. Predictions by selected equations slightly underestimated the field data used for validation, and their correlation was very high, exceeding 0.9 for coniferous forests. The coefficient of determination of the two-variable equations was smaller than that of the one-variable equation for broadleaved forests. It is suggested that canopy structure variables were not effective for broadleaved forests. The SV and TDB maps showed quite different frequency distributions. The ratio of the stem part of the broadleaved forest is smaller than that of the coniferous forest. This suggests that SV was relatively smaller than TDB for the case of broadleaved forests compared with coniferous forests, resulting in a more even spatial distribution of TDB than that of SV.

Keywords: stem volume; dry biomass; conifer; broadleaves; light detection and ranging (LiDAR); regression analysis; correlation coefficient

---

1. Introduction

Biomass and stem volume (SV) are important variables for forestry and carbon balance studies. The biomass and carbon stocks in forests are important indicators of their productive capacity, energy potential, and capacity to sequester carbon [1]. Biomass is a pool of atmospheric carbon fixed by plants and ranges widely based on tree size in large areas. Stem volume (SV) is basic piece of information for informing lumber production in forest management. The stem volume of a tree is the principal commercial product of forests as the stem contains a large proportion of the biomass of a tree [2].

Airborne laser technology was introduced for forest measurements in the early 1980s, and a laser profiler revealed that, in a large area, tree canopy height was measurable from the air [3,4]. Large-area forest inventory is a time-consuming task; however, biomass estimation using light detection and ranging (LiDAR) point data has become popular for creating wall-to-wall inventories. LiDAR-based inventories have been recognized as essential in providing more accurate estimates of biophysical properties than conventional methods. Although airborne laser observation and data processing is costly, it provides forest resource information over large areas with wall-to-wall coverage. Further, the use of laser data for forest inventories has shown promising results with improved accuracies [5]. Various studies have since been executed to evaluate the performance of airborne laser sensors for evaluating forest variables, such as SV and above-ground biomass (AGB) of pine [6], AGB of deciduous broadleaved [7] and evergreen coniferous [8] forests, and SVs of spruce and pine forests [9]. The improvement of LiDAR technology in the field of pulse density and accurate positioning [5] yields accurate small-footprint laser data, which are applicable for precise biomass mapping [10,11].

Various LiDAR variables have been examined and found to be useful for SV or biomass estimation by many scientists [10,12–15]. However, variable effectiveness and estimation accuracy differ as a function of footprint size [14], point density [15], scan angle [11], tree size, and canopy structure [16]. LiDAR data are also suitable for biophysical variable estimation, such as tree density and canopy height [17]. LiDAR data show strong coefficients of determination that mostly exceed 0.85 for logarithmic equations of various stand variables, including SV and AGB in a hardwood forest [18]. Separating areas with different forest types is essential to improve the accuracy of biomass mapping using LiDAR data [19], since individual analyses of different forest types improves the prediction accuracy of forest variables [20,21]. LiDAR-derived forest structure variables, such as canopy height, DBH, and AGB are used for large-area biomass mapping using high-resolution satellite data, since LiDAR data provides accurate stand information as reference data in a large area [22]. Above all, Næsset [5] pointed out that LiDAR-derived height variables, other than the maximum canopy height, was influenced little by pulse density even in the case of low-density data between 0.25 and 1.13 pulses m^{-2}. Thus, low-density LiDAR data could provide accurate canopy height information, and studying the performance of low-density small-footprint LiDAR data remains important and worth analyzing.

Tsuzuki et al. [23] pointed out that SV is theoretically proportional to the space between the canopy surface and the ground (hereafter, canopy space). The average canopy height is determined by the canopy space divided by the stand area. Thus, the canopy space and average canopy height are identical and the average canopy height will be useful for SV estimation. However, individual tree analysis has become popular for variables, such as tree height [24], AGB, and SV. Some recent research has focused on double-logarithmic relationships, especially for single tree analysis, between AGB or SV and variables that are derived from LiDAR data, including the average canopy height [12,25,26]. Alternatively, areal-based analysis produces better results than individual tree-based analysis for SV and AGB [26]. Although various studies have been conducted, most were stand-level studies. Therefore, it is important to evaluate the causes of the estimate variation in the stand-level analysis prior to operational use [27].

Most LiDAR-based studies separately analyzed the biomass of conifer, broadleaved, or mixed forests [6–10,12–18,25,28]. Therefore, the difference in LiDAR data performance in biomass estimation between evergreen coniferous and deciduous broadleaved forests [21] is one of interest in temperate zones, since evergreen coniferous and deciduous broadleaved forests are the dominant forest types. The performance of LiDAR data for biomass estimation is probably different among forest types [21]. Understanding the difference, and its cause in the estimation of SV and dry biomass, such as AGB and total dry biomass (TDB) by forest types, are essential to improve estimation methods. Although SV and TDB are indicators of biomass, they show tree biomass with different measures, volume of the stem part, or the dry weight of whole tree, respectively. Therefore, biomass of coniferous and broadleaved forests would be evaluated differently.

The aims of this study were to (1) identify LiDAR variables which have high correlations with SV, AGB, and TDB; (2) produce TDB and SV estimation models by multiple regression analysis using 38 LiDAR variables; and (3) validate the accuracy of TDB and SV estimation. The primary objective is identifying useful LiDAR variables for TDB and SV estimation in evergreen coniferous and deciduous broadleaved forests. Finding the difference in useful LiDAR variables between SV and TDB estimation is also an important objective. The resultant TDB and SV maps were compared to determine the relative difference of SV and TDB distributions caused by their definition.

2. Materials and Methods

2.1. Study Area

The study area is located in the Daihachiga river basin in Takayama City, Gifu in Central Japan between 36.16166°N, 137.32676°E (northeastern corner) and 36.13238°N, 137.44865°E (southwestern corner, Figures 1 and 2) in a cool temperate zone. The elevation ranges between 650 and 1600 m above sea level (ASL) with a steep topography and an average slope angle of 30°. The mapping extent covers a mountainous 8.3 km (east to west) by 2.0 km (north to south) area (36.14393°N, 137.40165°N at the center, Figure 2).

According to local information, most of the river basin was completely logged about 60–70 years ago after World War II and, therefore, the forests in this study site are considered relatively young. Planted forests of evergreen conifers, which include Japanese cedar (*Cryptomeria japonica* D. Don) and hinoki cypress (*Chamaecyparis obtusa* Sieb. et Zucc.), are dominant in the area below approximately 1000 m ASL. Some hinoki cypress stands are mixed with Japanese cedar. Planting Japanese cedar was most common around the 1950s to 1960s, because Japanese cedar is a fast-growing species for timber production used for the restoration of buildings after damage from World War II. Hinoki cypress was introduced around the 1970s to 1990s because of its valuable commercial quality [29]. Japanese cedar was also planted recently. Therefore, the ages of Japanese cedar and hinoki cypress stands are clearly different in the study area. Natural deciduous broadleaved forests dominate in areas above 1000 m ASL, and planted Japanese larch (*Lalix leptolepis* Gordon) forests exist in areas above 1200 m ASL. The dominant deciduous broadleaved species in this region are deciduous oak (*Quercus mongoloca* var. *grosseseratta* Rehder et Wilson), Japanese white birch (*Betula platyphylla* var. *japonica* Hara), Erman's birch (*Betula ermanii* Cham.), and the dominant species vary with elevation and successional stages of stands. Japanese larch was planted around the 1950s for a short period, and the tree size is similar among larch stands. Therefore, larch was not analyzed in this study. Models for deciduous broadleaved forests were used to map SV and TDB for larch stands.

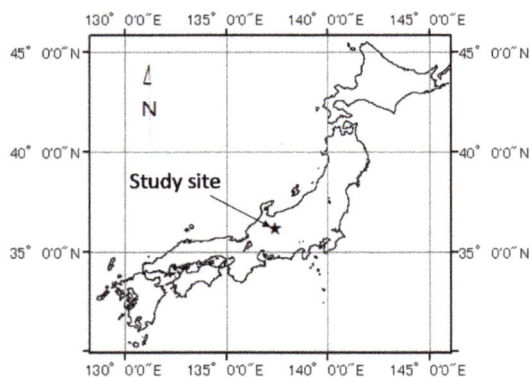

Figure 1. Location of the study site.

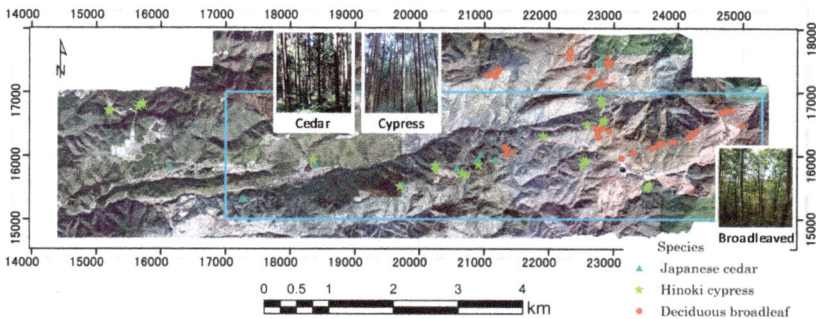

Figure 2. Plot location map. The orthophoto shows the common coverage of three light detection and ranging (LiDAR) data. The blue square area represents the location of the resulting maps of TDB and SV. The three photos show typical stands in the study site. The coordinate system of the map is Japan Plane Rectangular Coordinate System VII. Deciduous trees that had dropped leaves at the time of photo acquisition in 2012 appear in orange.

2.2. Sample Plots

Plot surveys were undertaken between 2010 and 2013 for deciduous broadleaved trees, Japanese cedar, and hinoki cypress. Sample plot areas were selected using aerial orthophotos and in situ field verification in advance of the field surveys for the following reasons: Since forest size distribution was uneven, it was difficult to find juvenile and old stands in the study area. Plots were selected along forest roads, but further than 10 m from the road side in order to facilitate the location identification on the aerial orthophotos, as well as to reduce access time to the plots. Plots were selected on graded slopes to minimize the effects of topographical accuracy differences in digital terrain models (DTM). Various height classes of stands were selected for the surveys, and circular plots were set in relatively homogeneous parts of stands. Plot radii varied between 5 and 17 m based on tree height and tree density in order to control the number of sample trees in a plot. Plot radius was determined to be large enough to include at least 40 sample trees. Stem diameter (cm) at 1.2 m DBH height and 4 cm minimum diameter of all trees was measured using calipers. The circular plot was divided into four quadrants for the convenience of tree identification. Since trees were very small and numerous, the DBH of all trees was measured in one quadrant of juvenile stands of which the top layer height was less than approximately 5 m. Tree height (H, m) was measured using a Vertex hypsometer (Haglöf, Avesta, Dalarnas, Sweden) for all trees for which DBH was measured.

A stake was set at the center of each plot, and plot center location was determined using a GPS receiver (either Mobile Mapper CX or Mobile Mapper 100 receivers, Thales, Arlington, VA, USA) with an external antenna by recording for at least 30 min by post-differential GPS (Figure 2). Plot coordinates were also checked on the aerial orthophotos (see the next sub-section). If the coordinate was uncertain based on the aerial orthophotos interpretation, coordinates were re-measured up to two times using the Mobile Mapper 100 receiver in the field until the location was confirmed on the aerial orthophotos.

During the four years, 12, 23, and 55 sample plots were measured for Japanese cedar, hinoki cypress, and broadleaved stands, respectively (Figure 2). Stem volume (SV, m^3) was calculated using DBH, H of sample trees, and the volume equations for Japanese cedar, hinoki cypress, deciduous broadleaved trees, and larch. These equations were modeled by the Nagoya Regional Forestry Office of the Japanese Forestry Agency [30–33]. Dry above- and below-ground biomass (AGB and BGB, respectively) of each tee were calculated using DBH, specific gravity of the wood for each species, and allometric equations for AGB and BGB [34]. SV (m^3 ha^{-1}), AGB (Mg ha^{-1}) and BGB (Mg ha^{-1}) of each plot were computed. Total dry biomass (Mg ha^{-1}, TDB) was computed by summing AGB and BGB. The survey results are summarized in Table 1.

Table 1. Summary of sample plot surveys.

Forest Type	No of Plots	Average DBH (cm)	Average Tree H (m)	Stem Volume (m^3 ha^{-1})	Above Ground Biomass (Mg ha^{-1})	Total Dry Biomass (Mg ha^{-1})
Japanese cedar *	4	1.1–9.4	1.8–5.8	2.4–80.9	1.9–67.5	2.6–86.1
	8	22.6–41.2	18.8–29.1	619.5–1068.3	273.8–384.6	325.7–467.1
Hinoki cypress *	21	14.6–34.3	9.6–21.4	106.6–555.7	98.8–281.0	110.4–444.9
Deciduous Broadleaved	5	1.1–25.7	1.7–21.3	6.2–408.7	4.9–263.0	7.2–317.6

* Planted tree species have been changed from cedar, cypress to cedar in conifer plantations.

2.3. Aerial Orthophotos and Forest Type Map

Three sets of aerial orthophotos with 50 cm pixels taken in June 2003, September 2008, and November 2012 (Figure 2), were supplied by the government of Gifu Prefecture and used as the reference for plot location, surveys, and location validation of the GPS measurements. For analysis, we used a forest-type map (Figure 3) [35], which was created by a decision-tree classification procedure using QuickBird images that were obtained in 2007, and a digital canopy height model based on LiDAR data, which was obtained in 2003. Forests were classified into three types in the map: evergreen coniferous forests (Japanese cedar and hinoki cypress), deciduous broadleaved forests, and larch forests. The forest type map was used to identify the forest-type distribution in biomass mapping. The forest-type map and orthophotos were geo-referenced to the Japan Plane Rectangular Coordinate System VII (JPRCS VII) with Japanese Geodetic Datum 2000 as raster images with 2.0 m pixels for the map and 0.5 m pixels for the photos.

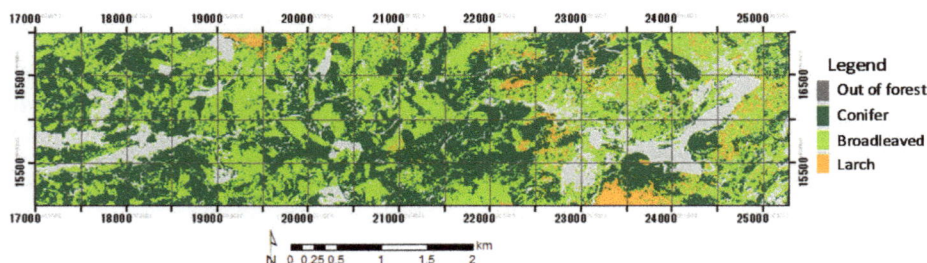

Figure 3. Forest type map of the mapping area.

2.4. Airborne LiDAR Data

The three airborne LiDAR datasets of the study area were obtained in October 2003, July 2005, and August 2011. The government of Gifu Prefecture supplied the 2003 LiDAR dataset. The geometric location of the data collected by airplanes or a helicopter was measured using a global positioning system and an inertial measurement unit. The point data were then geo-referenced to JPRCS VII. The footprint sizes of the three LiDAR datasets were between 0.2 and 0.4 m, and the point densities were 0.7, 1.8, and 1.0 pulses m^{-2} for the 2003, 2005, and 2011 LiDAR data, respectively. The LiDAR observations are summarized in Table 2.

Although the Gifu Prefecture government provided a 2 m raster DTM of the 2003 LiDAR data, the DTM was not accurate. For example, the terrain and canopy top showed the same elevation in very dense coniferous stands. Therefore we decided to produce a DTM using the three LiDAR datasets [36]. Elevations among the three LiDAR datasets were compared at six open areas, such as parking lots. The differences between the average of three, and each, LiDAR dataset were less than 10 cm. As a result, the elevation of the three LiDAR datasets was adjusted using these differences. The adjusted point data of the three LiDAR datasets were used in the following analysis including TDB and SV estimation.

Table 2. Summary of LiDAR observations.

Observation Date	Contractor	Scanner, Manufacturer	Beam Divergence (mrad)	Wave-Length (nm)	Flight Altitude Above Ground (m)	Foot-Print Size (m)	FOV (°)	Beam Density (pulse m^{-2})	Usage
October 2003	Kokusai Kogyo Co., Chiyoda, Tokyo, Japan	RAMS, EnerQuest, Denver, CO, USA	0.33	1064	2000 (Entire Gifu Prefecture)	-	±22	0.7	DTM production
25 July 2005	Nakanihon Air Service Co., Nagoya, Aichi, Japan	ALTM 2050DC, Optech, Vaughan, Ontario, Canada	0.19	1064	1200	0.24	±22	1.8	DTM production
28 August 2011	Nakanihon Air Service Co.	VQ-580 RIEGL, Horn, Horn, Austria	0.50	1064	600	0.30	±30	1.0	Biomass estimation, DTM production

The three LiDAR point data were combined for the production of the DTM. Raster data of slope and Laplacian with a 7 × 7 pixel window were produced using the DTM from the 2003 LiDAR dataset using ERDAS Imagine 2011 (Hexagon Geospatial, Madison, AL, USA) as a reference. Low points were removed from the combined point data using Terra Scan (Terra Solid, Helsinki, Uusimaa, Finland). Terra Scan provided the ground function to select points hitting the ground using two parameters, angle and distance. We assumed that proper parameters differed as a function of terrain. Therefore, we separated the point data into three groups of mild slopes, steep slopes, and ridges. Areas where the Laplacian was greater than 90 with a slope greater than 20° were classified as ridges. Areas other than ridges where the slope was between 0° and less than 20° were classified as mild slopes. The remaining areas were classified as steep slopes. The angle and distance parameters of the ground function were set the 18° and 0.5 m, 18° and 2.5 m, and 30° and 5 m for mild slopes, steep slopes, and ridges, respectively. Ground point data were then selected for the three groups and combined. A revised DTM was produced using the selected ground points by TIN of ArcGIS 10 (ESRI, Redlands, CA, USA) as a 2 m raster image using JPRCS VII [36].

Of the three LiDAR datasets, the dataset obtained in August 2011 was used for the biomass study. Gridding of the digital canopy height model reduces height accuracy [37]. Therefore, the digital canopy height of the 2011 LiDAR first return points was computed as the difference between the elevation of each point and the interpolated terrain elevation at the point location using the 2 m raster DTM and the bi-linear interpolation method. Various LiDAR variables, shown in Table 3 and as described in the next sub-section, were computed using the digital canopy height for each plot for statistical analysis. A raster LiDAR variable file with a 10 m pixel size, which included the variables in Table 3 as channels, was produced for the entire test site in the same way as the calculation for plots for mapping of SV and TDB.

Table 3. Variables computed from LiDAR point data.

Target Points	Variables					
All points	Average height	Standard deviation	Coefficient of variance	Maximum height	Canopy closure	
Points other than ground	Average height	Standard deviation	Coefficient of variance			
Points within canopy part	Average height	Standard deviation	Coefficient of variance	Height at every 10th percentiles	Height at every 10th part	Canopy closure at every 10th part

2.5. LiDAR Variables

Thirty-eight variables (Figure 4, Table 3), after Næsset [14], were calculated in order to evaluate the LiDAR-derived variables for the estimation of SV and TDB for planted evergreen coniferous forests (i.e., Japanese cedar and hinoki cypress forests) and natural deciduous broadleaved forests. Points were sampled from the area of each plot. Points over 0 m in the digital canopy height points were labeled as above-ground points. The canopy components were determined as follows: The average height and standard deviation (SD) of above-ground points were computed. The height which was two SDs below the average height was determined as the canopy bottom. Points above the canopy bottom were treated as returns from the canopy components. Average, SD, and coefficient of variance were computed for all points, above-ground points, and points within the canopy. Fortran programs were written and used for the generation of digital canopy height points and LiDAR variable computation.

Variables are denoted as follows: maximum canopy height of the plot, Hmax; canopy closure, CC; average of all points, THavr; SD of all points, THsd; coefficient of variance of all points, THcv; above-ground point average, AHavr; above-ground point SD, AHsd; above-ground point coefficient of variance, AHcv; average of canopy points, CHavr; SD of canopy points, CHsd; coefficient of variance of canopy points, CHcv; height for each percentile (e.g., 10 percentile, H10%); each tenth of the canopy height (e.g., four-tenths, Hd4); and canopy closure at this height (e.g., C4). These variables were also computed as a 10 m raster image for biomass mapping, as described previously.

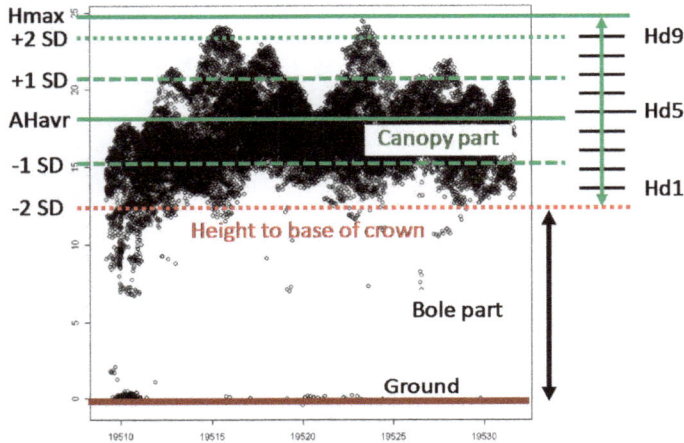

Figure 4. Graphical explanation of LiDAR variables. Hmax is the maximum canopy height, and AHavr and SD are the average height and the standard deviation of points above the ground in the plot, respectively. The height to base of the crown is defined as the height which is two SDs below AHavr. The canopy is defined as the part above the height to the base of the crown. Heights at every 10th percentile between 10% (H10%) and 90% (H90%) were computed. Heights (Hd1–9) and closures (C1–C9) at 10ths of the canopy heighs between one-tenth (d1) and nine-tenths (d9) were also computed.

2.6. Correlation Analysis

Sample plots were separated into three groups (A, B, and C) by systematic sampling based on TDB for evergreen conifer stands and deciduous broadleaved stands separately. For example, the plot with the smallest TDB among coniferous or broadleaved stands was placed in group A, the plot with the second-smallest TDB was placed in group B, the plot with the third-smallest TDB was placed in group C, and the plot with the fourth-smallest TDB was placed in group A. This process was repeated for other plots. Thus, each group was composed of similarly-sized forests. Product-moment correlation coefficients between SV and the LiDAR data variables were separately computed for evergreen coniferous and deciduous broadleaved forests for all sample plots and for each group of sample plots. Scores were given to variables to evaluate the suitability of variables for biomass estimation as follows: three points to the variable with the greatest correlation, two points to the variable with the second-greatest correlation, and one point to the variable with the third-greatest correlation in the sample sets described above. The points were summed for each variable to identify variables that were less affected by sample combinations for biomass estimation.

2.7. Regression Analysis

A multiple regression analysis by a step-up procedure using Bayesian information criterion for variable selection was applied to the evergreen coniferous and deciduous broadleaved datasets separately using JMP ver. 11 (SAS Institute Inc., Cary, NC, USA). Three combinations were tested for each of the three groups for coniferous and broadleaved forests. Samples of two groups, such as groups A and B, were used for modeling, and samples in the other group, such as group C, were used for validation. Regression models with one and two variables were built. If a canopy height variable was selected as the first variable, only variables other than height, such as canopy structure variables, like canopy closure or SD of canopy height, were used for the second variable selection. Coefficients of determination of equations were compared among the three groups, and the equation with the greatest coefficient of determination was selected as the best model for SV and TDB mapping.

Maps of SV and TDB were produced using the raster LiDAR variable file, forest type map, and selected models using a program we developed in Fortran. A model for deciduous broadleaved forests was used for larch forests based on the results of a comparison of estimates using models for conifer and broadleaved forests. Biomass distribution patterns were visually evaluated using the SV and TDB maps. Frequency histograms of TDB and SV were drawn, and their features were visually evaluated. A scattergram between TDB and SV of field plots was examined to infer the cause of the spatial pattern differences between the TDB and SV distribution maps.

ERDAS Imagine 2010 (Hexagon Geospatial, Madison, AL, USA) was used for raster data conversion, and ArcGIS 10 was used for map creation.

3. Results

3.1. Correlation Coefficients

Correlation coefficients greater than approximately 0.9 appeared in AHavr for SV, AGB, and TDB for conifer stands, with a score of 18 points (Table 4). Thus AHavr was the most effective variable for biomass estimation for this dataset. Alternatively, the mid-height between four- and six-tenths of the canopy of deciduous broadleaved forests (Hd4, Hd5, and Hd6) had correlation coefficients greater than approximately 0.85 for the three biomass variables with scores greater than 13 (Table 5). Although the mid-canopy height showed high correlation, the height ranged rather broadly. Tables 4 and 5 shows variables with higher points among LiDAR variables. Correlation coefficients were almost the same between LiDAR variables and AGB or TDB in Table 4. All correlation coefficients in Tables 4 and 5 were significant ($p < 0.01$).

Table 4. Correlation coefficients between LiDAR and stand variables—Evergreen conifer (n = 35).

LiDAR	Volume ($m^3\ ha^{-1}$)	AGB ($Mg\ ha^{-1}$)	TDB ($Mg\ ha^{-1}$)	Score *
Hd4	0.871	0.915	0.915	6
Hd5	0.876	0.912	0.912	8
Hd6	0.874	0.904	0.903	7
AHavr	0.894	0.915	0.913	18
H40%	0.886	0.900	0.896	7

* Score is an evaluation of correlation coefficients. Correlation coefficients that were the greatest, second, or third highest among the LiDAR variables were given 3, 2, and 1 points, respectively, for four cases of correlation calculations and summed. Since relationships between biomass and LiDAR-derived variables vary by samples, the general trend was evaluated by the score.

Table 5. Correlation coefficients between LiDAR and stand variables—Deciduous Broadleaved ($n = 55$) forests.

LiDAR	Volume (m³ ha⁻¹)	AGB (Mg ha⁻¹)	TDB (Mg ha⁻¹)	Score *
Hd4	0.912	0.854	0.852	16
Hd5	0.912	0.856	0.854	14
Hd6	0.909	0.857	0.854	13
AHavr	0.907	0.843	0.840	0
H40%	0.904	0.839	0.836	0

* Score is an evaluation of correlation coefficients. Correlation coefficients that were the greatest, second, or third highest among the LiDAR variables were given 3, 2, and 1 points, respectively, for four cases of correlation calculations and summed. Since relationships between biomass and LiDAR-derived variables vary by samples, the general trend was evaluated by the score.

3.2. Regression Analysis and Validation

The following equations were derived by regression analysis:
One-variable equations:
Evergreen coniferous forest:

$$SV = 43.0 \times AHavr - 171.3 \tag{1}$$

$$TDB = 18.3 \times AHavr - 1.3 \tag{2}$$

Deciduous broadleaved forest:

$$SV = 16.3 \times Hd5 - 53.6 \tag{3}$$

$$TDB = 12.0 \times Hd5 - 21.1 \tag{4}$$

The coefficients of determination adjusted for the degrees of freedom were 0.859, 0.896, 0.821, and 0.739 for Equations (1)–(4), respectively.
Two-variable equations:
Evergreen conifer forest:

$$SV = 55.2 \times AHavr + 825.4 \times CHcv - 482.4 \tag{5}$$

$$TDB = 18.5 \times AHavr - 216.0 \times C8 + 12.2 \tag{6}$$

Deciduous broadleaved forest:

$$SV = 17.9 \times Hd5 + 1281.4 \times CC - 1324.8 \tag{7}$$

$$TDB = 12.5 \times Hd5 - 107.0 \times C7 - 7.5 \tag{8}$$

where C7 and C8 (no units) are canopy closures at seven- and eight-tenths of the canopy height, CHcv is the coefficient of variance of the LiDAR points returning from canopies, and CC is the canopy closure (no units). The coefficient of C8 in Equation (6) was not significant ($p > 0.05$); however, the other coefficients were significant ($p < 0.05$). The coefficients of determination adjusted for the degrees of freedom were 0.903, 0.903, 0.849, and 0.762 for Equations (5)–(8), respectively. The numbers of samples were 22 for coniferous forests and 38 for deciduous broadleaved forests, and all eight equations were significant ($p < 0.0001$). High coefficients of determination were observed especially for Equations (2), (5), and (6) for SV and TDB of conifer stands. Although the coefficients of determination of the broadleaved forest were smaller than those of coniferous forest, the equations for SV estimation had a high coefficient of determination. These equations could provide accurate biomass estimates.

Validation results showed that predicted volume tended to be slightly underestimated based on the regression lines between field measurements and the predictions by the equations for SV and TDB. Slopes were between 0.928 and 0.971 (Figures 5 and 6). In particular, volume predictions tended to be slightly underestimated in dense Japanese cedar stands with high volume, and slightly overestimated in hinoki cypress stands with relatively low density. The models for conifer forests might underestimate the true Japanese cedar volume and overestimate the true hinoki cypress volume. Models should be developed for each species separately [20,21]. Japanese cedar trees had a multilayer structure because of intraspecies competition due to unthinned conditions. Understory trees were invisible from the air, since the upper canopy covered them. Therefore, the biomass estimate was relatively low, because the understory was not evaluated in the overall estimation using upper layer canopy information. On the other hand, almost all hinoki cypress stands had a uniform one-layer canopy and relatively low biomass, resulting in an overestimate using the canopy information. Thus, canopy structure influences the biomass estimates [38]. The standard errors (SEs) for volume validation (Figure 5) were 140.9 and 126.8 m^3 ha^{-1} for one- and two-variable equations for coniferous forests, respectively. The SEs were 32.3 and 37.1 m^3 ha^{-1} for one- and two-variable equations for broadleaved forests, respectively. For TDB (Figure 6), the SEs were 49.0 and 47.6 Mg ha^{-1} for one- and two-variable equations for coniferous forests, respectively, and 35.8 and 36.0 Mg ha^{-1} for one- and two-variable equations for broadleaved forests, respectively.

In the validation results, SEs of the two-variable equations were worse or almost equal to those of one-variable equations for SV and TDB of broadleaved forests. Samples for modeling and validation were different, and the results suggest the following: The models were built to minimize residuals among the modeling plots by the least square method. The no-height variable was set in order to reduce canopy structure difference. However, the canopy shape of broadleaved trees is irregular and differs tree by tree. Therefore, the canopy structure was different among the sample plots and between modeling and validation sample groups. Although the second variable was set to reduce residuals in the modeling group, it was not effective for broadleaved forests.

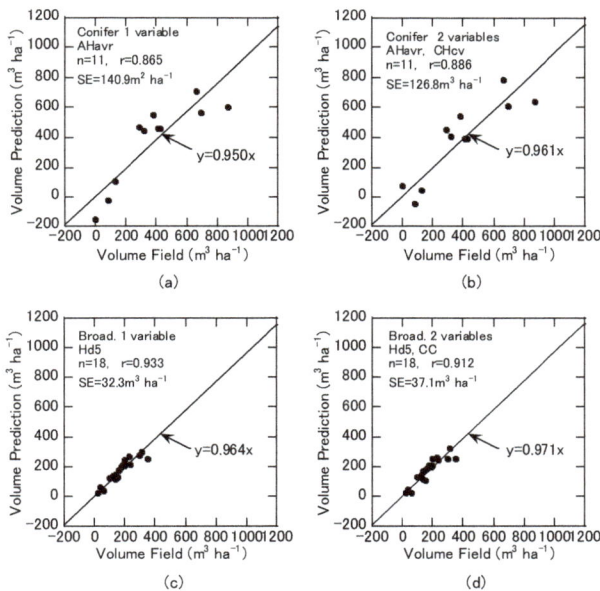

Figure 5. Validation results of SV prediction based on comparison of predicted and field SV. (**a**) One- and (**b**) two-variable models for evergreen coniferous forests, and (**c**) one- and (**d**) two-variable models for deciduous broadleaved forests.

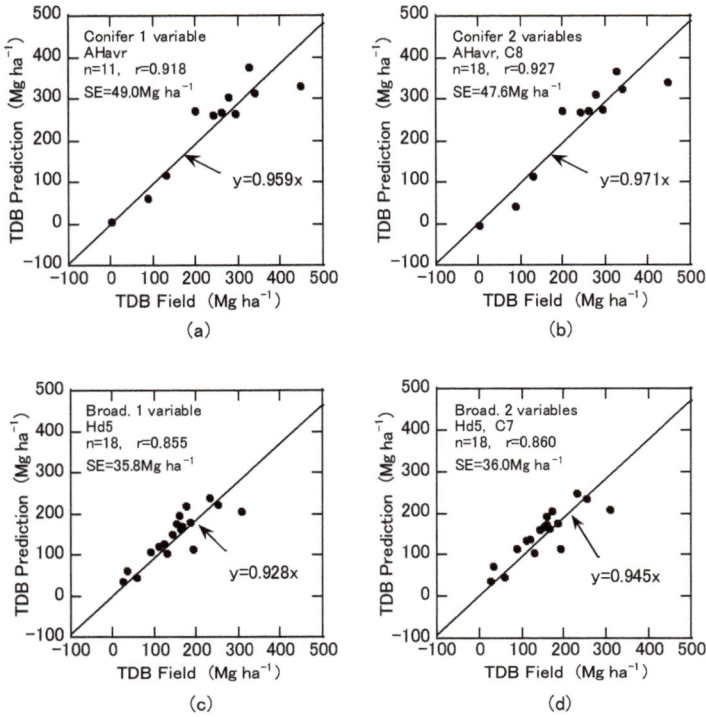

Figure 6. Validation results of TDB prediction based on the comparison of predicted and field TDB. (**a**) One- and (**b**) two-variable models for evergreen conifer forests, and (**c**) one- and (**d**) two-variable models for deciduous broadleaved forests.

3.3. Biomass Distribution

According to the plot survey, the SVs of broadleaved and conifer stands reached 400 and 1000 m^3 ha^{-1}, respectively (Table 1). Forest owners have not cared for their forests for decades, and many mature Japanese cedar stands were not thinned. Japanese cedar is relatively older than hinoki cypress and is a fast-growing species. Dense, mature Japanese cedar stands with high SV exist in the study site.

Figure 7 shows forests with high SV in the mapping area. A small village was located to the left (west) of the mapping area, and planted evergreen coniferous forests were common (Figure 3). Areas with high SV in the west mainly included mature Japanese cedar forests. However, there were no villages in the eastern two-thirds of the mapping area, and broadleaved and young coniferous forests dominated. Therefore, the SVs of these forests were small, less than 500 m^3 ha^{-1}, with a few exceptions (Figure 7), which included areas with high-SV coniferous forests along forest roads in the eastern part of the mapping area. SV was rather small in large portions of the study site, and SV was consistently low in hinoki cypress and deciduous broadleaved forests (Table 1). Regarding TDB, distribution (Figure 8) was slightly different from that of SV. The maximum of TDBs were similar among Japanese cedar, hinoki cypress, and deciduous broadleaved forests (Table 1), and TDB was more evenly distributed than SV in the study site.

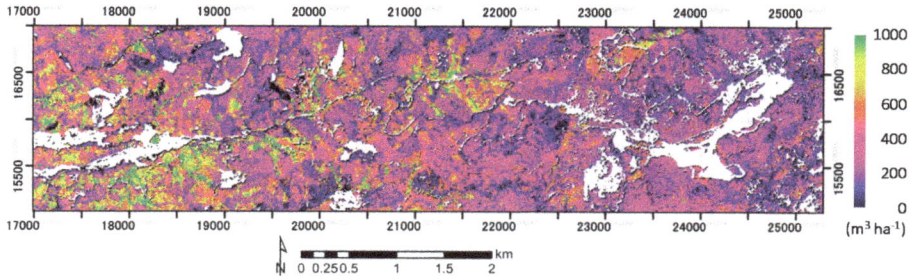

Figure 7. Distribution of SV using Equations (5) and (7). The white areas are non-forested.

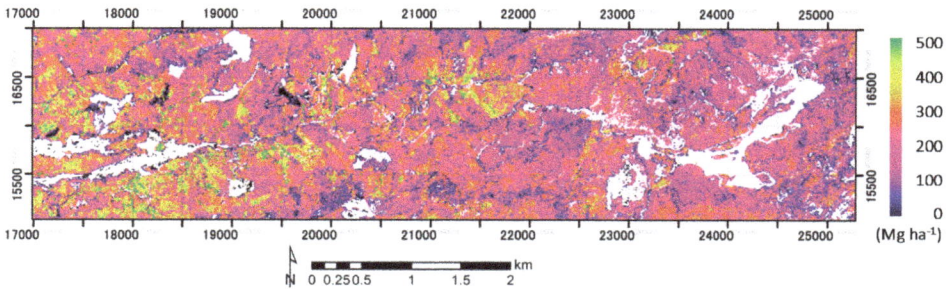

Figure 8. Distribution of TDB using Equations (2) and (8). The white areas are non-forested.

Frequency distributions were different between SV and TDB; the distribution of SV was right skewed; however, that of TDB was close to a normal distribution (Figure 9). The averages were 265.6 and 174.6 Mg ha^{-1} for TDB and 280.5 and 224.3 m^3 ha^{-1} for SV for evergreen coniferous and deciduous broadleaved forests, respectively. Predicted SV and TDB were compared (Figure 10). The regression line in Figure 10 shows a relationship between predicted SV and predicted TDB of deciduous broadleaved forests. The relationship was different between broadleaved and conifer forests, with conifer stands having relatively greater SV than TDB as compared with broadleaved stands. The ratio of stem part of broadleaved forests are smaller than that of coniferous forests, and the ratio may be greater in mature than juvenile stands of coniferous forests. The difference between coniferous and deciduous broadleaved forests was larger in high-stock conifer stands; however, the reason was unclear.

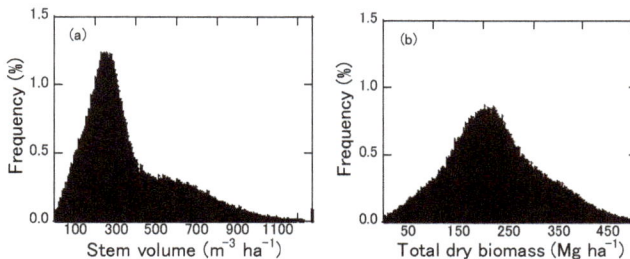

Figure 9. Frequency distributions of (**a**) SV and (**b**) TDB maps.

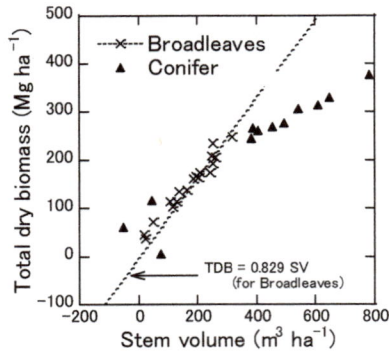

Figure 10. Comparison of predicted SV and TDB at validation plots.

4. Discussion

Sample size can impact statistical analyses. Therefore, sample selection is very important. We paid attention to the following components: (1) large numbers of samples are favorable for deriving stable results; (2) the impact that combining sample plots bears on statistical analysis; (3) the plot area should be sufficiently large to include enough number of trees which shows stand structure; and (4) minimizing the influence of DTM accuracy is important to evaluate the performance of LiDAR point data for TDB and SV estimation. We planned to measure plots as much as possible within a limited time. In order to reduce edge effects [5] and survey time, variable plot sizes were chosen, which included more than 40 trees. Survey areas were identified along forest roads using the aerial orthophotos. Futher, because of uneven age class distribution and small forest patch size in the study site, more forests in various tree sizes were selected. Sample plots were selected on graded slopes to reduce the influence of uncertain DTM accuracy.

The range of tree size and the combination of samples bear an impact on analyses. Therefore, forests with sizes ranging from very small to very great were surveyed to generalize results across coniferous and broadleaved forests in the study site. Selecting one plot from every three plots based on the order of TDB size, and combining them as three groups, made the tree size range and distribution similar among the three groups. If the sample size is small, statistical results can be greatly affected. We tried to maximize the number of samples by combining two groups for the regression analysis. The remaining plots were used to show the performance of the derived model independently from the regression analysis.

4.1. Correlations

The results showed that effective LiDAR variables differed by forest type (Tables 4 and 5). As for coniferous forests, AHavr showed the greatest score of the height variables by following other height variables, whereas only three mid-canopy heights showed similar scores for broadleaved forests. The difference between coniferous and deciduous forests is probably caused by different canopy structures. Regarding stand variables, AGB and TDB showed greater correlation coefficients than SV for coniferous forests; however, SV showed the best correlation for broadleaved forests. The specific gravity for dry biomass calculation was similar between Japanese cedar and hinoki cypress. The dry weight per unit volume was, therefore, almost identical between these species. Regarding parts of the tree used for biomass calculation, only the stem was included for SV, whereas the whole tree was included for TDB. This likely caused the observed differences in the correlation coefficients for SV and TDB (Tables 4 and 5). Additionally, broadleaved species have a wide range of specific gravity, and species composition changed widely among sample plots. The average canopy height is identical to the canopy space, which shows a linear relationship with SV [23], and probably TDB. However, uneven specific gravity made TDB distribution uneven and caused worse correlation coefficients for AGB and TDB than for SV. Correlation coefficients were almost the same between AGB and TDB. Therefore, we suggest that the estimation of TDB could be as reliable as that of AGB, and AGB was, therefore, not analyzed thereafter.

4.2. Regression Analysis

Some studies reported coefficients of determination for SV of evergreen conifer stands that ranged between 0.887 and 0.97 [9,10,13,15]. The coefficient of determination of Equation (5) was 0.903 for SV of evergreen conifer stands, which is similar to those of the previous studies. Lim et al. [18] reported 0.931 as the coefficient of determination for the logarithmic estimation model for SV of broadleaved forest, and our study produced a value of 0.849 using the linear Equation (7). Although our value was smaller, logarithmic models tend to show greater coefficients of determination [10]. Therefore, our results may still be robust.

For the second variable, we selected no-height variables. The two-variable equations revealed that canopy structure variables, such as CHcv, CC, and C7, would be effective second variables for the estimation of SV and TDB. Since C8 was non-significant, the significance of the second variable was not high. When we applied an ordinary stepwise multiple regression analysis, height variables were mostly selected for the first and second variables. Two height variables could provide duplicate information for biomass estimation. Næsset [9], however, reported an SV estimation model with multiple height variables. In our cases, SEs in all one-variable equations were highly correlated with LiDAR height variables. The variation of variables, such as DBH, H, SV, TDB, and tree density, increases with stand age owing to both the uneven growth of individual trees and management operations, such as thinning. The size variation would be a function of age and, thus, canopy height. Therefore, LiDAR height variables probably correlated with variation in biomass among stands, related to the SEs in one-variable equations. Reducing SE was the greatest task to improve the accuracy of volume and biomass estimates, and the second height variable had an important role in the two-variable equations for reducing SE. Selecting two height variables could improve model accuracy; even if the first variable showed high correlation, the second variable would be meaningful without overfitting.

The validation results showed that there was substantial underestimation of SV for small stands, especially in one-variable models (Figure 5a). Stem volume and TDB are almost zero at an average canopy height of 1.2 m or less, since DBH, which is measured at that height, is zero. Using equations that pass through the origin are probably better to reduce estimation errors for small forests than using ordinary equations for biomass estimation. The large negative error was reduced by using the two-variable Equation (5) (Figure 5b). Correlation coefficients slightly improved by adding a second variable to the validation dataset. However, SEs became greater for SV and TDB of deciduous broadleaved forests. This indicated that the canopy height strongly affected the estimates, while the canopy structure variables did not affect the estimates significantly for deciduous broadleaved forests.

Standard errors of coniferous forest estimates were greater than those of deciduous broadleaved forests for SV and TDB (Figures 5 and 6). There were three possible reasons for this. First, although the canopy shape of Japanese cedar was different from that of hinoki cypress, and because of a lack of field plots of different age classes of each species and their planting history, we combined these data for the purpose of modeling. Therefore, the models were not best adapted to either Japanese cedar or hinoki cypress. Second, Japanese cedar and hinoki cypress forests were man-made. The thinning history, however, was very different among stands and, thus, the relationship between canopy height and SV or TDB was affected by different management histories. Finally, because broadleaved forests were native, the forests could maintain the greatest biomass and canopy height in the young stage. Canopy height and the variation of SV or TDB of broadleaved forests may be relatively smaller than those of coniferous forests. Validation of the cause is important to improve models.

4.3. Biomass Distribution

The spatial distributions of SV (Figure 7) and TDB (Figure 8) were quite different. TDB was more homogenously distributed than SV, and their frequency distributions were different, as shown in Figure 9. Figures 7 and 8 may, however, give a different impression to users regarding the biomass distribution. This is likely due to the difference in the proportion of stem to the entire tree between coniferous and broadleaved trees. The regression line in Figure 10 relates to the stem ratio of deciduous broadleaved trees. Coniferous forests showed a different trend from broadleaved forests.

The equations used for the calculation of SV [30–33] and TDB [34] for individual trees were produced in different ways. Equations were produced for each DBH size class for major tree species using numerous sample trees for SV. A common equation for TDB for all species was, however, produced. Additionally, differences in the modeling process suffers from the reliability of their samples. Although LiDAR height variables reflect spatial information and were similar to volume, TDB differs from volume because it reflects weight. This difference also influences the spatial distribution of the two biomass variables. Since high SV coniferous stands had relatively small TDB, the peak was broader in TDB than that in SV (Figure 9), which caused differences in the distribution patterns between SV and TDB (Figures 7 and 8).

SV and TDB are variables that are used in forestry [2]. The former is used for the trade of lumber, and the latter for the trade of biomass. TDB is also used in studies that evaluate ecosystem ecology and carbon production to help mitigate the effects of global warming [1,2]. However, different measures can yield different estimates, as shown in Figures 7 and 8. This study revealed that TDB would be a better measure than SV for conifer biomass mapping, and SV would be better than TDB for deciduous broadleaf biomass mapping owing to their higher model correlation coefficients and validation results compared to another variable (Figures 5 and 6).

5. Conclusions

Although this study builds on previous similar research [10,12–15], comparisons of coniferous and broadleaved forests or SV and TDB estimates, such as those presented herein, are scarce [21]. TDB and SV showed higher coefficients of determination in the coniferous forest models and the broadleaved forest models, respectively. The results suggest that TDB and SV are suitable variables for biomass mapping for coniferous and broadleaved forests, respectively, from the view point of correlation. However, the reason why a different biomass variable showed a better performance for coniferous or deciduous broadleaved forests was unclear. It is important to confirm this evidence in other forests and to identify the underlying causes in order to understand the meaning of LiDAR variables selected in statistical models. Stem, branch, and root ratios, and specific gravity used in the calculation of TDB, are likely the causes for the difference. Since statistical analyses cannot easily show deterministic mechanisms and root causes, utilizing any geometric tree model [39] would be necessary in future studies.

It has been recognized that stand-level analysis requires abundant plot data to produce empirical models for biomass estimates. Further, these models are often not transferable to other areas [38]. If DBH is measured or estimated, SV or AGB can be estimated using DBH and stem count density [40] or tree height [41]. Using equations for AGB estimation by single-tree-based approaches may be better than the stand-based approaches [41]. Our results will be transferrable to forests of the same species with a similar structure and the results will not be universal among various forests, as pointed out by other studies [5,38]. Forest structure varies greatly and is impacted, for example, by planting rates or thinning practices. Estimation errors could be larger in forests after thinning, since thinning changes the gap and tree distribution in various ways. This can resulted in changes in the relationship between biomass and LiDAR variables [38]. Although the first variable was almost the same in each forest type, the second variable was different. Standard errors of the two-variable equations were greater than those of the one-variable equations. These facts indicate that structural differences among the sample plot groups influences the second variable selection. The difference in stand structure, including canopy height and variation, is especially influenced by thinning and must, therefore, be analyzed and included in prediction models to improve biomass estimates.

Forest plots were selected on graded slopes in order to reduce the influence of topography in this study. It was probably successful, since the correlation coefficient between the dominant tree height and CHavr was very high ($r = 0.987$). Steep and complex topography reduces the accuracy of biomass estimates using LiDAR-derived variables, since tree size is more variable on uneven terrain than graded slopes. However, topographic influence remains unclear in the biomass estimation. A precise DTM is necessary to reveal the influence of topography on stand-level biomass estimation using LiDAR data. Since forests exist on steep and complex mountain slopes in many parts of the world, the topographic effect on biomass estimation is an important issue.

Acknowledgments: This study was supported by JSPS KAKENHI (grant number JP22248017, "Development of validation method for the scaling up of carbon flux simulation using ecological process models", Grant-in-Aid for Scientific Research A). The 2003 LiDAR data and three aerial orthophotos were provided by the Gifu Prefectural Government. The authors greatly appreciate field survey support from Mr. Kenji Kurumado (Takayama Research Center of Gifu University), Naoko Fukuda, and Hiroto Kawai (former Awaya Laboratory staff), and Siqinbilige Wang, Nabuti Alatan, and Weilisi (former Awaya Laboratory students). We give special thanks to the editor and the four anonymous reviewers for their valuable comments and suggestions on our manuscript.

Author Contributions: Yoshio Awaya designed the research and performed the field surveys, laser data processing, and statistical analysis; Tomoaki Takahashi provided mentorship for laser data processing and statistical analysis.

Conflicts of Interest: There was no conflict of interest between the two authors.

References

1. Food and Agriculture Organization of the United Nations (FAO). *Global Forest Resources Assessment 2015 How Are the World's Forests Changing?* 2nd ed.; FAO: Rome, Italy, 2016; pp. 1–44.
2. West, P.W. *Tree and Forest Measurement*, 2nd ed.; Springer: Berlin, Germany, 2009; pp. 22–63.
3. Arp, H.; Griesbach, J.C.; Burns, J.P. Mapping in Tropical Forests; A new approach using the laser APR. *PE&RS* **1982**, *48*, 91–100.
4. Aldred, A.H.; Bonner, G.M. *Application of Airborne Lasers to Forest Surveys (Inst. Information Report PI-X-51)*; Petawawa National Forestry: Petawawa, ON, Canada, 1985; pp. 1–62.
5. Næsset, E. Area-Based Inventory in Norway—From Innovation to An Operational Reality. In *Forestry Applications of Airborne Laser Scanning*; Maltamo, M., Næsset, E., Vauhkonen, J., Eds.; Springer: Dordrecht, The Netherlands, 2014; Volume 27, pp. 215–240.
6. Nelson, R.; Krabill, W.; Tonelli, J. Estimating forest biomass and volume using airborne laser data. *Remote Sens. Environ.* **1988**, *24*, 247–267. [CrossRef]
7. Lefsky, M.A.; Harding, D.; Cohen, W.B.; Parker, G.; Shugart, H.H. Surface lidar remote sensing of basal area and biomass in deciduous forests of eastern Maryland, USA. *Remote Sens. Environ.* **1999**, *67*, 83–98. [CrossRef]
8. Lefsky, M.A.; Cohen, W.B.; Acker, S.A.; Parker, G.C.; Spies, T.A.; Harding, D. Lidar remote sensing of the canopy structure and biophysical properties of Douglas-fir western hemlock forests. *Remote Sens. Environ.* **1999**, *70*, 339–361. [CrossRef]
9. Næsset, E. Estimating timber volume of forest stands using airborne laser scanner data. *Remote Sens. Environ.* **1997**, *61*, 246–253. [CrossRef]
10. Means, J.E.; Acker, S.A.; Fitt, B.J.; Renslow, M.; Emerson, L.; Hendrix, C.J. Predicting forest stand characteristics with airborne scanning lidar. *PE&RS* **2000**, *66*, 1367–1371.
11. Holmgren, J.; Nilsson, M.; Olsson, H. Estimation of tree height and stem volume on plots using airborne laser scanning. *For. Sci.* **2003**, *49*, 419–428.
12. Næsset, E.; Økland, T. Estimating tree height and tree crown properties using airborne scanning laser in a boreal nature reserve. *Remote Sens. Environ.* **2002**, *79*, 105–115. [CrossRef]
13. Næsset, E. Predicting forest stand characteristics with airborne scanning laser using a practical two-stage procedure and field data. *Remote Sens. Environ.* **2002**, *80*, 88–99. [CrossRef]

14. Næsset, E. Effects of different flying altitudes on biophysical stand properties estimated from canopy height and density measured with a small foot-print airborne scanning laser. *Remote Sens. Environ.* **2004**, *91*, 243–255. [CrossRef]
15. Magnusson, M.; Fransson, J.E.S.; Holmgren, J. Effects on estimation accuracy of forest variables using different pulse density of laser data. *For. Sci.* **2007**, *53*, 619–626.
16. Takahashi, T.; Yamamoto, K.; Senda, Y.; Tsuzuku, M. Estimating individual tree heights of sugi (*Cryptomeria japonica D. Don*) plantations in mountainous areas using small-footprint airborne LiDAR. *J. For. Res.* **2005**, *10*, 135–142. [CrossRef]
17. McCombs, J.W.; Roberts, S.D.; Evans, D.L. Influence of fusing LiDAR and multispectral imagery on remotely sensed estimates of stand density and mean tree height in a managed loblolly pine plantation. *For. Res.* **2003**, *49*, 457–466.
18. Lim, K.; Treitz, P.; Baldwin, K.; Morrison, I.; Green, J. Lidar remote sensing of biophysical properties of tolerant northern hardwood forests. *Can. J. Remote Sens.* **2003**, *29*, 658–678. [CrossRef]
19. Zhao, K.; Popescu, S.; Nelson, R. Lidar remote sensing of forest biomass: A scale-invariant estimation approach using airborne lasers. *Remote Sens. Environ.* **2009**, *113*, 182–196. [CrossRef]
20. He, Q.S.; Cao, C.X.; Chen, E.X.; Sun, G.Q.; Ling, F.L.; Pang, Y.; Zhang, H.; Ni, W.J.; Xu, M.; Li, Z.-Y. Forest stand biomass estimation using ALOS PALSAR data based on LiDAR-derived prior knowledge in the Qilian Mountain, western China. *Int. J. Remote Sens.* **2012**, *33*, 710–729. [CrossRef]
21. Cao, L.; Coops, N.C.; Hermosilla, T.; Innes, J.; Dai, J.; She, G. Using small-footprint discrete and full-waveform airborne lidar metrics to estimate total biomass and biomass components in subtropical forests. *Remote Sens.* **2014**, *6*, 7110–7135. [CrossRef]
22. Mora, B.; Wulder, M.A.; White, J.C.; Hobart, G. Modeling Stand Height, Volume, and Biomass from Very High Spatial Resolution Satellite Imagery and Samples of Airborne LiDAR. *Remote Sens.* **2013**, *5*, 2308–2326. [CrossRef]
23. Tsuzuki, H.; Kusakabe, T.; Sueda, T. Long-range estimation of standing timber stock in western boreal forest of Canada using airborne laser altimetry. *J. Jpn. For. Soc.* **2006**, *88*, 103–113. (In Japanese with English Summary) [CrossRef]
24. Kwak, D.; Lee, W.; Lee, J.; Biging, G.S.; Gong, P. Detection of individual trees and estimation of tree height using LiDAR data. *J. For. Res.* **2007**, *12*, 425–434. [CrossRef]
25. Takahashi, T.; Yamamoto, K.; Senda, Y.; Tsuzuku, M. Predicting individual stem volumes of sugi (*Cryptomeria japonica D. Don*) plantations in mountainous areas using small-footprint airborne LiDAR. *J. For. Res.* **2005**, *10*, 305–312. [CrossRef]
26. Kankare, V.; Vastaranta, M.; Holopainen, M.; Räty, M.; Yu, X.; Hyyppä, J.; Hyyppä, H.; Alho, P.; Viitala, R. Retrieval of forest aboveground biomass and stem volume with airborne scanning LiDAR. *Remote Sens.* **2013**, *5*, 2257–2274. [CrossRef]
27. Breidenbach, J.; McRoberts, R.E.; Astrupa, R. Empirical coverage of model-based variance estimators for remote sensing assisted estimation of stand-level timber volume. *Remote Sens. Environ.* **2016**, *173*, 274–281. [CrossRef] [PubMed]
28. Næsset, E.; Gobakken, T. Estimating forest growth using canopy metrics derived from airborne laser scanner data. *Remote Sens. Environ.* **2005**, *96*, 453–465. [CrossRef]
29. Forestry Agency of Japan. *Annual Report on Forest and Forestry in Japan Fiscal Year 2013*; Forestry Agency of Japan: Tokyo, Japan, 2014; p. 223. (In Japanese with English Summary)
30. Forestry Agency of Japan. *Description of Stem Volume Table Development for Manmade Sugi Cedar Stands in Nagoya Regional Forestry Office*; Forestry Agency of Japan: Tokyo, Japan, 1959; p. 17. (In Japanese)
31. Forestry Agency of Japan. *Description of Stem Volume Table Development for Manmade Hinoki Cypress Stands in Nagoya Regional Forestry Office*; Forestry Agency of Japan: Tokyo, Japan, 1959; p. 14. (In Japanese)
32. Forestry Agency of Japan. *Description of Stem Volume Table Development for Deciduous Broadleaved Stands in Nagoya Regional Forestry Office*; Forestry Agency of Japan: Tokyo, Japan, 1959; p. 19. (In Japanese with English Summary)
33. Forestry Agency of Japan. *Description of Stem Volume Table Development for Larch Stands in Nagoya Regional Forestry Office*; Forestry Agency of Japan: Tokyo, Japan, 1963; p. 29. (In Japanese with English Summary)

34. Komiyama, A.; Nakagawa, M.; Kato, S. Common allometric relationships for estimating tree biomasses in cool temperate forests of Japan. *J. Jpn. For. Soc.* **2011**, *93*, 220–225. (In Japanese with English Summary) [CrossRef]

35. Fukuda, N.; Awaya, Y.; Kojima, T. Classification of forest vegetation types using LiDAR data and Quickbird images—Case study of the Daihachiga river basin in Takayama city. *J. JASS* **2012**, *28*, 115–122. (In Japanese with English Summary)

36. Fukuda, N.; Awaya, Y. Investigation of DTM generation using LiDAR data—A case in Daihachiga river basin. *Chubu For. Res.* **2013**, *61*, 107–108. (In Japanese)

37. Gaveau, D.L.A.; Hill, R.A. Quantifying canopy height underestimation by laser pulse penetration in small-footprint airborne laser scanning data. *Can. J. Remote Sens.* **2003**, *29*, 650–657. [CrossRef]

38. Rosette, J.; Suárez, J.; Nelson, R.; Los, S.; Cook, B.; North, P. Lidar Remote Sensing for Biomass Assessment. In *Remote Sensing of Biomass—Principles and Applications*; Fatoyinbo, T., Ed.; InTech: Rijeka, Croatia, 2012; pp. 3–26.

39. Ko, C.; Gunho Sohn, G.; Remmel, T.K.; Miller, J. Hybrid Ensemble Classification of Tree Genera Using Airborne LiDAR Data. *Remote Sens.* **2014**, *6*, 11225–11243. [CrossRef]

40. Yao, T.; Yang, X.; Zhao, F.; Wang, Z.; Zhang, Q.; Jupp, D.; Lovell, J.; Culvenor, D.; Newnham, G.; Ni-Meister, W. Measuring forest structure and biomass in New England forest stands using echidna ground-based lidar. *Remote Sens. Environ.* **2011**, *115*, 2965–2974. [CrossRef]

41. Maltamo, K.; Eerikäinenn, J.; Pitkänen, J.; Hyyppä, M. Vehmas Estimation of timber volume and stem density based on scanning laser altimetry and expected tree size distribution functions. *Remote Sens. Environ.* **2004**, *90*, 319–330. [CrossRef]

MDPI

St. Alban-Anlage 66

4052 Basel

Switzerland

Tel. +41 61 683 77 34

Fax +41 61 302 89 18

www.mdpi.com

Remote Sensing Editorial Office

E-mail: remotesensing@mdpi.com

www.mdpi.com/journal/remotesensing

www.ingramcontent.com/pod-product-compliance
Lightning Source LLC
Chambersburg PA
CBHW051725210326
41597CB00032B/5603